W9-DFK-731

SELF-ORGANIZING NATURAL INTELLIGENCE

Self-Organizing Natural Intelligence

Issues of Knowing, Meaning, and Complexity

by

MYRNA ESTEP
Indiana University, Bloomington, U.S.A.

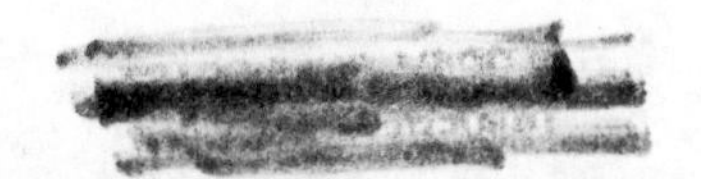

A C.I.P. Catalogue record for this book is available from the Library of Congress.

ISBN-10 1-4020-5275-8 (HB)
ISBN-13 978-1-4020-5275-0 (HB)
ISBN-10 1-4020-5299-5 (e-book)
ISBN-13 978-1-4020-5299-6 (e-book)

Published by Springer,
P.O. Box 17, 3300 AA Dordrecht, The Netherlands.

www.springer.com

Printed on acid-free paper

Dedication

To the memory of my mother,
Mary Magdalene Stanley Estep,
November 24, 1917 –
May 5, 2004

Table of Contents

Introduction

Having been born and raised in the rural coal-filled mountains of southern West Virginia during the nineteen forties, my brothers, sisters and I learned early to respect the earth and other living things. Our parents taught us to tend large areas of land, to plant them thick with crops every year, and take care of our animals. Along with cows, chickens, and our crops, they often spoke of the earth itself as a living thing, and that's the way I grew to see and understand it. The very dirt under our feet was viewed as a precious thing, not something to disparage or belittle, let alone poison or wear out. It was that dirt, after all, that helped create what we saw as the natural miracle of food to eat.

But it was not only our job to plant, harvest then help with canning and storing food in our cellar each year. We also had to keep wild animals from helping themselves too much to our crops and livestock. I suppose it was from my earliest years that I learned the tricks of wildlife, along with the occasional off-the-beaten-path human, to get young plants and the occasional chicken to eat. Almost from necessity, I grew up with a healthy respect for natural intelligence.[1]

It was only quite a few years later when I attended a great university in the Midwest, however, that I learned animals didn't have any intelligence.[2] They didn't have any intelligence, it was claimed, because they have no language abilities.[3] Animals can neither speak nor write any recognized language because they do not have the vocal apparatus, language areas of the brain, and in most cases do not have fingers suited for writing. Hence, so the argument went, they cannot reason. Reasoning in language was the

---

[1] Unless otherwise indicated, I use "natural intelligence" and "intelligence" interchangeably as that intelligence naturally found in humans and animals. These are in contrast to "artificial intelligence" which is deliberately designed by humans for specific purposes. These terms are more precisely specified and defined later.

[2] I use the term "animal" to refer to nonhuman animals, though obviously humans are animals as well.

[3] The term "language" is taken to refer to any alphanumeric system, with rules of grammar. "Natural language" is taken to refer to languages which are historically given with no explicit rules laid down from the start which govern their use. Such rules continually change. This is in contrast to "artificial language" which is taken to refer to languages that are essentially simple, with rules explicitly set forth. See Nordenstam 1972.

essence of intelligence, the experts claimed, and scores on IQ tests, which animals cannot even take, eventually became circularly definitive of that intelligence. Thus, according to experts, it followed that animals had none.

The Continuing Influence of Behaviorism

At the time, those disciplines that study and conduct research into intelligence were under the influence of the movement known as behaviorism, led by B.F. Skinner. Skinner and his adherents held that all behavior, human and animal, was explainable as responses to external stimuli. Following the classical science model, simple, linear causal chains were taken to explain everything a human or animal does.

Based upon inherited neural mechanisms, behaviorism is a mechanical determinist theory whose origins can actually be traced back many centuries, at least to Descartes. According to Descartes, animals were merely reactive mechanical organisms, like wound-up clocks. They could not feel pain and they certainly possessed no intelligence. They had no intelligence, he claimed, because they had no soul, which meant they also had no free will.

On the other hand, he held that humans do have souls and hence also have freedom of will because they have the ability to reason. Reason, he argued, is the highest achievement of mankind and it is reason and the freedom of our wills that he held distinguishes us from animals. Though our view of animal pain has changed since Descartes, we still largely view them the same way he did, as reactive creatures with no real intelligence.

We also still largely view intelligence the way he did. As the basic S-R theory developed, the more important behaviors of human beings came to be viewed as those that are acquired. Since humans are equipped with a vocal apparatus and language centers in the brain, language could be explained as an acquired, conditioned response to external stimuli. Animals, on the other hand, were and still are largely viewed as creatures equipped solely with reflexes and instinctive behavior. We're all familiar with Pavlov's dogs, taught by classical conditioning to respond in certain ways to external stimuli.

But over many years, even with extraordinary *ad hoc* changes to the basic theory, later passing under other names, it became clear to some researchers that unbiased observation shows that basic stimulus-response schemes do not work. One of the most important things to recognize about early behaviorism is that both humans and animals were and still are largely conceived by some to be passive receptors of external stimuli. Jettisoning earlier concepts such as freedom and will, as well as other references to internal motivation, the aim was to focus solely upon the environment and external observable forces acting upon both humans and animals. We and our animal friends came to be viewed as reactive organisms, not in control of ourselves; we become whatever external stimuli make us to be.

But innumerable clinical and other studies accumulated irrefutable evidence of failures of S-R schemes to explain even ordinary human and animal behavior. Among many other things, they leave out self-directiveness found in intentional and sometimes prescient, anticipatory behavior involved even in simple everyday problem-solving. They also entirely leave out exploratory, playful, as well as improvised, innovative, and creative (yes—oh that dreaded of concepts to the behaviorists—even creative) activity of both humans and animals. Though later versions of behaviorism were not quite so strict, they nonetheless maintained the basic reactive scheme to explain all behavior.

The basic S-R model remains influential in the behavioral sciences though it is no longer found in its most stark form advocated by Skinner. Still found in intelligence theories by those who call themselves "materialists," "eliminative materialists," "naturalists," and "neural network theorists," among others, the underlying S-R scheme was modified placing more control in neural mechanisms and chemistry of the brain. In place of *single* causal chains whole networks of interconnected causal chains were put in place, acknowledging multiple determinants as well as multiple resultants. The capacity for thought or reason, viewed as exclusively an aspect of specific language portions of the brain, came to be seen as the controlling feature of intelligent behavior.

Myths of the Representational, Top-Down, Linguistic Mind

Among the most persistent myths about intelligence is that language is a necessary condition to think or to act intelligently. Representationalism, the picture of mind or intelligence itself as a large set of symbolic alphanumeric representations of the world with rules for their manipulation, arose as a prominent view along with the computational theory of mind.

This myth is a natural consequence of centuries of influence by twin scholarly movements known as *nominalism* and *conceptualism* that arose long before behaviorism. Basically, nominalism is the view that all we can know of what we may call "reality" is the language we use to describe it. There are no facts out there, objectively existing independent of language speakers. All we have, according to nominalists, are the language labels we use to describe or name our experience. There is always a language representational interface between us and reality, if there is any such thing.

We have no direct or immediate contact with reality; we have nothing of reality itself.

Similarly, conceptualists claim that all we have and can know are concepts "in our minds." There is nothing beyond those, they say, that is real. Though possibly not realizing the historical sources, these twin movements have led some even recently to issue stark pronouncements about "making our own reality."

It is important to stress this essence of nominalism and conceptualism due to their continuing influence upon our view of intelligence. Both doctrines hold that there is nothing cognitive that is not mediated by language. We do not have any *immediate* contact with the world or reality, they claim. We can only have *mediated* contact through language *about* our experience or by way of representations in the brain.

Following this line of argument, however, still others who eventually became known as postmodernists argued even more strongly that nothing in our language can correspond with anything outside of ourselves. Indeed, they still argue that we can make our language do just about whatever we want it to, *hence we can make reality any way we want it to be.* This line of argument extending from earliest nominalism and conceptualism has led to a tradition bereft of moorings tied to concepts such as *fact, evidence,* and *truth*.

Others known as realists argued that though we may only have our language or representations, they must nonetheless correspond to something outside ourselves that we can put to test and verify. But testing and verification must always be done with instruments defined by linguistic means. Any other claims to know must be excluded from the domain of intelligence.

The effect of these combined arguments by both postmodernists and realists has been to entirely remove the locus of intelligence from the person to the mechanics of language and objective representations of reason. Later additional arguments placed the rules governing the mechanics of language and reason in specific areas of the brain. Those parts of the brain became the machine's central processor and the claimed true locus of all intelligence.

Over many centuries, especially since Descartes, the influence of mechanism along with the twin movements of nominalism and conceptualism led to the mechanics of artificial computer languages, which paradoxically became the paradigm for the mechanics, the essence, of *natural* intelligence.

With the development of the neurosciences, especially Hebb's notion of the cell assembly, more attention eventually focused upon the overall dynamics of the brain instead of behaviorist explanations that did not work in any case. Neural systems became important in the development of computer models of the mind.

Additionally, it should be mentioned that the shift from behaviorism to neurology helped to better explain certain diseases such as autism that confounded (and still confounds, to some degree) the experts. Nonetheless, global theories of the brain tending toward global theories of intelligence followed the same representationalism found in nominalism and conceptualism.

Today, representationalism and the computational theory of mind are pervasive in fields such as psychology—especially in intelligence research and IQ studies, and also, obviously, in artificial intelligence. These are determinist, top-down, verbal,[4] knowledge-based approaches to mind and intelligence, still adhering to an underlying though extended mechanical S-R model.

Perhaps the most highly influential view today is that intelligence is found in those neural centers of the brain where purported "grammar genes" are the controlling feature. The science of genetics has been used to add yet another layer of determinism to the model. It is a *genetic* determinist, logico-linguistic, top-down model driving the train of intelligence research in the U.S. But it is an approach that has proven to be a manifest failure to capture even the most rudimentary aspects of any natural intelligence system.

Among other things, this approach leaves entire categories of human as well as animal cognition out of the intelligence picture altogether. It leaves out entire facets of complex sensorimotor awareness, control, and intentional behavior evident even in simple tasks because of the prior assumption that these have nothing or very little to do with intelligence.

The arguments upholding this view are based not only upon false assumptions about the science of genetics but also false assumptions about human cognition generally. Moreover, they are based upon ignoring reams of documented evidence of human and animal intelligent behavior found in everyday experience that is unrelated to language. At the most fundamental level, they are based upon the false assumption that intentional doings, *knowing how,* are reducible to *knowledge that* (or just *"knowledge"*),[5] the kind of knowing we can put in declarative sentences

[4] The word "verbal" means "by linguistic means." It refers to communication by language, either spoken or written.

[5] The term "procedural", as in "procedural knowledge", is sometimes used to refer to *knowing how*. I will use the phrase "knowledge that" and the term "knowledge" equivalently to refer to declarative assertions, or claims to know, in language. In the

and encode into machines. This assumption underlies virtually all existing theory of intelligence, as well as the standardized tests used to measure it, even while extensive evidence shows that it is false. Indeed, the evidence shows that *knowing how* and *knowledge that* are two very different kinds of intelligence, not one.

Knowing How and Sensorimotor Intelligence

This astonishing neglect of the intelligence of *knowing how* has not only diminished and thwarted our understanding of basic performances or practical tasks such as knowing how to tie one's shoes or drive a car, it entirely leaves out whole facets involved in understanding what everyone would agree is highly intelligent behavior such as knowing how to prove complex mathematical theorems or perform surgery. Indeed, evidence shows that the intelligence of knowing how expands across *all* our intelligence. It is the most fundamental kind of intelligence, found in the most practical or procedural physical tasks to the highest levels of human thought and creative endeavors of the mind. Yet it is missing almost entirely in predominant theories of intelligence.

To some degree there is an irony involved in the fact that the very top-down, logico-linguistic and knowledge-based serial and additive approach to intelligence so pervasive in intelligence research cannot address even basic intentional doings. It especially cannot address sensorimotor performances found in those who are largely visually-oriented (as opposed to verbal), nor does it capture the *know how* involved in fundamental tasks of basic math and logic. It leaves out an entire panoply of indexical signs of intelligence traversing the entire spectrum of intentional doings.

In contrast to much of the research directed to human intelligence, many researchers in artificial intelligence have more recently turned to more mathematical and less logical approaches to understanding and simulating human and animal intelligence. Based in part on the development of high-speed computers and advanced imaging techniques, great strides have been made by designing sensory architectures for biologically inspired approaches, especially in robotics. This was necessary to get away from the Good Old Fashioned Artificial Intelligence (GOFAI) approach emphasizing the knowledge-based, top-down, serial approach to intelligence founded upon so many false assumptions.

---

philosophical literature "knowledge that" is used to emphasize the declarative sentence or proposition following "that". "Knowing how" was earlier called "practical intelligence" by the ancients (though they did not use that notion to mean the same thing we do today when we generally use that phrase).

Yet even the more recent mathematical approaches are still inadequate. For example, though neural network (connectionist) approaches are promising in efforts to simulate actual perceptual and intentional sensorimotor performances, they are fraught with some fundamental flaws. Those flaws largely center on some of the same basic classical science assumptions that simple causal chains of the top-down, linear, functional-block oriented strategies are sufficient to describe and explain the emergent dynamics of actual intelligence.

Moreover, these approaches still largely rely upon the old representationalist, top-down *knowledge that* assumptions and fail to capture the unique, context-sensitive smoothness, timing, and immediacy of human and animal sensory and sensorimotor awareness of *knowing how* clearly evident in intelligent behavior spanning *all* intelligence.

In other areas, while some recent animal intelligence researchers affirm that language is not necessary for thought (Marc Hauser 2000), they still largely rely upon those same representationalist models. They still follow what amounts to the same top-down computational model. Moreover, their efforts are premised as well upon faulty classical science reductionist, linear assumptions, striving for simple, direct causal chains that cannot account for emergent intelligence phenomena found in actual experience.

Among other things, extensive empirical research demonstrates the need for a broad theory of *signs* by which both animals and humans exhibit as well as *disclose* their cognition in the shapes and patterns of what they do. For example, as things now stand, major theories of intelligence do not include accounts of relatively ordinary stealth patterns used by both humans and animals to elude predators. They also neglect more complex and interesting deadly offensive patterns such as aggressive mimicry found in both humans and animals.

Likewise, these theories do not touch upon highly intelligent improvisational elements found in the likes of both defensive and offensive combat behavior as well as insurgency and counter-insurgency strategies. They have made few attempts to tie their theories of intelligence to the concepts and strategies commonly debated in and around intelligence agencies in the capitals of the world and often deployed by military as well as criminal and terrorist groups.

Toward Signs and Self-Organization

A broad theory of signs would include a set of classifications, clearly defined, of the self-organizing and emerging patterns and shapes human and animal cognition takes in the world extending far beyond written and spoken natural and artificial language behavior. Moreover, it would extend beyond patterns of individuals to include patterns of teams, networks, and coordinated groups or "cells" of individuals who share common objectives on many levels. Yet to find efforts to set forth such a theory of signs, one must look back many decades in the annals of philosophy.

Additionally, none of the predominant verbal-based theories of intelligence today address the *self-organizing dynamics* of intelligence found in the actual experience of living sentient beings, let alone groups of such beings. The predominant theories are static, reductionist, nondynamic, and may as well be describing programmed, unmoving automatons. The actual living, breathing, moving, brilliant, sweating, angry, laughing, planning, violent, coordinating, improvising, and loving activity found in the intelligent behavior of living beings is largely if not entirely missing.

To summarize, right now intelligence is largely viewed as a one-dimensional, top-down, language-and number-based special *single* ability that some have more of than others. It is viewed as given largely at birth in one's genetic inheritance, and cannot be modified to any significant degree by education, training, or experience.

Moreover, the predominant view is that intelligence is found in one or two parts of the brain but not in others. It is definitely not found in other parts of the body or in anything the body does that is *unrelated* to the use of language, logic, and number. It is definitely not found in lower animals or at least certainly not in the superior ways it is found in us (or at least in some of us, so the argument goes) due to language centers of our brains. Most importantly, it is held to be entirely measured by standardized IQ tests.

Unlike just about everything else in the natural world, intelligence is viewed as this special single ability that does not emerge from innumerable interactions among those components making up a life. On the contrary, this inborn central processor is that special ability that anoints riches and material success on those who are privileged to be born with much of it. Like Venus issuing forth from the head of Zeus, intelligence is viewed as issuing forth from certain parts of a superior brain that is master of all it surveys. And like a goddess that imparts favors on all who worship her, intelligence assures the one who has a lot of it much success and power in this world.

It is this almost *super*natural and strictly hierarchical view of intelligence that has formed our view of humanity itself, as well as our relation with the rest of the animal kingdom. It is a view that has also wrought sometimes miserable and often very inhumane consequences to the world. Cut off from the natural world in many ways, it is little wonder that it is this view—more than just about anything else—that not only gave rise to what C.P. Snow called the "Two Cultures," but to some of the most pernicious social and economic divisions found in our history.

The world needs a different way of looking at intelligence

I have approached natural intelligence as a multi- and interdisciplinary phenomenon. I view intelligence as very much a part of the natural world and hence as a living thing, an emerging richly textured set of patterns that are highly complex, dynamic, self-organizing, and adaptive.

In both scope and content, this book draws upon the behavioral sciences, the neurosciences, philosophy, as well as computer sciences and engineering to address the above and many more fundamental issues about the nature of intelligence. Above all, I wish to broaden and re-carve the universe of intelligence without the biases of historical accident and whims of power, based solely upon the facts of intelligence found in the natural world.

Myrna Estep
Silverton, Colorado

List of Abbreviations

AI	Artificial Intelligence
AIP	Anterior, intra parietal region
AL	Artificial Life
ASL	American Sign Language
CNS	Central Nervous System
ERP	Event Related Potential
fMRI	Functional Magnetic Resonance Imaging
GOFAI	Good Old Fashioned Artificial Intelligence
LGN	Lateral Geniculate Nucleus
MIP	Medial intra parietal region
M1	Primary Motor Cortex
MT	Middle Temporal
MT+	Middle Temporal complex
MST	Medial Superior Temporal
MSTl	Medial Superior Temporal lateral
MSTd	Medial Superior Temporal dorsal
PF	Performative Intelligence
QL	Qualitative Intelligence
QN	Quantitative Intelligence
rCBF	Rate of Cerebral Blood Flow
RTC	Reticulo-Thalamo-Cortical
SDT	Signal Detection Theory
SIGGS	Set, Information, Graph, General Systems
SOFM	Self-Organizing Feature Map
TBC	The Bell Curve
TSS	Test Score Semantics
UG	Universal Grammar
VIP	Ventral intra parietal region
WISC	Wechsler Intelligence Scale for Children
WAIS	Wechsler Adult Intelligence Scale-
WAIS-R	Wechsler Adult Intelligence Scale-Revised [WAIS-R],

List of Figures and Tables

Acknowledgements

A book is almost never the result of a single person's efforts, and this one is no exception. I am fortunate to have learned much from parents who knew first-hand how to survive the coal-filled mountains of West Virginia and the Great Depression as well, then later raising five children. It is to them that I am indebted beyond measure.

More than anyone else, I owe both a personal as well as professional debt to my husband, Professor Richard Schoenig. He made it possible for me to have the time, comfort and leisure to write this. I will always be grateful for his invaluable friendship, his enormously broad-ranging knowledge, understanding, wisdom, advice, and steadfast support.

Over the years, I have also been very fortunate to meet individuals either in person or in their works from whom I was able to learn some of life's most insightful, enduring and endearing lessons. I am especially indebted to Professors Elizabeth Steiner and George Maccia; Sir Roger Penrose; Professor Stuart Kauffman; Ian Stewart; and Jack Cohen, whose works I heavily relied upon in this book. If their lessons were not mastered sufficiently to be reflected in my own work, then that is my failure. It is certainly not theirs.

Thanks are also due to Professor Ted Frick at Indiana University; Professor Alwyn Scott of the Mathematics Department at University of Arizona; and Professor Burton Vorhees of the Mathematics Department at Athabasa University, Canada. I am grateful to all of them for constructive comments and suggestions on earlier versions of my manuscript. And I will always be grateful to the faculty and administration of Indiana University in Bloomington who years ago made it possible for me to study in such a naturally beautiful place and eventually develop as a scholar.

Finally, I owe my sincerest gratitude to Robbert van Berckellaer and Cynthia de Jong at Springer Verlag in Dordrecht, The Netherlands for their support and assistance; and to the Victorian mining town of Silverton, Colorado for its warm friendliness and natural beauty.

Figure and Photo Acknowledgements

I wish to thank *Washington University School of Medicine,* its Neuroscience program, and especially its Departments of Anatomy and Neurobiology and Susan Danker for permission to reproduce photos appearing in Fig. 4.2. Washington University School of Medicine http://thalamus.wustl.edu/

I wish to thank the *University of Western Ontario Medical School*, in particular Professor Tutis Vilis of the Departments of Physiology and Pharmacology for permission to reproduce photos appearing in Fig. 4.1. and Fig. 4.3.

I also wish to thank *North Carolina State University,* in particular Professor Christopher Healey of the Computer Science Department for permission to reproduce the List of Some Preattentive Features in Table 4.1, and Fig. 4.7. Triesman's feature integration model.

1 The Problem of Intelligence

In common parlance, the word "intelligence" has both descriptive and normative senses. In its descriptive sense, it is used as a noun to mean anything from secret information gathered about a purported enemy to the capacity of individuals to reason and solve problems. Intelligence agencies gather information they call "intelligence" about real or suspected enemies so that their government may be in a more powerful position to deal with the supposed enemy. When individuals use their minds to reason about problems and end up solving them, we think of that behavior as exhibiting intelligence.

The power to reason, especially to reason well, brings with it an increased capacity to effectively solve a wide variety of problems. It is that capacity that also bestows even greater knowledge and power on the individual who has it. Thus in common language, "information", "knowledge", "reason", and "problem-solving" are clearly tied to these senses of "intelligence". And so is the concept of power.

These senses of "intelligence" are even more fundamentally tied to the explicit use of language and the rules of logic, including the use of number.[1] They are tied to a notion of intelligence that is exhibited specifically in human language behavior, including reading, writing, and speaking. Though some research in lower animal use of language and logic has expanded the notion of intelligence to other animals in addition to us, the concept is still largely confined to human beings.

This sense of intelligence is sometimes referred to as *natural* or *real* intelligence to distinguish it from artificial intelligence. The term "real" is not adequate to make the distinction, however, because insofar as they exist, both kinds are real. I use "natural intelligence" and "intelligence" interchangeably as that kind naturally occurring in humans and animals. These are in contrast to "artificial intelligence" which is deliberately designed by humans for specific purposes.

In spite of a wide range of intelligence research programs covering diverging theories, the accepted usage of the term "intelligence" is still

[1] Though language use and number use are not identical, their expressions fall under the concept "alphanumeric".

largely defined in terms of scores on Intelligence Quotient (IQ) tests. The problem space of natural intelligence has been carved such that a single score, labeled "*g*" (for general intelligence), is taken to define "intelligence".

Moreover, the descriptive use of the term leads even ordinary laypersons to conceive of intelligence as a *single* thing found in the human being, specifically in certain parts of the cerebral cortex of the brain. It enables a person to solve problems or achieve many things. That is especially so if the person has a lot of this thing called "intelligence." Both the descriptive sense and advocates of the "*g*" concept lend support to the view that intelligence is this single thing, a mental faculty observed in a person's use of language and logic, a quantifiable superior use of specific portions of the brain.

In this sense, the ordinary layperson has much in common with early American psychologists who chose to ignore findings–and warnings—of European scientists who had carefully studied human intelligence over many decades. Alfred Binet (1857–1911) and later Jean Piaget (1896–1980) knew that intelligence was not and could not be a single reified, scaleable thing found "in the head." The phenomenon of natural intelligence, of which human intelligence is one kind, is far more complex than that.

Unlike his American counterparts, for example, Binet knew that testable intelligence, even using his own instruments, is a very small part of a much larger natural phenomenon. And Piaget later warned that the word "intelligence" "is nothing but a collective term, used to designate a considerable number of processes and mechanisms whose significance becomes clear only if they are analyzed singly and in the order in which they develop"(Piaget 1971, p. 40). They both understood that human intelligence should be conceived as functionally integrated throughout everything a person intentionally[2] does, not confined to the use of language and related symbolic tools, or to certain parts of the brain.

Anticipating a view confirmed by research almost a century later (though still debated), Binet devised a program to assist special education children using a set of physical exercises "designed to improve, by transfer to mental functioning, the will, attention, and discipline" necessary for studying academic subjects. At one point, he warned against those who claim that intelligence is a fixed quantity, a quantity that cannot be

[2] The concept of intentionality has sometimes recently been defined to mean that one must know what one is doing, where "know" is essentially limited to verbal knowledge. I am not using intentionality in that limited sense. It is expanded here to include any deliberativeness in behavior, including mental and physical deliberate acts.

increased (Binet 1909). He demanded "We must protest and react against this brutal pessimism; we must try to demonstrate that it is founded upon nothing."

But many American scientists largely ignored Binet and Piaget. In spite of Binet's admonitions, they took his scores as measures of what they claimed was this single entity called "intelligence" that they also held was inherited and fixed for life. More egregiously, they ignored his warnings that his tests measured a very small slice of human intelligence. They proceeded to commit the twin fallacies of reification and hereditarianism, as well as a fundamental category mistake, based upon the false assumption that intelligence is a single thing, a fixed, inherited, scaleable quantity that cannot be increased (or at best can be increased only minimally), found in specific portions of the brain.

Moreover, placing enormous unqualified faith in test instruments and methods, many of these American psychologists held that these scores "marked people and groups for an inevitable station in life" (Gould 1981, p. 157). In spite of often overwhelming differences in the quality of life of individuals and groups, they held that those differences were the direct result of inherited and *unalterable* differences in intelligence. In a word, inherited intelligence was conceived as the great hierarchical natural social class "divider" among individuals and groups.

Much has already been written about these early American psychologists, such as H.H. Goddard, Lewis M. Terman, and R.M. Yerkes, and the disastrous effects of social policies based upon their views of intelligence and very naïve understanding of genetics. Though their view of intelligence as a single general capacity for conceptualization and problem solving was later challenged by Thurstone and Guilford, who argued for a number of factors or components of intelligence, it is the former view that has been largely adopted in the United States.

This book is not about this earlier unfortunate period in the history of intelligence research and misguided social policies based upon it. However, with the publication of *The Bell Curve* (Herrnstein and Murray 1994), some of the same fallacies and issues involving reification and especially hereditarianism were repeated—and continue to be debated today. These issues have become increasingly important over the years since that publication due to the prevalent acceptance of those authors' view of intelligence and the ready acceptance of public policies based upon it.

It is a genetic determinist view echoing earlier American psychologists Goddard, Terman, and Yerkes that intelligence is that single thing, now called "*g*", a fixed, inherited, scaleable quantity that cannot be increased. The acceptance of this view has also unfortunately been bolstered in part by the growth of interest in and misinformation surrounding the science of genetics as well as the neurosciences.

Though some of the same *known* faulty assumptions and fallacious arguments for reification and hereditarianism persist—and indeed, seem to proliferate greatly under more politically conservative administrations, it is the genetic determinist view of intelligence that appears to prevail.

In efforts to correct that trend, my objective is to turn to more *fundamental* theoretical, philosophical, empirical, and methodological research issues about intelligence in general and natural intelligence in particular. Along the way, I will revisit many of those basic arguments, concepts, and assumptions held by those who view human intelligence along the same lines as our earlier hereditarian American psychologists.

1.1 Some of the Basic Issues

Recurring fundamental issues about the nature of intelligence found in human beings seem to revolve around three questions:

1. Is intelligence a single entity or is it many?
2. Is intelligence inherited or acquired?
3. To what extent does the cultural context or environment influence the development of intelligence, if at all?

Though many prevailing theories take positions on these issues, there are still other significant questions that are often not asked in the intelligence research community.

4. What are those intentional patterns, outside the scope of language behavior, by which humans exhibit or disclose their intelligence?
5. Indeed, to what extent does the domain of human intelligence overlap with animal intelligence?
6. Fundamentally, what is the range, depth, and scope of intelligence?

Of course, the first two questions may pose false choices. Moreover, it may be misleading or outright false to think of intelligence as a single or multiple "entity." This is fundamentally a category mistake since intelligence is a process or group of processes, not an entity. This mistake has in turn led some theorists to argue that intelligence is found in specific molecules and locations in the brain.

A more scientific approach is to look upon intelligence found in a person or animal as a group of capabilities or processes bound together to form a single intelligent being. We may speak of the intelligence of that being as though it is a single thing while actually referring to many capabilities or processes. It may also be the case that some facets of intelligence or those having a direct bearing upon it, are acquired or emerge over time while others are clearly genetically inherited.

The Single and Multiple Capacity Views

But our natural language use of such terms is not a very good guide. The issues that fundamentally divide many if not most intelligence scientists clearly fall into two diverging groups. There are those scientists who hold the view that intelligence is a single capacity that every human being possesses to a greater or lesser extent. However it is defined, these scientists also hold that standardized verbal instruments that consist of linguistic and logico-mathematical tests can measure this capacity.

On the other hand, there are other scientists who hold that intelligence is composed of a much wider set of competences. Moreover, they argue that many if not most of those are not measurable by standard verbal methods.

The former group argues for what is called the "*g*-theory" of intelligence, where the letter "*g*" stands for general intelligence. It is conceived as the single determinant of all things intelligent located in specific parts of the brain. The *g*-theory proponents have been largely successful in promoting a simple, direct causal chain view of intelligence that originates from a single cause and is measurable by standard IQ tests. The latter group of scientists, on the other hand, argues for the multiple intelligence or "MI-theory." These multiple intelligences include but extend beyond a person's verbal and mathematical abilities. They are found integrated throughout intentional human activity in the world.

However the above questions are parsed, other issues flow from them. If intelligence is conceived to be a single, inherited property located in certain parts of the brain, then it may be argued that explanations of intelligence are reducible to explanations of portions of the central nervous system. On this view, intelligence may be reducible to certain neuronal clusters found in the brain. This is a genetic determinist position that has been gaining substantial ground in some research communities over at least the last 15 years. With this view, it is held that questions about the influence of context or environment on intelligence can be largely ignored.

On the other hand, if intelligence is conceived to be many capabilities, some of which are found in one's interactions and transactions with objects in one's environment, then questions about the influence of the context or environment on intelligence must be asked. Moreover, the view of intelligence-as-many also raises the issue of whether or not measures of performance on standard IQ tests are viable and complete measures of these many capabilities making up intelligence.

Standard IQ tests are paper and pencil tests that measure linguistic *cum* logico-mathematical capabilities. Yet there are intentional, clearly intelligent human activities that are unrelated to the use of language and logic. Recognizing this fact leads to yet another issue implied but not stated above in question 4: Is intelligence found in *any* human activity unrelated to the use of language and logic? If the answer to that question is yes, then standard IQ tests do not measure it.

Where Are the Facts of Intelligence Found?

A more fundamental question underlying all the above particular questions is: "Where are the facts of intelligence to be found?" Are they found in measures of the central nervous system, in language and logic performance scores on standard IQ tests? Or are the facts of intelligence found distributed throughout one's *knowing how* to interact and transact with and transform objects in one's context or environment? Are the facts of intelligence found in one's genetic profile, neuronal firings, or scores on standard IQ tests? Or are they found in an individual *knowing how* to live his or her life?

If intelligence is conceived of and researched as a single thing occupying a specific controlling place (or distributed among certain places) in the brain, a question I will ask is the following: "How successful has that view been in explaining the *facts* of intelligence as found in actual human and animal experience?" If intelligent beings are endowed with many intelligences what is the evidence for the many? How are they related to one another? More to the point, we might ask, by virtue of what are they all bound together to form a coherent, intelligent single being?

Though Gardner's (1983) theory of multiple intelligences is the best known among several of the multiple theories, there are reasons to raise issues with his classification of kinds of intelligence. I will assess many arguments both for and against the single theory view. But I will also take issue with proponents of multiple intelligence theory as well. At minimum, the intelligence research community in general has left undone the substantial theoretical and experimental work necessary to show the scope of intelligence we have in common with the rest of the animal kingdom. Indeed, from my own point of view, it is unfortunate that much of the research currently underway is not even asking that question.

For clarification, my use of the phrase "natural intelligence" is not identical or equivalent with Gardner's use of the term "naturalist" in "naturalist intelligence" (Gardner 1998). He has identified at least eight kinds of intelligences, of which naturalist intelligence is one. My use of "natural intelligence" is far broader and includes all the kinds of intelligence Gardner identifies as well as much more.

It will also become clear that by "intelligence research community" I include other disciplines besides psychology. Professionals from other disciplines currently addressing the nature of intelligence generally include but are not limited to neuroscientists, biologists, geneticists, biochemists, computer scientists, mathematical physicists, and philosophers who have written extensively about the mind, brain, and theories of consciousness.

Unfortunately, though the study of intelligence, mind, and even the brain extends back as early as the Greeks, one finds few cross-bred studies that call upon serious efforts in these disciplines. And though the study of intelligence is well represented both historically and across the academic curriculum, it is currently very much claimed as the province of psychology, especially in the United States.

Regardless of the discipline in which one finds intelligence research, however, major problems follow from varying positions taken on the above issues. In chapters that follow I will address each of these in turn.

1.2 The Faulty Sciences of Intelligence

Among other problems, some intelligence research largely adheres to what many in other scientific fields view as faulty scientific method. If accurate, this may be due as much to the sheer complexity of the problem of intelligence as it is to any failure in scientific method. But the complexity of a research problem is not resolved by trying to define it out of existence.

Nor is it resolved by relying solely upon known classical data collection procedures and methods to the neglect of theory and concept formation that may expand searches to the poorly understood or to the unknown. The latter practice in particular has much in common with the man who lost his keys in the woods but persisted in looking for them under a lamppost since it was only there that he had any light.[3] Among other things, the futility of such an effort should be obvious.

[3] This story is attributed to E. Steiner whose work in systems theory and educational theory is well known. See *Methodology of Theory Building*, Educology Research Associates, Sydney, Australia, 1988.

1.2.1 The Anti-Theory Bias

Coming largely out of the American behavioral science tradition, much intelligence research is decidedly *anti*-theory and *anti*-concept formation. Some researchers who study intelligence tend to hold the position that theory and concepts somehow arise naturally out of data collection. This is an erroneous understanding of inductive processes inherited from earlier periods of Western science. In the early history of American science, for example, theorizing was often viewed as mere speculating. Research methodology was largely restricted to verification procedures. Data collection was thought to be not just central, but the *whole work* of research (Steiner 1988).

Though that view has changed to a large extent, especially in natural sciences such as physics, the behavioral and certain of the life sciences still lag behind. For example, commenting recently on a colleague's work on intelligence, noted neuroscientist Antonio R. Damasio states, "There is an enormous resistance against theory in neuroscience and biology" (Rothstein 2004). His own research on the brain and intelligence has met the same anti-theory resistance as have others in the field.

A Misleading Heritage of Inductivism

Historically, the preoccupation with data collection and its interpretation is part of the heritage of inductivism that includes not only a misunderstanding of induction and scientific inference generally, but also a misunderstanding of the nature of human learning. Among other things, though there are at least two distinct concepts of induction, only one was generally recognized. Moreover, also along misconceived inductivist lines, human learning was and is still taken by many to be a passive experience of associations. These are points made by many, including Medawar (1964, 1969), and ones I will more fully address in later sections.

An anti-theory and anti-concept formation bias based on the heritage of inductivism is well documented in certain of those sciences that study the nature of intelligence (Steiner 1988). But this bias has been around much longer in the history of Western science generally before the study of intelligence became a part of psychology. It has been around long enough for the behavioral sciences to have learned something of the problems created by such a view. For example, even in 1861, Darwin noted (Darwin and Seward 1903):

> "About thirty years ago there was much talk that geologists ought only to observe and not theorize; and I well remember someone saying that at this rate a

man might as well go into a gravel-pit and count the pebbles and describe the colors. How odd it is that anyone should not see that all observation must be for or against some view if it is to be of any service."

Indeed, it is common to find a similar anti-theory bias as well as a clear bias *for* data collection to the exclusion of theory in some prevailing intelligence research efforts. This is so even where, other than asserting correlation coefficients among data, the relation between data collected and cognitive functions is not made evident or even clearly identified. Good recent examples of this are reaction time studies of Jensen (Jensen 1998), but examples are evident in research on cognition generally.

Confusing Cause and Correlation

First, though the point is amply argued elsewhere (Gould 1981), the assumption that a correlation somehow implies a *causal* relation is not warranted. The empirical fact that there is a correlation among sets of data does not imply anything at all about a causal relation among the sets. Second, as Gould notes, "It is not even true that *intense* correlations are more likely to represent cause than weak ones" (Gould 1981, pp. 242–243). That is, the strength of a correlation does not tell us anything about the nature of the cause. Yet the assumption that it does underlies basic arguments by proponents of the *g*-theory.

The *inference* to a cause must come from somewhere other than the fact of a correlation, even a strong one. Likewise, no such inference may be warranted at all. We might show any number of positive and negative correlations among a variety of data sets and still fail to find a cause anywhere, much less be warranted to make an inference to one. This may be so because there is no necessary connection among the sets of data; at best only probabilistic claims may be made about correlations and none at all about cause.

These issues point to larger problems of causality, statistical probability, verification and validation. These are certainly crucial in experimental science. But defining the entire scientific enterprise in terms of data collection and correlations, and delineating the object of research solely or primarily in terms of those procedures is to thwart the goal of scientific method. The aim of scientific method is usually taken to be the generation of scientific knowledge, the generation of *theory* consisting of causal laws and law-like statements permitting prediction and control.

Even the most precise data collection and verification procedures have *underlying theory and concepts* on which they depend for their validity. Indeed they depend upon these even to begin to do anything. Those who collect data must know what to select and what not to select as data. Yet in intelligence research, underlying theory and concepts are often not only unquestioned, they are also apparently often unrecognized. Underlying implied concepts and theoretical structures are left unexamined for logical and other problems, leading sometimes to wholesale fallacious inferences based upon those same unexamined assumptions and concepts.

No matter how good the data, it is only as good as the assumptions and concepts it rests upon.

1.2.2 Invalid Reductionism

Reductionism in general is the view that scientific understanding of a phenomenon can be gotten from analysis of its parts. And it has been remarkably successful in the natural sciences such as physics. There are very large differences, however, between reductionist strategies in the physical sciences and those sciences directed to the study of living things.

A rather pervasive underlying fallacy often found in intelligence research is a kind of reductionism in which a researcher projects into structures or phenomena of a lower order the characteristics of structures of a higher order such as intelligence.

It is given that all extant scientific inquiry into the natural world, including intelligence, posits kinds of hierarchical structures. For example, human societies are composed of individuals; individuals of cells; and cells are composed of molecules, and molecules of atoms. A viable question to then ask is whether or not an explanation of components at one level of the hierarchy can be reduced to explanations of components at a lower level.

Strict reductionists in particular hold that descriptions of higher level processes and structures can only be explained in terms of descriptions of some lower level processes and structures. Descriptions of higher level phenomena such as intelligence must be "reduced" to descriptions and explanations of lower level processes such as neural activity.

One must also keep in mind that the term "reduction" here means establishing a clear deductive relationship among the sets of description. A *valid* reduction absolutely requires establishing such a deductive relationship. The descriptions of higher level intelligence must be validly deduced from descriptions of lower level activity.

Coming out of the discipline of physics, reductionists are generally determinists. As applied to an understanding of intelligence, however, this is basically the idea that if you want to explain some higher level behavior or a complex problem, such as our intelligence, the answer is found in understanding the building blocks that make up that behavior. Those building blocks are then taken to completely causally "determine" that higher level behavior such as intelligence. The idea is that if the higher level behavior is reducible to these lower level building blocks, then one has a causal explanation for the higher level behavior. One can then explain the higher level behavior by composing it from the lower level.

In the case of intelligence, some argue the causal determinants, the building blocks, are neurons. Others argue that the building blocks are genes. Since DNA is the more basic of the two, and reductionists would presumably want to ultimately reduce intelligence to our DNA sequences, we should look at that argument first.

The Faulty Genetic Argument

The question then becomes: Can the intelligence of individuals be reduced to the behavior of neural cells of their brain? Can an individual's intelligence be reduced to their DNA? A further question might then be: Can the biology of the brain, including intelligence, be reduced to physics?

Though some argue for just such reductions, their arguments are often filled with logical, conceptual, methodological and empirical error. Such wholesale material reductionist arguments found in intelligence research are fraught with logical inconsistencies, methodological problems, and unwarranted claims not backed by the data.

In this reductionist genetic deterministic point of view, proteins unleashed by genes are claimed to cause or control (determine) behavior. There is an assumption that somehow genes act on their own, that they autonomously turn on and off the synthesis of particular proteins that eventually "cause" our intelligence.

But genes do not actually work this way. As Sapolksy explains (Sapolsky 2000), more than 95% of DNA is "non-coding." That means that it does not act on its own. The regulation of on- and off-switches of our DNA actually comes in some instances from chemical messages from other (non-DNA) parts of the cell; others come from other cells in the body; in still other cases, genes are turned off or on by environmental factors. There is in fact an interaction between genes and the environment.

Empirically, genetic scientists know that there is no direct causal (deductive) chain from our genes to the features that appear at the level of the entire organism. They know that the way we are nurtured by our families, societies, and culture at large either reinforces or retards what we are genetically naturally given. It's not "all in the genes," as Sapolsky explains, and the study of genetics will never "gobble up every subject from medicine to sociology."

Additionally, neurons act chemically and electrically to perform and permit certain brain functions. In the scientific theory of neural activity, there are concepts and theory appropriate to describe neural activity at that level. But among those concepts and theory describing the activity of neurons one will not find concepts and theory describing individuals solving a calculus problem, understanding a poem, or falling in love. It is not neurons that solve math problems, understand poems, or fall in love; it is individuals. Individuals are not DNA made flesh.

Moreover, from the empirical fact that every time you solve an equation an imaging technique can show certain neural activity in your brain, it does not follow that the neural activity *is* the solving of the equation. No matter how many established correlations there are between neural activity in your brain and your intelligent behavior, you are the one who solved the equation. Indeed, you are the one who chose the equation in the first place. It was not the mechanics of neural clusters that did it for you.

As Cohen explains (Cohen 1994) human DNA space is not a map of human space because, among other things, there is no unique correspondence between the two spaces. There is no way to assign to each sequence in DNA space a unique animal that it "codes for." Cohen further explains, "Biological development is a complicated set of transactions between the DNA 'program' and its host organism, neither alone can construct a creature and neither alone holds all the secrets, not even implicitly."

I would say that biological development is a complicated set of transactions between the DNA program, the host organism, as well as the environment in which the organism finds itself. The point is there is an individual in a context in addition to the neural networks in the individual's brain.

A Neo-Darwinist Influence

In spite of a rather widely held view that there is "a mapping from genes to character, from a genome to a phenome," popularized by some neo-Darwinists, this simple reductionist view of our biology is wrong for a whole host of reasons, some of which I will discuss in later sections. For now, it is sufficient to recognize that it is another instance of the fallacy of projecting into structures or phenomena of a lower order the characteristics of structures of a higher order. Even more specifically, those genetic reductionist theories that argue that a human's genes completely specify the living person are factually wrong.

It should be mentioned that a variation on this kind of reductionist fallacy is also found in behaviorist stimulus-response theories of cognition that still manage to find their way in some popular theories of learning. That is, one finds the method of reducing higher level cognitive processes to lower-level conditioned associations among stimulus events. Indeed, some prevailing S-R theories actually deny that there is such a thing as intelligence, claiming instead that we are all bundles of environmentally conditioned responses, on all levels, to stimuli.

Though perhaps not widely accepted, this point of view on intelligence is still found in some academic research circles. Historically, it is nowhere better represented than in the classic works of B. F. Skinner, but it is also found in the works of the philosopher W. V. O. Quine and some of his present-day eliminative materialist followers.

In their defense, reductionist efforts in general are built in part on the understandable supposition that one should not use complex interpretations of data if simpler interpretations are possible. But the fact is that almost without exception such simple "reduced" interpretations of intelligence *do not even fit the data* of intelligence research. This is especially the case in investigations of higher level human cognition, as in logico-mathematical reasoning. It is also the case in investigations of animal cognitive processes and is found across the board in cognition research generally (Sebeok and Rosenthal 1981).

Among other things, what is missing in reductionist theories of intelligence is any coherent account or explanation of how intelligence develops from interactions of all those components at various levels of the building blocks. A central issue is "How does intelligence develop from the level of genes or neurons, and from the interactions, transactions, and reactions to the environment in which the body and brain housing those genes and neurons finds itself?"

More to the point, however, just how far down in the hierarchy should a reduction go? If a reductionist explanation is taken to the level of neurons and DNA, why not to the level of atoms or even subatomic particles? The actions of our brain mass depend on cell biology and chemistry; chemistry depends upon quantum mechanics; quantum mechanics depend upon the laws of physics. The combinatorial effects alone amount to at least an immense number[4] of required calculations, making the reductionist task quite literally impossible. It is, to quote Stewart (1995), a "reductionist nightmare."

[4] An immense number, $\Im = 10^{110}$. In contrast to a finite number of items that can be put on a list and examined, for an immense number of items (though countable) this is not possible. There would not be sufficient memory capacity in any computer that could ever be built to store an immense number of items. See Walter M. Elsasser, *Atoms and Organism; A New Approach to Theoretical Biology*, Princeton University Press, 1966.

Even at higher levels and given the absence of context in many intelligence research studies, one does not find coherent explanations or accounts of relations between sensorimotor functions and the development of thought itself. That is, one does not find explanatory accounts of relations between perception and intelligence or like accounts of structural differences and relations between higher level thought processes, such as abstract conceptual structures, and intentional sensorimotor behavior. Such explanations are daunting and require considerably more than simple reductions that do not work and may be impossible in any case.

1.2.3 Neglect of Emerging Intelligence

The hierarchical building blocks reductionist view of intelligence obviously has many problems. Among those problems is the fact that there is enormous empirical evidence showing that natural intelligence generally is a kind of emergent phenomenon.[5] If nothing else, it is one of nature's enormously complex set of patterns of certain kinds revealed and disclosed sometimes in very complicated, subtle and interesting ways in action and thought by members of the animal kingdom. Evidence shows that it is a phenomenon that emerges from an "ocean of complexity"[6] and is not a *direct* consequence of simplicities of natural laws at lower levels.

Intelligence emerges at higher structural levels made possible by the interactive activity of sometimes immense numbers of elements across multiple domains and large numbers of structures, and relations that interact and combine in highly complex ways at lower levels, progressively leading to higher ones, still interacting and combining all along the way in even more highly complex ways. Emergent phenomena cannot be understood by breaking them up into constituent parts, at lower levels of development, and analyzing those parts independently in isolation.

That is because the properties of emergent phenomena of interest, in this case intelligence, are properties of the *interactions between the parts*, rather than being properties of the parts themselves. The model for emerging intelligence must be organizational, not isolated parts or "building blocks." The interaction-based properties are not there *in the parts* when they are studied independently. The DNA of intelligent beings such as ourselves and our nearest relatives in the animal kingdom, chimpanzees, cannot be studied independently to find our intelligence.

[5] One only has to consult numerous publications on this topic such as Piaget's earlier work, especially his 1950. Also see Alwyn Scott 1994.

[6] The phrase is Ian Stewart's. See his 1995.

Again, our DNA is not us. Our genes are not us (Cohen and Stewart 1994) nor are they our intelligence.

Methodologically, the empirical facts supporting the view that intelligence is emergent also indicate the need for a different mathematical and scientific approach to study it. This is a matter I will discuss in much greater detail in a later chapter. Mathematically, emergent phenomena are *nonlinear*. That means that they are phenomena that do not obey the mathematical superposition principle that requires a general solution as a function of a certain finite number of particular solutions (Saaty and Bram 1964). That is, where φ is an operator, it is said to be linear if the effect of operating on the sum of two entities, for example functions, is equal to the sum of the effects of operating on them separately: $\varphi(f + g) = \varphi(f) + \varphi(g)$. Where this equivalence does not hold, the operator is nonlinear.

Inadequacies of the Classical Linear Approach

Yet that is precisely the way intelligence is treated by most researchers today. The *g*-theorists in particular treat what they call general intelligence as a function of a finite number of particular solutions. General intelligence, they claim, is an overall score that is a function of a battery of psychometric tests. They hold that all that needs to be understood about intelligence can be gotten by understanding the quantifiable parts in isolation, such as verbal and math scores, then composing or adding together our understanding of the parts. Any future research, such as implications for employment demands, would extend the inquiry by simply adding more to the parts in isolation (Jensen 1998).

Though the classical linear and reductionist approach in the sciences of intelligence is easier, and certainly more expedient and cost-effective, at best it reveals a small portion of the reality it portends to study. With this approach, complex cause can be expressed as a convenient sum of simple components. Then the combined effect becomes the sum of the effects from each component of the total cause.

Yet this approach leaves us a severely truncated, distorted and blind view of the actual scope and contents of intelligence. The empirical facts of emergent intelligence phenomena in a broader domain of actual experience, are not generally recognized by reductionists.

However, emergent phenomena are certainly not outside the low-level laws of nature. They interact with, cooperate with, build on, or follow from them in complicated ways we have yet to understand. That understanding will require the development of new conceptual, theoretical, and methodological tools beyond current linear mean-field theory, data collection and simple verification procedures.

Researching intelligence as an emergent phenomenon means that it can only be understood by examining its many hierarchical layers and their interactions, instead of reducing it to one or a few building blocks. It means trying to understand the fundamental mechanisms of self-organizing complexity found in dynamic living beings that give rise to emergent phenomena. Focusing upon the nature of self-organization in intelligence will alone require a fundamental redirection to our traditional ways of viewing intelligence as issuing forth from a centralized genetic linguistic source or as imposed by external "civilizing" forces and authority.

It also means that the problem space of intelligence must be re-carved permitting a broader scope of inquiry allowing nonlinear analyses as well as syntheses. The scientific community must, among other things, construct intelligence from the relevant physics *and* biochemistry *and* electrophysiology *and* neuronal assemblies *and* cultural configurations *and* mental states[7] *and* analysis and synthesis of interactions among all these. This is a much larger and more daunting task than simple reductionist efforts, involving all relevant levels and their interactions involved in a complex emergent phenomenon.

To briefly summarize, though reductionism in intelligence research is expedient, it has purchased that expediency by carving much of the actual space of intelligence out of the picture. Generally, attempted reductions fail by defining the scope of the reduced phenomena too narrowly or by otherwise making inflated claims that cannot be logically supported or even backed up empirically by the data. Moreover, these arguments are often unable to provide adequate, fallacy-free scientific explanations of even the simplest intelligent acts.

1.2.4 Neglect of Theory Construction and Concept Formation

Theory construction is now often recognized as necessary even in the behavioral sciences. However, little consideration has been given concerning how one goes about it. How does one go about constructing theory or forming a concept? How does one even begin? More importantly, how does one recognize existing *implied* theory that may be hidden beneath all that data collection? Accepted ways of doing things sometimes blind us to questions we should be asking about the structures lying just beneath the surface.

The reticence to address these tasks among researchers in diverse fields studying intelligence is still pervasive. Biology, psychology, the neuro-

[7] This point was made by Alwyn Scott in his 1994, p. 160, though he was referring to consciousness.

sciences, education, computer science, artificial intelligence, and philosophy, among others, all contribute explicit or implied concepts, theory, and method to rational inquiry into intelligence. Given the multidisciplinary and interdisciplinary nature of such inquiry and research, there is a need to address those fundamental structures and methods having a direct bearing upon our understanding of the nature of intelligence.

More pointedly, there is a need to address alternative theory and concepts more appropriate to the *facts* of intelligence. In later sections, I will show that the task of theory construction requires that one first peel away layers of data collection efforts and their underlying assumptions and interpretations to gain access to underlying *implied* theory and concepts. One must first be able to recognize theory especially where it may be partially hidden among other structures and methods of science. Only then can those underlying structures be critically assessed for their adequacy. If they are shown to be inadequate, then existing structures may be emended or extended.

Mechanism and Organicism

Many intelligence research efforts, for example, adhere to an underlying mechanical (machine) theory or model. They adhere to this model even while they do not explicitly acknowledge doing so. Because it remains largely hidden beneath data collection efforts, it is not critically examined or questioned to determine its adequacy. For that very reason, it is part of an underlying structure that entails fallacies, such as begging the very question at hand.

In a mechanistic or machine model of anything, for example, the emphasis is on *non-modifiable* parts that are the determining factors in what that thing does. A machine is an object that consists of parts that act in predetermined ways. The content and form of the parts, the way the parts are combined, determine the function of the machine.[8] It is a powerful model for deriving one-way causal chains in efforts to explain the behavior of a thing, or at least some of it.

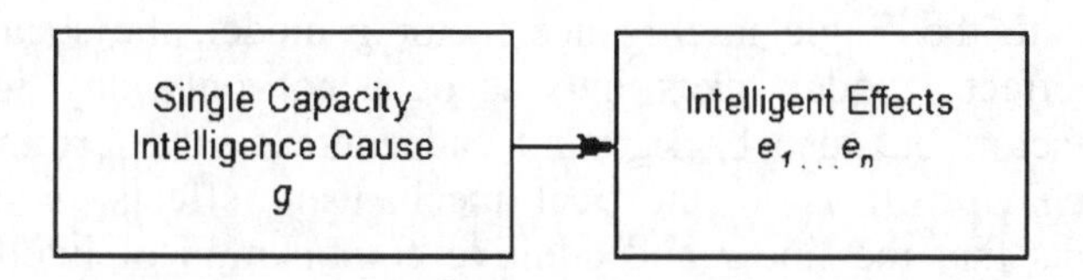

Fig. 1.1. Single Capacity Mechanistic Effects Model

[8] Generally, I use the term "machine" as identical with the concept "algorithm".

The usual intelligence researcher takes an experimentalist's standpoint in which intelligence is reduced to a single capacity or factor (or to a very small number of such factors) such as *g*, and the effects are taken as linear and additive. This is represented in the graph above. It is a non-statistical mechanistic effects model in which all context or background variables are controlled so as not to affect theory or measurement of intelligence.

On the above model, it is the single intelligence factor *g* that is taken to determine all intelligent behavior. It is this model or point of view that posits a central all-purpose "controller" or genetic central processor in charge of all things intelligent. This all-purpose controller is viewed as the single general capacity, the intelligent cause that determines all intelligent effects. It is this mechanistic point of view, for example, that focuses upon tests of how quickly the machine can perform, as in reaction time measures on intelligence tests, to determine intelligence levels. These are actually measures of just how quickly that central processor, the intelligence factor *g*, works.

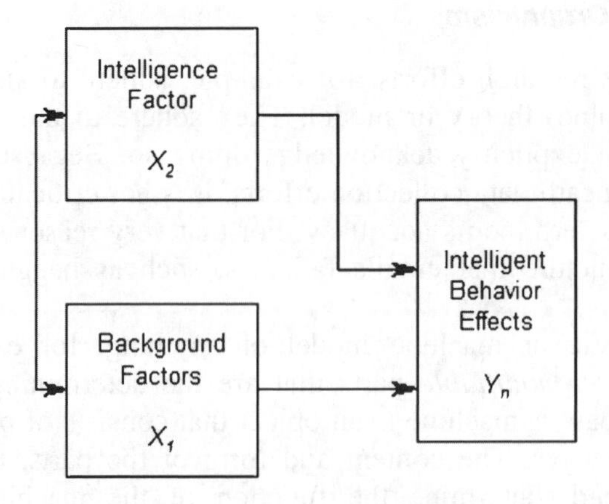

Fig. 1.2. Sociological Mechanistic Effects Model

In contrast to the single intelligence factor *g* model, the sociological mechanistic effects model takes into account not only any inherited intelligence factor, but also background and context as determinants of intelligent behavior. It is a statistical mechanistic effects model that basically posits that the linear and additive combination of intelligence, background and context factors determine all intelligent behavior.

Yet the use of such models in intelligence research assumes *as given* on one level *what one seeks to prove* on another. This is the fallacy of begging the question. In intelligence research based upon either of the above non-statistical or statistical mechanical models, it is intelligence itself that is assumed to be a non-modifiable machine and the determining factor for much of the rest of human behavior.

This uncritical assumption in turn is based in part on prior unquestioned assumptions that the components of intelligence, the parts, are discrete, atomistic, linear and additive. In sum, these models view intelligence as a linear, nondynamic system. Intelligence is viewed as given from the start, not emerging from interactions of components of the organism. It is viewed as a central processor acting upon the organism to produce kinds of observed intelligent behavior.

The mechanistic view is in stark contrast to the organismic or self-organizing view. In some sciences when researchers collect data and construct theories about living things, many have recognized the need to use appropriate concepts and underlying models to direct their inquiry. These are models in which states of affairs are represented like an organism, a dynamic self-organizing living thing.

In contrast to the machine model, the organismic point of view stresses the organization of the parts of the organism and the interactions among them, especially in contexts. It does not focus upon the properties of the parts in isolation but their organization and information input, intake, action upon, output, and transfer in environments. The dynamics of the entire organism in an environment are viewed as nonlinear, self-organizing complexity.

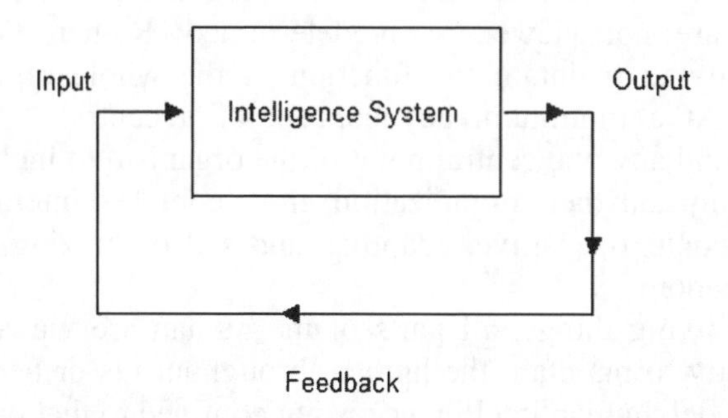

Fig. 1.3. Simple Feedback Organismic System Model

Early attempts to model an organism's interactions with an environment displayed the usual simple feedback model as in the above graph. Biologists focused upon negative feedback because that was identified as the mechanism used by living things to stabilize physiological processes, preventing extreme oscillations in those processes.

For example, regulation of body temperature in warm-blooded animals is by receptors in the hypothalamus. Blood temperatures are maintained within a certain range not only for comfort but the health and life of the animal. If external temperatures rise above or drop below certain thresholds, physiological responses by the animal will act to stabilize body temperature by bringing it back into a certain range. The animal acquires temperature information that elicits a negative response to counteract the information, seeking to reestablish stability. As helpful as negative feedback is, however, it is a limitation on the use of information by an organism.

In contrast to negative feedback, positive feedback reinforces change in a system in the same direction as initial information indicates. Unless checked by negative feedback, positive feedback can result in a snowballing effect resulting in wild oscillations and eventual destruction of a system. Viewing the world's ecosystem as a whole, unchecked rising global temperatures or population growth may be examples of this.

Nonetheless, taken together these two measures of information are still limitations of the feedback model to adequately characterize intelligence. In a later chapter I will present a more fully developed self-organizing complexity theory model demonstrating information-extensions beyond the simple feedback model. The information extensions are necessary for an adequate theory models approach to the science of natural intelligence.

Unlike the mechanistic models above, in an organism the parts are not assumed to have non-alterable natures and fixed actions. The functions of an organism are not viewed as predetermined. Rather, the parts act interdependently to maintain the function of the whole organism in an environment. At a minimum, any concept of a central "controller" is expanded beyond any one central point of the organism to include all parts of the organism and their organization, their complex interactions. It is conceived as context-sensitive, adaptive, and self-organizing, a nonlinear living phenomenon.

Like other living things, all parts of the human are viewed as acting interdependently to maintain the human throughout his or her life. On an organismic model, human intelligence is not confined to that part of human behavior when he or she is engaging in verbal behavior but expanded to include human intentional behavior as humans act, react, interact, transact, transform and are transformed by what they experience in the environment and in themselves.

Narrowing the Intelligence Domain to Suit Tools at Hand

In some ways, some researchers have excluded the broader domain of intentional behavior so as to render the object of their research, intelligence, amenable to the tools they have at hand. These are tools to measure those parts they can submit to IQ tests. That is, these researchers have attempted to carve the universe of intelligence to suit their available data collection and verification procedures instead of carving it based on available facts. They have tried to define much of the problem space of intelligence out of existence.

Narrowing the object of research to available test and measurement tools has the added benefit that one can also derive one-way causal chains of behavior from these procedures, thus affirming a chosen self-fulfilling research strategy. Such a focus excludes what does not fit in chosen underlying, largely unexamined mechanistic models and supporting concepts and methods.

More to the point, without addressing fundamental theory, concepts, and assumptions such as these, it is abundantly clear that wholesale fallacies and other logical and methodological problems result.

And conclusions based upon fallacies, no matter how much they are abetted with collected data, cannot be valid.

1.2.5 Unexamined Assumptions, Concepts, and Fallacies

Some intelligence theories are built upon unquestioned if not unrecognized assumptions, underlying theory, models and concepts such as those above. In some cases, where assumptions have been challenged, the answering arguments have been superficial and fallacious as well.

Almost without exception, for example, proponents of *g*-theory have largely based their research on the following implicit, largely unquestioned or unproven claims:

1. That intelligence is only *human* intelligence; all other kinds of apparent animal intelligence can be explained as instinctual or "hard-wired";
2. That intelligence is a *single* all-purpose general capacity, a largely inherited mental ability confined to certain parts of the brain;
3. That intelligence is embedded solely in the use of language and logic; it is not found in anything the human does that is *unrelated* to the use of language and logic;
4. That general intelligence is operationally defined in terms of performances on standard IQ tests;

The scope of the domain of intelligence is taken to be verbal, substantially if not entirely omitting visual and sensorimotor intentional behavior unless they are defined in terms of language performances as on IQ tests. More generally, these researchers have narrowed the scope of intelligence to the ability to verbally generalize. This point of view does not take into account enormous evidence that intelligence generally and reason in particular may not need language at all.[9]

Ironically, even the actual use of language itself, the actual *doing* of language speaking, reading, and writing, goes far beyond the scope of language as measured by standard IQ tests. Such tests focus almost exclusively upon declarative and description functions of language, leaving out the vast majority of forms and functions actually found in human language behavior and experience.

For example, human beings use language indexically to either literally or figuratively point to objects in their experience. Examples of indexical use include using words such as "this" and "that" when speaking of something the person can point to or of some idea or abstraction in their minds. Moreover, language speakers use language demonstratively in their own behavior to disclose their own knowing. This is often done quite indirectly, for example, with the use of physical gestures and figurative language such as metaphors. The role of gesture and metaphor in higher cognitive efforts such as theorizing itself, or the quite figurative and imagist notion of elegance, as in mathematical proofs, are nowhere captured by intelligence tests.

The Scope of Cognition

More fundamentally, the intelligence research community has not clarified and defined the scope of *cognition* that is usually taken to formally demarcate the scope of intelligence. Though most ordinary dictionaries define "cognition" in terms of *knowing*, the history of its treatment in the behavioral sciences has been to eliminate references to the word "knowing" altogether while substituting limited kinds of language behavior. The cognitive has been largely taken to be isomorphic with the linguistic.

However, there is no widespread agreement that cognition excludes behavior that is unrelated to language and logic, or that it either excludes or includes such things as artistic activity, exploratory behavior in the environment, or in the realm of ideas, socializing, or even some of the more creative endeavors of the mind.

---

[9] One has only to consult reams of evidence provided by the autistic research community, among others. I recommend Grandin 1994.

Moreover, perhaps due to professional disagreement as well as insufficient data, the research community has not always clearly defined behavior that is inherited, either as reflex or instinctive action, and distinguished these from behavior that is acquired. Nor has it clearly defined the structures of perception and their relation to instinctual, reflex, and acquired behavior. Given these definitional gaps alone, the scope and depth of intelligence is left poorly understood.

Additionally, questions related to consciousness and awareness[10] generally are missing almost entirely from intelligence research. Related concepts such as "thinking" and "mental act" are also left ambiguous or are arbitrarily reduced to language representations in the mind. In part, these concepts are related to issues in the entire spectrum of perception research, pattern recognition studies, and research into kinds of sensorimotor awareness related to intentional behavior. But they all rest on poorly defined boundaries that are fundamentally necessary to any serious science of intelligence.

However, given some broadly agreed upon parameters, numerous studies of higher primates alone show that human intelligence is only *one kind* of natural intelligence (Goodall 1990; Allen et al. 1997). At a bare minimum, the intelligent use of tools found in the animal kingdom, especially in chimpanzee behavior, is evidence of enormous overlap with our own intelligence. Human intelligence does not exhaust the category of *all* intelligence.

Though human intelligence has been studied and researched possibly more than any other kind, there is a wide consensus of opinion in the scientific as well as lay community that we do not have a clear understanding of and wide agreement on the content or scope of what has been studied. Aside from *g*-theorists, most researchers agree that we do not have an adequate definition of intelligence generally, including human intelligence.

In spite of arguments by advocates of certain single theories of (human) intelligence, profound arguments and evidence show that our current conceptions of intelligence in general and human intelligence in particular are entirely too narrow. The past decade alone has shown that these

[10] The word "awareness" is unfortunately often thought related to some mystical or mysterious ability. I am certainly not using that term with any such meaning. Indeed, my notion of "immediate awareness" as developed in previous publications and included here is supported by empirical scientific evidence demonstrated in clinical trials. The reader should note that significant questions related to levels and degrees of awareness have become public with unfortunate cases of persons in what is termed "persistent vegetative state." With these cases our unfortunate lack of understanding of awareness and consciousness became apparent.

advocates have been forced to repeatedly revise, sometimes in *ad hoc* manner, some of the most cherished assumptions and principles of their theories.

For example, as already noted, proponents of *g*-theory assert that intelligence is a single capacity that can be measured and operationally defined with standard IQ tests. The basic argument is that since we know that many different measures of mental ability are correlated positively with each other, hence, based upon a variety of multivariate tests, it follows that there is an underlying general factor, *g*, called general intelligence (Murray 1995).

It is worth mentioning, however, that what such standard tests actually give is a *value*, called the "Intelligence Quotient," or "IQ". As one example, the Stanford-Binet test provides a procedure for observing where a person ranks on certain tests relative to others, where other factors are ruled out. Since performance on IQ tests is given a value, intelligence is sometimes called a "variable." But strictly speaking, intelligence is not a variable. Variables are symbols, say *x*, taking on one of a set of values ranging from low to high, say from 60 to 150.

In this as well as many other instances, otherwise careful researchers sometimes take a symbol *of* something for *the thing* symbolized. They commit what is called the fallacy of reification, which leads them to commit still other, sometimes subtle, fallacies as well. In this case, researchers have claimed that the symbol, *x*, the variable that is actually a label for an average of many performances on tests, *is* intelligence. Perhaps even more egregiously, the flip side of this fallacy is to assume that because a researcher has put a name or label on one of a set of values, that they have solved a concrete problem.

More specifically, "intelligence" is operationally defined by many within the scope of language and logic as scores (variables) on standard IQ tests. From such scores, it is then held to be a concrete, specific mental activity; it takes place "in the mind," "in the brain." It is not found in the rest of the body or in anything the person does with their bodies unrelated to the use of language and logic.

This is an unquestioned and largely unchallenged assumption by some intelligence researchers: intelligence *is* the use of language, including number. This assumption has been followed by a proliferation of studies on linguistics, especially recursive features of human language use, genetic origins of language, and the dismissal of lower animal intelligence as merely instinctual or a matter of imitation or cueing.

In spite of reams of multivariate test results and multiple regression analyses, however, those theories have barely, if at all, touched an understanding of *natural* intelligence. Minimally, this is so because such tests are aimed at symbolic artifacts of a narrow slice of human intentional behavior, not the broader domain of intentional behavior itself.

It is also because lower animals have largely been left out, along with the entire scope of intentional behavior humans have in common with animals that is largely unrelated to language. Though there is evidence that certain higher primates have some language ability, even assuming the narrow definition of "intelligence," they would simply not qualify even to take the tests. On some theories, in the entire natural universe, this rarefied thing called "intelligence" is held to be found only in human beings.

1.2.6 A Bankrupt Theory of Knowing in the Sciences of Intelligence

Yet even for human beings, there was a time in our evolutionary development when language use, in the sense of alphanumeric symbolic expressions, was non-existent. Millions of years ago, human beings had yet to evolve to the point that we were language users. One of our earlier ancestors, *Australopithecus*, did not have language ability. It is not even certain that our immediate ancestor, *Homo erectus* had language ability, in spite of evidence in skull remains of Broca's area of their brains. On some theories of intelligence, it would have to follow that we (or our ancestors) were not then intelligent beings, though there is no escaping the fact that even in a more primitive state, our ancestors were nonetheless excellent problem solvers of a certain kind.

More problematically for those theories, from the point of view of adequate scientific explanation of intelligence, it appears from much accumulated evidence that Neanderthals had at least rudimentary language ability. Yet they became extinct approximately 30,000 years ago. Even though our ancestor *Australopithecus* and perhaps *Homo erectus* did not have language ability, in order to survive, they were nonetheless able to *reason*, in some more primitive sense, even without language and without knowing the rules of logic or how to use numbers.

That is, without knowing in the only sense of "knowing" that some intelligence theorists would accept. Another unquestioned and implicit assumption by some verbal theorists of intelligence, for example the *g*-theorists (among others), is that *to know a rule is to know how to speak or write with it and to know how, where, when, and in what right proportion, to apply it.*

Kinds of Knowing and the Intellectualist Legend

But there are at least two senses of "know" here: there is *knowledge* of a rule and there is *knowing how*, where, when, and in what right proportion to apply it. Like other theories of intelligence, the *g*-theorist view of human knowing contains false assumptions about the relation between *knowledge* of rules and *knowing how* to apply or use them. It assumes what Ryle (1949) referred to as the "intellectualist legend" to explain the move from knowing a rule to knowing how to apply it. It is one thing to know a rule, expressed in a statement or sentence; it is quite another to know how to apply or use a rule in any rational action, practice or performance.

According to the legend, originally derived from Descartes, actual intentional doings with the body are not part of intelligence. Only the rules stated in language describing or prescribing the doings are part of intelligence. Intelligence involves only mental or language-dependent *knowledge* with the use of alphanumeric natural or artificial languages. According to the legend, *knowing that* rule or prescription is both necessary and sufficient to *know how* to apply it.

Moreover, among other things, the *g*-theorist view of knowing omits entirely any consideration of *knowing how* to do something or *knowing what to do* in the absence of any explicit language rule. For the *g*-theorist, strictly speaking, there is no such kind of knowing. Though allowing for learning from experience, for g-theorists there must always be an explicit language-based rule, a kind of mental script that one can read and understand that drives the train of all the person intelligently does.

However, these assumptions were shown to be false decades ago, a finding confirmed many times by later philosophers and scientists who studied the problem (Maccia 1987, 1989; Scheffler 1965). Ryle proved once and for all that there are at least two kinds of intelligence; there are two kinds of knowing. There is *knowledge that* (or "knowledge") and *knowing how*. They are not reducible to one another. Indeed, he showed that "*knowing how" names a different kind of intelligence* altogether from the traditionally recognized *knowledge that*. But many intelligence theorists, particularly *g*-theory proponents, appear to be unaware of the distinction and its profound implications for their own theories.

The theory of knowledge, or epistemology, assumed by *g*-theorists as well as many other intelligence researchers is bankrupt from the start. It is bankrupt because it fundamentally assumes what is false: that there is only one kind of knowing, *knowledge that,* expressed verbally. And it further assumes that *knowing how* (intentional doing) is reducible to *knowledge*. They assume a view of knowledge that can be subsumed safely within the scope of certain forms and functions of language and logic, including number.

Unlike the intelligence of *knowledge that* which can be found evidenced in artifacts of written and spoken language, the intelligence of *knowing how* is found in actual intentional doings that are largely unrelated to language. *Knowing how* traverses the entire spectrum of knowing, the entire spectrum of intelligence. It is the most fundamental kind of intelligence, found in all intelligent behavior.

That is, facts of *knowing how* are found in doing, not in linguistic artifacts. Knowing a rule, in the sense of *knowledge that* (such as knowing some prescription or algorithm), may be necessary in the case of knowing how to do some performances or tasks, but it is far from sufficient to know how to use it, to apply it. This spans a continuum of intelligence from simple tasks such as knowing how to tie one's shoes to knowing how to prove theorems or perform any surgery.

It should be mentioned here that though doing mathematics is thought by many to be an entirely language-based intelligence activity, the actual relation between number (and the doing of mathematics generally) and language is very far from clear. This is so in spite of a prevalent view, especially among nominalism-inspired intelligence theorists, that reduces number and numerosity in general, such as knowing how to count, to language. This is an issue to which I will return much later and show that such a reduction is not possible and is not consistent with the evidence.

1.2.7 A Missing Distinction between Rule-governed and Rule-bound Intelligence

Moreover, there are kinds of *knowing how* requiring no such rule at all. Indeed, our fundamental understanding of the relation between rules and intelligence in general is not clear. The very notion of "rule" itself is not clear and it may be beneficial for us to spend a little time trying to clarify it.

In one formal sense, the notion of "rule" may be thought equivalent to the mathematical notion of "function". The idea of a function is that a value depends upon some argument to which a procedure, recipe or algorithm has been applied. In a relevant sense, we may use the notion of function here to mean that one clearly rule-governed act of intelligence depends on some other clearly rule-governed act of intelligence because a procedure or rule has been applied. *A function is a way to turn one act of intelligence into another by following a definite procedure or rule.*

Thus in one familiar sense we have an algorithmic or functional notion of *knowing*, of intelligence. It is a notion of knowing set forth in a rule (or set of rules) in the form of step-by-step operations. Performing these step-by-step operations leads to the performance of some overall intelligent function or task. This may be thought of as the *rule-governed* notion of intelligence, at least in its practical sense. This notion is effectively captured in the use of any existing algorithm, procedure, or recipe used to accomplish some task. Additionally, this notion of intelligence is fully consistent with the single capacity theory and many multiple theories as well.

However, there is another sense of knowing which cannot be captured by this sense of function and algorithm. There are kinds of intelligent performances for which we do not have a *prior* rule, procedure or algorithm that, when successfully followed, leads to the accomplishment of a given intelligent task. In other words, for such intelligent performances there is no existing overall algorithm or rule or set of rules characterizing the intelligent performance itself.

Nonetheless, intelligent beings act intentionally and quite rationally without such rules. Moreover, such acts are not exceptional but, when examined closely, can be found in everyday activity. Such activity is clearly intentional, patterned and intelligent even in the absence of a prior rule in part because the intelligent being is making or generating the rule as they carry out the activity. Because the activity is not random, there *is* some algorithm or set of rules involved in the actual activity. How they are generating such patterned intentional intelligent behavior is another question which, for the time being I will set aside. However, this issue points to a set of problems that any fully developed theory of natural intelligence must address.

This intelligence or knowing is not governed by some prior procedure or algorithm. In this sense, such intelligence may be said to be rule-*bound*, but not rule-*governed*. Rule-bound intentional behavior is found in virtually everything we do because much of our everyday activity on many levels is not rule-governed. There are no over-all algorithms characterizing much of our intentional, clearly intelligent activity. Needless to say, innovative, improvisational, exploratory, playful and creative behaviors are also kinds of rule-bound intentional behavior. At its most fundamental levels, *knowing how* is rule-bound instead of rule-governed.

I will have much more to say about the distinction between rule-governed and rule-bound later to more fully explain the concept of emergent self-organization in intelligence. I will also expand upon the relation between these and other kinds of intelligence. Prevailing theories of intelligence, however, have nothing to say about this distinction, nor the kinds of intelligence characterized by it.

1.2.8 Neglect of Multiple Signs and Disclosure of Intelligence

There are many kinds of performance-embedded *knowing how* just as there are many kinds of language-embedded *knowledge that.* The highly complex relations between them must be examined in any responsible and comprehensive theory of intelligence that addresses the involvement of the brain and the rest of the body. Among other things, what is needed is a thorough analysis and assessment of how organisms acquire and act upon information. This will be a recurring question throughout.

Signals, Cues, and Clues

The biological literature has experimentally shown that information can flow within groups or even within individuals themselves *via* two distinct pathways: signals and cues (Lloyd 1983; Seeley 1989). In general, *signals* are stimuli shaped by natural selection. An example is a chemical trail left by ants leading to food sources. A *cue* is a stimulus conveying information only incidentally such as a deer trail through the woods (Camazine et al. 2001, p. 21).

To expand this to include verbal pathways found in language-speaking primates, we must include the concept of *clue* as well. In general, clues are facts capable of being described in language.

Of course the term "information" is defined technically in its mechanical sense in terms of uncertainty of occurrences at categories, including signals, cues, and clues. Thus there is a prior need to view an organism's acquisition and action upon information spanning verbal, visual, and sensorimotor categories of intelligence.

I have stressed that many intelligence theorists emphasize the brain, particularly language centers, the Broca and Wernicke areas, to the apparent exclusion of all else in their concept of intelligence. Minimally, the behavioral, neurological, epistemological, and anthropological evidence shows that such exclusion is not warranted.

Moreover, it does not appear warranted from a genetic point of view either. Scientifically, we know that very little of the overall scheme for embryonic development is special to the brain. Although thousands of genes are involved in brain development, a large number of them are shared with or have close counterparts in genes that guide the development of *the rest of the body* (Marcus 2003).

But on many intelligence theories, other than being a host or carrier for the brain, the rest of the body has little to do with intelligence. In contrast, extensive research into *sign* use, expanding far beyond just symbol use by both humans and lower animals shows that symbolic alphanumeric expressions (spoken or written) are only *one* means or one way by which we express or represent *one kind* of knowing in our intelligent behavior (Neurath et al. 1955).[11] It is only one way that humans (and some lower primates) acquire and act upon information.

Just as lower animals, human beings express, exhibit or disclose their knowing by what they do. What they do is often a sign or pattern of signs ("signature") of their intelligence. The concept of sign is taken to include the notions of signal, cue, and clue, mentioned above, but expands beyond these to include patterns of doings that may involve any combination of these.

Viewing intelligence as a self-organizing system, signs of that intelligence arise from multiple interactions among visual, verbal, and sensorimotor components of the human or lower animal. Those are the components of knowing systems; they are the categories of ways humans and lower animals acquire and act upon information that flows to, through and from them in terms of signals, cues, clues, and patterns of these.

Beyond the use of symbols, human beings use other signs exhibiting or disclosing their intentional doings. Indeed, the doings themselves also function as signs. Humans also exhibit or disclose their knowing, their intelligence, with patterns or forms in their own minds and patterns or forms in the way they move and touch with their own bodies. Humans can move and touch intelligently, clumsily or stupidly, or fatally. We also exhibit or disclose knowing, our intelligence, with the use of artifacts beyond symbols such as *icons,* that is with public images such as diagrams, schemas, drawings, and paintings.

Exhibiting and Disclosing Intelligence

It is important to stress the differences between *exhibiting* intelligence and *disclosing* intelligence. Prevailing single- and multiple capacity theories of intelligence emphasize exhibition and almost entirely neglect disclosure. To exhibit is to present for others to see. It is to show outwardly, to display; to demonstrate openly. It is a public showing. All other things equal, there is nothing hidden or intended to be hidden about exhibitions. IQ tests are designed for takers to exhibit what they know. They are intended to measure that intelligence that is exhibited, specifically in language and number.

[11] Obviously, the concepts "sign" and "symbol" are not identical or equivalent. I use Morris' definition of "sign" which in general is the broader concept, with "symbol" largely confined to alphanumeric indices.

However, most of our intelligence is actually *disclosed* by what we do. It is disclosed usually in ordinary things we do everyday, in what we do, how we do what we do, and in the timing, sensitivity, and smoothness with which we do what we do. The disclosure of something is the act or process of revealing or uncovering. To disclose something is to expose it to view, as by removing a cover.[12] We remove a cover on a major kind of our intelligence by our doing, our *knowing how*.

For example, just as lower animals, human beings disclose a major kind of their intelligence by making intelligent moves with their bodies when avoiding predators in the wild or in the streets. They may know how to engage in stealth patterns of movement that they may only have imagined as a possible way of avoiding becoming a victim.

More to the point, human beings disclose their intelligence when they perceive even very well-hidden predators before becoming their target. Evidence I present later shows that human sensorimotor awareness embedded within the cognitive structures of *knowing how* permits "homing in" on indicators of predators before subjects are even aware *that* they are doing so.

More often than not, moreover, humans perceive the indicators of the predator yet if asked cannot *say* what those indicators are. Even more interesting, humans will often verbally deny *what tests show they have accurately perceived.* Their language reports of their own knowing, their own intelligence, correlate negatively with what they actually know. Language is not always a valid guide to intelligence, to *knowing*. Sometimes it misses the facts of intelligence altogether.

There is a real sense in which this kind of accurate human perceiving, a kind of knowing, is a function of intelligence that extends beyond human language in the sense that it extends beyond the human ability to talk about it, to describe it (De Becker 1997). It is a part of our intelligence that we are unfortunately often taught to ignore by a tradition dominated and to some extent blinded by an almost exclusive focus upon language and its rules.

More to the point, it is a tradition that largely ignores the overlap of our intelligence with that of lower animals who would not for a moment ignore the signs, the signals and cues, of a predator in their midst. Yet human beings ignore such indicators some of the time because we are taught to ignore nonlinguistic signs within ourselves and around us.

This kind of sensorimotor-emergent knowing is what I have earlier referred to as *immediate awareness*. It underlies and is embedded within

---

[12] See any standard American English dictionary such as *The American Heritage College Dictionary*, Third Edition, New York: Houghton Mifflin, 1993.

our *knowing how*. It is not equivalent to *knowing how*, but *knowing how* depends upon it to work. Without immediate awareness, we would never have survived as a species. Moreover, we would also not be able to function intelligently in the world, especially that part of our intelligent behavior removed from paper and pencil tests of our intelligence.

From the moment of our birth if not before, we are immersed within constantly changing unique particulars such as shapes, smells, forms, colors, shades, edges, and surfaces that remain beneath the surface of our verbalizable knowledge that they are there. Our sensorimotor awareness, our *immediate* awareness, interacts, reacts and transacts with this myriad of unique particulars enabling us to *know how* to do many things, including surviving by detecting hidden predators.

We manifest, exhibit, and disclose immediate awareness in what we *know how* to do.

1.2.9 Mechanical "Hard-Wired" and Natural Intelligence: Absent the Difference

For all the above reasons and far more, human intelligence operationally defined as the single capacity denoted by "*g*" and in terms of IQ tests cannot be identical or equivalent with *natural* intelligence. But proponents of this view argue that what others call "natural intelligence" is really just hard-wired, instinctual behavior.

Among other fallacies committed, their arguments are circular. They define intelligence in terms of language performance on IQ tests which, *ipso facto*, are beyond the capability of lower animals. They then conclude from their own definition, not from evidence, that all lower animal behavior is not intelligent but instinctual. At best, animal behavior appearing to be intelligence is held to be the result of imitation or cueing.

Instinctual, "hard-wired" behavior is inborn, patterned behavior characteristic of a species in response to specific environmental stimuli. Instinctual behavior is not intentional, but involuntary reflex, and usually *single*-pathed, determined by natural selection. For example, in response to specific environmental stimuli, say, the sight of a predator, an animal may immediately take flight. Intentional behavior, on the other hand, is deliberate and purposeful. It is planned behavior, not an instinctual or reflex response. It often involves multiple pathways to an envisioned (though not necessarily *visible*) goal, not just a single path.

However, the intelligence of our non-language-speaking ancestor, *Homo erectus*, is distinguished in part from *Australopithecus* by the fact that *H. erectus* designed tools for specific purposes. Such tools designed for specific tasks were later shown to require a multistage process going far beyond simple imitation, a limited process. Clearly, our ancestors had the intelligence of *knowing how* even in the absence of language ability.

Among numerous animal experiments, the famous "dangling banana" experiments by Wolfgang Köhler (1973) demonstrate the same kind of *knowing how* intelligence. In many efforts by chimpanzees to reach a banana suspended overhead on a string out of arm's reach, they tried multiple pathways. The only tools at their disposal were empty boxes and a stick. The chimps initially did not see the boxes as tools they could use to reach the banana, but eventually tried multiple ways of stacking the boxes to do exactly that. In spite of some failures, they persisted in trying to stack the boxes in the right configuration so as to climb on top of them then use the stick to knock the banana free. They eventually succeeded.

All criticisms to the contrary, the chimpanzee behavior exhibited planning, insight, and persistence. They tried multiple paths to reach the banana, not one. Thus their efforts were clearly not reducible to instinctual or reflex response. Moreover, they demonstrated an ability to use an indirect if not novel approach to reach a goal. They also later demonstrated an ability to transfer what they learned to new situations.

But even where such animal behavior may *appear* to be quite intelligent, it cannot be, according to some theorists of intelligence. Human intelligence is the only intelligence in their theory because it is formed and informed with language.

The self-correcting, clearly purposeful and intentional *problem-solving* behavior of the chimpanzees is not accepted as intelligent behavior by these theorists. Ignoring empirical evidence in favor of a self-fulfilling definition and uncritically held assumptions renders their position inevitable.

Yet regardless of where one comes out on the instinctual *versus* intentional/intelligent behavior debate, proponents of such intelligence theories commit other kinds of fallacies and circularity as well. Though I will address these issues later, "intelligence" is ill defined in part because it is test designers who define it. Those efforts have resulted not only in circularity but ambiguity and unwarranted narrowness in our over-all understanding of intelligence.

1.3 Requirements for a New Science of Intelligence

The above problems with the current sciences of intelligence largely map requirements for a new approach. Based on arguments in later sections, the neural, genetic, and linguistic *cum* logico-mathematical reductionist arguments are rejected. Rather, it is a fundamental principle here that the facts of intelligence are found in the broadest domains of actual human and animal experience. The facts of intelligence are found in what humans and animals do, specifically what they know how to do.

From this principle, other requirements for a science of intelligence follow. Throughout, it is those facts of intelligence found in a broad domain of actual human and animal experience that drive the requirements for a new scientific approach and methods.

These methods are not limited to the classical approach with existing data collection restricted to narrow portions of the domain of human language experience. At minimum, the classical approach is too narrow because it recognizes only one kind of intelligence, one kind of knowing, while excluding at least two other major fundamental cognitive categories. These major categories of intelligence, of knowing, show up in both human and animal intentional behavior.

1.3.1 A Broader Theory of Knowing

Knowing how and *immediate awareness* are kinds of intelligence found in conjunction with one another in intentional behavior. Because of their dynamic, self-organizing structures spanning sensory (including somatosensory) and sensorimotor systems, they require an entirely different scientific approach from that used to measure largely verbal intelligence.

Knowledge That, Knowing How, Immediate Awareness

Minimally, verbal intelligence is largely a public matter because in principle it can to a large degree be manifested in public, alphanumeric language structures that are separate from the person. Those language structures are available to anyone to inspect. On the other hand, *knowing how* in conjunction with *immediate awareness* is somehow manifested *in the person* or animal and evidenced in what they do. It is manifested in, among other things, what they do, how they do it, and the manner, sensitivity, timing, and seamless quality with which they do what they know how to do.

This distinction between verbal intelligence, *knowledge that*, and the intelligence of *knowing how* is obviously not just a cognitive distinction. It is also a distinction *between where one looks for the facts* of what a human or animal knows. For *knowledge that* intelligence, one looks to language structures for facts related to whether or not one knows. For *knowing how* intelligence, one looks to actual structures and patterns of doing for facts to determine whether or not one knows how. Among other things, it is this distinction that entails a like distinction between rule-governed and rule-bound intelligence.

Among other things, rule-governed intelligence is what we call a recursively enumerable set. That means it is a set of things we can count. It is a computable (machine model) set, on the standard digital computer, a set of problems or function instances that can be defined over discrete, countable domains. Verbal intelligence as measured by standard intelligence tests is clearly rule-governed intelligence. Those intelligence problems require "yes" or "no" decisions. As such, there are classical effective algorithms that can be used to address these problems. Standard IQ test formats are extensions of these algorithms.

On the other hand, *knowing how* intelligence is largely rule-bound intelligence. As such, it requires a different approach to its computability, if it is entirely computable at all. Knowing how problems are those that can not be addressed with classical algorithms requiring "yes" or "no" decisions. Minimally, this kind of intelligence requires an approach that generates dynamic self-organizing patterns of interactions among very large numbers of components or elements in the way something is done. Knowing how requires that we look at the dynamic patterns in the actual doing of something, requiring an altogether different model for this kind of intelligence.

Fundamental properties of intelligence are found in *knowing.* This means that the requirements for a new scientific approach include a more complete theory and classification of kinds of knowing, in addition to a more complete theory of the emergence, manifestation and disclosure of knowing in actual experience. These will also provide a more comprehensive view of the relation between human and animal intelligence, and, as well, the relation between these and artificial intelligence.

Single capacity theories of intelligence restrict the scope of intelligence to verbal (linguistic) activity. However, arguments and evidence in the following chapters demonstrate the need to extend the scope of intelligence to kinds of intentional behavior *unrelated* to the use of language. I have already made reference above to such kinds of behavior found within a broader domain of actual human and animal experience.

Technically, the term "knowing" refers to a cognitive relation between subject(s) and object(s). It is used here in place of "knowl*edge*" since that term is usually limited to language-based declarative sentence knowing. Arguments and evidence support the broader scope of the concept of "knowing" to include at least two additional major cognitive categories not currently addressed in current intelligence theories. An adequate theory of *knowing*, not limited to "*knowledge*", will provide the theoretical bridge linking human and lower animal intelligence.

These two additional categories of cognition, of knowing, are interrelated in highly complex ways that account for the dynamic emergence of natural intelligence. As noted, evidence for that emergence requires a different scientific approach than the classical approach currently used in most intelligence research. The dynamic nature of *knowing how* in conjunction with *immediate awareness* shows that they require a different mathematical and methodological strategy.

Briefly, the classical approach defines problems over discrete rather than continuous domains. As already noted, it is reductionist, linear, additive, and focuses almost exclusively upon existing data collection and verification procedures. It focuses on *knowledge* representations in symbolic language behavior and locates natural intelligence primarily in only one area of the brain, language centers of the cerebral cortex.

Knowing how, on the other hand, refers to the actual dynamic shapes and patterns of intentional doing in the world that can be unrelated to language behavior. Even when we become language speakers, we must know how to use words; know how to read and write; we must know how to determine what is relevant in a context; know how to recognize and discriminate among the mouthings, vocalizations, tones, and signings of language tokens.

Knowing how in conjunction with immediate awareness is at the foundation of all our intelligence. Because of their dynamic structures they require definition over *continuous*, rather than discrete domains, and can be shown to be self-organizing, nonlinear, adaptable, flexible, and even self-complicating intelligence phenomena.

A research focus upon *knowing how* in conjunction with immediate awareness extends the location of intelligence beyond a few language parts of the brain to include much (if not all) of the rest of the brain and body. The major structures of intelligence are viewed as emergent, nonlinear phenomena, evidenced by a broad range of multiple signs and "signatures" in both human and animal behavior.

Almost without exception, however, the current intelligence research strategies narrowly define intelligence so that it is measurable by classical standard linear instruments and is limited to verbal behavior. By their very self-imposed limitations, existing methodological approaches cannot measure at least two major categories of cognition found in the larger domain of natural intelligence evidenced in human and animal behavior.

A more complete and exhaustive classification of kinds of intelligence, however, shows the need to include these two additional categories. Addressing these entails not only extensions beyond the classical approach as noted above, especially recognizing the distinction between rule-governed and rule-bound intelligence, it also requires extensions beyond the classical computational approach to intelligence research. These distinctions, among others, have already been implemented in more advanced sciences and are also required in the sciences studying intelligence.

The intelligence of *immediate awareness* involves a radical departure from the entrenched traditional view that all human intelligence is reducible to acts of classification. For many centuries, the human mind has been modeled and viewed as a representation-filled classification machine. By its very nature, however, classification involves kinds of symbolic representations and the use of rules of logic. The act of classification is the act of identifying instances of a class or category of things by comparing its properties with other members.

However, evidence shows that immediate awareness proceeds by the cognitive use of *nonlogical* indexicals. Nonlogical indexicals are indicators found in the cognitive structures of *knowing how* and immediate awareness. As such, they overreach the scope of symbolic representation; they overreach the scope of verbal intelligence. The objects of this knowing are not class objects because they are not language objects and are not gotten by logical operations of comparing properties. As such, they are unique individuals, not class objects.

Nonetheless, our scientific understanding of the intelligence of immediate awareness is possible because it is publicly accessible with a more informed view of the shapes and patterns of intentional doings found in human and animal behavior. That accessibility is made possible by extending our understanding of how intelligence is exhibited and disclosed, that the full spectrum of intelligence is comprised of multiple signs, including signals, cues, and clues, and their relations. These multiple signs are not random doings, but intentional. The fact of their intentionality means that these doings show *patterns*, and are part of the highly complex patterns of natural intelligence.

1.3.2 A Broader Theory of Signs of Intelligence

A broader domain of human and animal experience within which we must look for facts of natural intelligence leads to a broader view of the signs by which animals and humans exhibit and disclose their knowing. Intelligence is not narrowed to that slice of the domain of experience that includes verbal behavior. We must extend beyond that narrow slice and look more at the multiple *ways* human beings and animals exhibit and disclose what they know. What is needed is a broader theory of signs to include rule-bound knowing, exhibited and disclosed by multiple signs, patterns, and interactions among these.

Neither the full scope of intelligence itself nor the dynamic growth of intelligence can be explained solely within the domain of verbal intelligence structures. These must include accounts of flexible task-centered problem-solving skills that are, among other things, sensitive to changing environments. These skills are part of the intelligence of *knowing how* and immediate awareness.

Toward Three-Dimensional Signs and Patterns

Research into the full scope of intelligence will require extending beyond the largely (though not exclusively) one- or two-dimensional symbolic domain of verbal intelligence to include the broader realm of three-dimensional patterns of sign-making, sign-exhibiting, and sign-disclosure of dynamic intentional doings. This extension includes a continuum of cognition spanning from the verbal to nonverbal sensorimotor and associational categories obviously tied to sensory and motor categories.

Three-dimensional signs such as signals and cues, include gestures (as with hands), but also full-body doings such as tasks and other performances. Some studies of human gestural patterns have tied those to language use, especially speech, showing that spoken languages themselves are not limited to one- or two-dimensional symbols (as alphanumeric letters or numbers), words and rules of grammar (McNeill 1992). Thus, those researchers sought to show that even spoken languages, let alone sign languages used by hearing impaired persons, must be considered from the broader perspective of three-dimensional referential functions of gestures.

Moreover, those studies follow a long line of research directed to the relation between gesture and thought, seeking explanatory links between "inner forms" of thought and "outer forms" of expression that may or may not take language form.

These lines of inquiry must be continued. At present, for example, I mentioned earlier that the research community does not have an explanation of the relation between mathematics and language. Some research suggests that the actual doing of mathematics appears to rely more on those inner abstract forms such as images than on any alphanumeric representations "in the mind." Yet we have little understanding of the relationship between those and the alphanumeric expressions on paper mathematicians produce.

We have little understanding of the role of imagery in general in human intelligence. Indeed, there are disputes about the meaning of "image" and whether or not we even experience such things, though the scientific evidence to date supports the existence of mental imagery and its effective role, for example, in mental practice of motor movement. Some researchers who acknowledge there are such things want to limit them to visual images while others want to expand the concept to include other sense images. These include the auditory, olfactory and kinesthetic senses.

At the foundation of such inquiries is that recurring central question: How do humans and animals acquire and act upon information? That question leads back to the need to extend intelligence beyond one- or two-dimensional cognitive categories to include three-dimensional sensory and motor domains and to establish sign categories of dynamic, self-organizing systems. These sign categories, including but not limited to signals and cues, span all sensorimotor capacities, including visual, auditory, olfactory, gustatory, and somatosensory categories (including touching, moving, and proprioception).

It should be stressed that neither a person nor an animal *knows how* just by doing something; it is the way they do it—the timing, sensitivity and smoothness in their patterns of doing—that shows whether or not they know how. This fact requires very close scrutiny of the structures of immediate awareness which make knowing how possible. It is the structures of immediate awareness which provide the foundation for an extended theory of signs by which natural intelligence is exhibited, disclosed, and revealed.

Elsewhere (Estep 2003), I have given meaning to "immediate awareness" as consisting of a complex hierarchy (or stacked set of "sheets") of primitive cognitive relations. Those primitive relations consist of the preattentive and attentive phases of neural activity, along with the sensory and sensorimotor systems. This involves the somatosensory system, specifically moving and touching,[13] but also includes imagining, the use of images and abstract, especially spatial forms in the mind.

It is interesting to point out that in contrast to much of the human intelligence research community, computer science areas of robotics have been making strong gains based upon biological models of sensory and motor systems of humans and animals. Since the early years of artificial intelligence, patterned on what was then taken to be human intelligence, that community has changed from largely top-down, knowledge- and logic-based serial and linear approaches to more massively parallel and distributive nonlinear approaches to understanding intelligence. They have looked more at the ways living biological systems are organized so as to permit the generation of kinds of intelligent behavior in what that living system does.

[13] Technically, the sensory system includes the somatosensory system, but I have explicitly mentioned all these here to make certain there is no ambiguity as to which systems are involved in immediate awareness.

1.3.3 Methods of Nonlinear Science: The Emergence of Self-Organizing Dynamical Intelligence

The classical scientific approach to the study of natural intelligence can be safely described as static and nondynamic. Largely on the model of an inorganic machine, it views intelligence as static and nondynamic and employs methods that are static and nondynamic as well. It is a top-down, logic- and knowledge-based, serial processing approach. Among other things, classical methods assume that effects of intelligence are additive and linear, especially as measured by standard IQ tests. The dynamic self-organizing complexity, directiveness, flexibility, adaptivity, and even the rhythm of intelligence are missing altogether on the classical view.

With a broader domain of both human and animal experience from which the facts and data of intelligence are found, there is a clear requirement for the sciences of intelligence to expand theory models and methods beyond static linear and additive models.

Where intelligence is viewed as a living thing on the model of an organism, it exhibits *dynamic* self-organizing complexity and emergent properties and relations that must be accounted for. Living systems of organized complexity such as natural intelligence require *geometric* configuration theory models that permit analysis of properties and relations of entire large ensembles of coupled elements in nonlinear dynamic interaction with one another and their environment. That organismic theory model must incorporate the strengths of mechanical models, but it must go beyond them to permit inquiry and configuration analyses that in turn permit representations of that self-organized complexity.

Recall that classical models and methods in intelligence research are mechanistic, either statistical or non-statistical. By their very structures, they are limited to either representations of *organized simplicity*, limited to very few factors, or representing *unorganized complexity* in the form of average combinations of factors. While these models are directed to setting forth simple, direct causal chains that are based upon fallacious assumptions in any case, the aim of a self-organizing complexity theory models approach is to capture the pattern-generating rules of natural intelligent agents in the context of their actual experience.

Self-Organization

Self-organization refers to kinds of pattern-formation processes found in both physical and biological systems. Patterns in self-organizing systems emerge at global levels from large numbers of interactions among lower level components of those systems. Additionally, just to be technically precise, the rules or algorithms that characterize those lower level interactions are executed based on local information in a self-organizing system without reference to global patterns (Camazine et al. 2001). Patterns emerge from the internal dynamic processes of a self-organizing system; they are not imposed on the system from any external source. That is, a self-organizing system is just that: it organizes itself without any instructions from outside itself.

Our concern here is with self-organization in living things, but we must also briefly compare properties of self-organization in physical (inorganic, mechanical) systems with those in biological systems. Self-organization in physical systems involves large numbers of components that are inorganic. For example, the grains of sand in a desert or chemicals in a reaction experiment. Compared to biological components, these inorganic components are relatively simple. They obey physical laws and their large number of interactions produces predictable deterministic patterns.

Biological systems also obey the laws of physics, but the mechanisms of self-organization in living things involve a much greater level of complexity precisely because the interacting components are living. Moreover, the rules governing the interactions of large numbers of living components also differ from physical systems in part because they are they are influenced by genetically controlled properties not found in physical systems (Camazine et al. 2001).

Though incomplete in crucial respects, the difference between mechanical physical self-organizing systems and biological self-organizing systems is driven home with the following:

> ". . the subunits in biological systems acquire information about the local properties of the system and behave according to particular genetic programs that have been subjected to natural selection. This adds an extra dimension to self-organization in biological systems, because in these systems selection can finely tune the rules of interaction" (Camazine et al. 2001, p. 13).

Fine tuning the rules of interaction is not an option available to self-organizing physical systems. However, this description is incomplete at best. In biological systems interactions between components are minimally based upon information transfer as signals or cues. As noted earlier, signals are stimuli shaped by natural selection specifically to convey information, while cues are stimuli that convey information only incidentally.

However, complex social environments may be intentionally designed with cues to deliberately, not incidentally, convey information. Moreover, with sufficient genetic engineering, signals originally shaped by natural selection can also be altered. They can be reshaped by human intervention.

Because human beings are also biological systems that are self-aware, they can and do use their intelligence to alter the rules of self-organization themselves.

The defining characteristic of self-organizing systems such as natural intelligence is that the organization emerges from multiple (in some cases immense numbers of) interactions among their components. Moreover, development or growth of natural intelligence does not arise from independent modules but is an emergent consequence of large numbers of interactions across many domains.

Thus, as with an organism, we must view natural intelligence, *knowing,* as a large population of simpler components which through time works upwards, synthetically constructing larger aggregates of rule-*governed* or rule-*bound* objects. These objects interact *nonlinearly* with one another and with their environment—in support of the overall life-like dynamics and emergent patterns and qualities of a natural knowing system. The self-organizing organismic theory model approach is a bottom-up, highly distributed and massively parallel view of knowing, of natural intelligence.

Theory Models Approach to Intelligence Inquiry

As earlier noted this view of intelligence as a living thing requires a mathematical and scientific approach very different from the classical top-down, verbal rule-governed, logic-based, linear and additive approaches. Those approaches are largely anti-theory and anti-concept formation; are either reductive or, as indicated by the primary value attached to data collection and verification procedures, inductive.

Given the enormous variety and range of intelligence experience, a theory of natural intelligence must incorporate the major mechanisms by which intelligent beings, both human and animal, acquire and act upon information.

What is needed is an integrated mathematical configuration model permitting characterizations of organized complexity. Again, these characterizations are based on nonlinear assumptions. Such characterizations can be provided by an integrated organismic theory model, formed from the integration of *set* theory, *information* theory *graph* theory and *general dy*namical system theory. This theory model will be referred to with the acronym "SIGGS" throughout the remaining chapters.

Set Theory

Set theory is of course necessary to demarcate the scope of the natural intelligence universe based upon logic and facts found within the domain of animal and human experience. Though I have outlined as well as referred to this above, in the next chapter I will endeavor to more precisely carve the problem space of natural intelligence according to principles of set theory.

Information Theory

Information theory is necessary to adequately characterize how organisms, including humans, acquire and act upon information. The term "information" used here is not identical to the ordinary language use which often refers to content or meaning. Looking upon an organism as an intelligence system, information theory provides a way to give meaning to the categorization of components of the system, connections of that system and its environment, and the uncertainty of occurrences at those categories.

Every system has information in the sense that occurrences of its components or affect relations or both can be classified according to categories. The added condition of uncertainty of occurrences at categories is necessary to develop information properties on the system and its environment (or *negasystem*). The concept "information" can be mathematically defined in our theory model in terms of uncertainty of occurrences at cognitive categories.

Though we have sorted three distinct pathways of information transfer that organisms use, including signals, cues and clues, we must rigorously clarify and classify the categories of occurrences of that information. This classification must span the entire cognitive repertoire of intentional behavior, including interacting components of verbal (by linguistic means), visual and sensorimotor categories.

Graph Theory and Dynamical Systems Theory

Graph theory is necessary to adequately characterize connections among components in a group. With digraph theory, an intelligence system group becomes a set of points; system affect relations become sets of directed lines; and digraph properties of a system result when certain condition are placed on its affect relations or its group. Given the large number of connections and relations among components at local levels in self-organizing systems, emerging in kinds of global dynamics, digraph theory provides powerful mathematical tools to characterize those connections and relations (Steiner 1988).

Graph theory and tools of dynamical systems theory are necessary to describe the integrated behavior of intelligence that is coordinating the actions of many, possibly an immense number, of components. The domain of experience is characterized by a very high degree of variation at all levels. At the brain level alone, there is extraordinary variation in neuronal chemistry, neural network structure, synaptic strength, to name just a few. Individual humans have highly variable personal histories, genetic influences, bodily responses, and motivations. As one noted neuroscientist has stated regarding the brain (Edelman 2004), this variation must not be dismissed as noise but seen as fundamental. Patterns nonetheless emerge from enormous variation, even under unpredictable circumstances.

The usefulness of dynamical systems theory as a basic mathematical and methodological framework for natural intelligence is that even though we may not know all the details of the order of connections or interrelations among possibly an immense number of primitive relations, we can nevertheless build a theory model to explain *fundamental*, generic properties of natural intelligence.

In part, this can be done with the use of random Boolean networks within the context of phase space, a space of possibilities. This will permit viewing a large population of fundamental objects and relations in highly complex dynamic interactions. The properties of Boolean networks have been thoroughly studied and techniques have been developed for determining the dynamical properties of specific kinds of networks (Forrest and Miller 1991). Thus they are of tremendous value in studying those fundamental properties of complex dynamical systems such as natural intelligence. I will return to Boolean networks and their usefulness in intelligence theory in a later chapter.

From a Symbol-based View to a Geometric View of Natural Intelligence

The nonlinear theory model approach permits reorienting the perspective on intelligence from a classical top-down *symbol*-based view to a *geometric*-based performance of intelligent agents in a fitness landscape. By turning to a phase space of intelligence possibilities, utilizing a broad array of variables from major categories of knowing, the very dynamics of a natural intelligence agent can be evaluated as that agent works his or her way through the landscape.

The geometric orientation to the study of natural intelligence can also be greatly effected with the use of high-speed computers, particularly for the study of *knowing how* and immediate awareness. This requires the use of highly parallel distributed processing and connectionist models capable of simulating rule-*bound* know*ing*, as opposed to limiting simulations to rule-*governed, symbol*-based know*ledge*. Such computer-based research programs have already proven highly useful in the study of gait analysis (Simon 2004), part of the intelligence of *knowing how*.

For a study of a natural intelligence system, where the computer is used as an instrument permitting experimentation of a kind in a hypothetical universe, what is needed is an approach to computation permitting a focus upon on-going, dynamic interactive *knowing* behavior—rather than a focus upon final results. That is, what is needed is a computational architecture of an intelligent system permitting a natural method of knowing behavior *generation*.

This natural method must reflect the distributed and parallel structures of human knowing systems. These natural intelligence systems include a hierarchy of the above mentioned populations of simple components constructing aggregates of simple rule-governed or rule-bound objects interacting nonlinearly with one another and with their environment to produce emergent structures of intelligence.

At each level, the primitives must be identified and rules governing or bounding their behavior under conditions at that level must be specified. Primitive knowing behavior, our species of knowing how in conjunction with immediate awareness, must be organized in the architecture of the artificial system similarly with their natural counterparts. From this organization, emergent properties of knowing, of natural intelligence, arise.

These considerations also apply to an understanding of the relation between natural and artificial intelligence. The classical approach to artificial intelligence has been premised upon a discrete, top-down view of natural intelligence as symbol-based, rule-governed *knowledge that.* That approach, the usual Artificial Intelligence (AI) approach, is a serial processing strategy, with problems defined over the natural numbers, integers, rationals, or domains encodable in the integers, requiring a great deal of elaborate programming and *prior* know*ledge* engineering. It is an approach built upon a centralized control structure with access to large sets of *predefined* data structures, operating with algorithms defined by mathematical formulas and discrete procedures.

1.4 Summary

The above should make clear that I am approaching the intelligence domain as *multidimensional.* The scope of intelligence is extended beyond the single capacity theory to reflect the facts of natural intelligence found in human and animal experience. Natural intelligence is an emergent consequence of interactions across multiple domains. An extended domain of intelligence includes many kinds of knowing beyond the traditional language-based propositional knowing *that* and beyond applied knowing *that* in action or behavior.

Because of the dynamical nature of these extended kinds of knowing, I also argue for an extension of our theoretical methods including mathematical models for defining the problems of that domain. For reasons cited above and more to be explored, nonlinear methods and models are superior to classical methods and models in intelligence research. These include the following:

1. Nonlinear theory models approach to natural intelligence permits a more complete and exhaustive classification of kinds of knowing; it does not limit the scope of cognition to only one kind or category of knowing, verbal (linguistic) intelligence.

2. Nonlinear theory models approach incorporates the strengths of classical statistical and nonstatistical models and methods, permitting characterizations of organized simplicity and unorganized complexity, but goes beyond these to permit characterizations of organized complexity.

3. Nonlinear theory models approach is not reductionist as are classical models.

4. Nonlinear theory models approach permits a dynamic configurational analysis of the natural intelligence of an entire organism in its interactions, transactions, interactions, and reactions in an environment.

Borrowing a point made elsewhere (Langton 1989), the traditional classical approach to intelligence research has from the beginning embraced and continues to embrace underlying theory, assumptions, and methodology that bear little or no demonstrable relationship to the method by which intelligence is actually generated in natural systems. It is an approach which has focused upon outputting intelligent *solutions*, *knowledge that,* rather than intelligent behavior, *knowing*, thus missing most of the domain of intelligence altogether.

The use of nonlinear theory modeling in natural intelligence research can profoundly enlarge our understanding of natural intelligence. It can change the current single or limited multiple capacity view of intelligence as a top-down, knowledge-based rule-governed non-dynamic phenomenon to one of the dynamic emergence of self-organizing intelligence and growth in possibilities.

More to the point, nonlinear theory modeling can provide us with a more realistic view of the nature of actual intelligence found in human and animal experience.

2 The Universe of Intelligence

The universe of intelligence includes both natural and artificial kinds.[1] Natural intelligence occurs in living things in the world, if not also in the broader natural universe. Artificial intelligence is generally held to be that which is found in machines intentionally designed to simulate the logical forms of human reason. The concept "machine" is logically equivalent to the concept "algorithm" and it is broadly construed to include software.

The distinction between natural and artificial intelligence however, should not be seen as mutually exclusive. There is an overlap between the two, though that is not generally recognized or acknowledged in those sciences and disciplines of engineering that concern themselves with either kind. We will take a look at the area of overlap between the two later.

Moreover, some computer scientists have endeavored to map natural intelligence onto machines, creating artificial intelligence or, as some call it (depending upon the engineering method), artificial life. They have endeavored to do all this without first getting clear on the nature of intelligence as such and *natural* intelligence in particular. We will also look at those issues as well. For now, we have to get clear on more fundamental distinctions to allow us to proceed with the maximum of clarity. Some readers may find the following section a bit elementary. If so, please feel free to skip to later sections.

2.1 Carving the Problem Space

When rational inquiry is directed to anything, it starts within what is called a "universe of discourse." One can think of this in some of the same ways we think about the natural universe, only a universe of discourse is specifically directed to a certain subject matter of inquiry or research. It is comprised of highly specialized language about a given subject. It contains concepts, distinctions, definitions, conjectures, explicit as well as implicit

[1] I use the terms "intelligence" and "natural intelligence" interchangeably and specifically use the term "artificial" when referring to that which is intentionally designed by humans for specific purposes.

assumptions, arguments, and methods. The natural universe, on the other hand, contains everything there is.

A universe of discourse contains everything that is known or rationally conjectured about a subject matter at hand. It is within that universe, containing concepts, distinctions, definitions, assumptions, arguments, and methods that the inquirer's or scientist's question or problem must be posed. However, because this universe is mostly in the background, the rational inquirer may not explicitly speak or write much about it, or he or she may limit what they have to say about it to just a few things.

Usually, an inquirer or scientist will start out with a few agreed upon definitions or concepts as well as a few assumptions to clarify what the problem or question is that they are posing. They will leave the rest unstated. Nonetheless the entire universe of discourse is there as a set of *prior* agreed upon concepts, distinctions, assumptions or conditional assertions that are held to be true, if only provisionally, and which permit the inquirer to ask rational questions or pose rational problems to be solved. Those problems must be solved within that universe.

Most of the background universe of discourse of any particular subject matter is never explicitly examined by the one doing inquiry. That background is there in principle as a position upon which a given inquirer may "rest" while he or she sets about to extend the scope of that universe by addressing unanswered questions or unsolved problems. The existing concepts, distinctions, definitions, assumptions, arguments, and methods are *assumed* by the inquirer or scientist in order to proceed.

Of course, the individual scientist over many years may and indeed should examine much of that background in the course of their education, training, or research. But it is in fact impossible in any given lifetime to examine all of it, to ensure that the assumptions are all true or that the concepts, definitions, distinctions, and arguments are all valid, or that the evidence supporting all the arguments holds up. For that, we tend to rely upon the public, verifiable work of generations of scientists that came before or will come after all of us.

2.1.1 Rational Inquiry and Ideology: The Differences

It is worth mentioning at this point that rational inquiry is that inquiry that critically assesses, in terms of reason and evidence, any concept, definition, assumption, distinction, argument, and method used to solve any rational problem or to pose and answer any rational question. Though it is acknowledged that a given inquirer or scientist cannot in his or her lifetime personally critically assess all those that exist in any given universe of discourse, in principle *any* of those assumptions, definitions, concepts, and methods are subject to critical scrutiny by reason and evidence.

There are *no* sacred assumptions, definitions, concepts, and methods that are beyond question in rational inquiry. If any one of these does not meet the standards that reason and evidence require, they may be changed or jettisoned for alternatives. Where those assumptions, definitions, concepts, and methods are changed or jettisoned, however, the rational inquirer must give powerful arguments for doing so. In turn, those arguments must withstand the same critical scrutiny by reason and evidence.

Another way of putting this in perspective is to say that any rational inquiry—by that I mean any question or problem posed to be addressed in terms of reason and evidence—actually starts as a conditional or hypothetical. "*If* such and such is true or valid, then. . ." The "*If such and such is true or valid*" part of the conditional is often not actually stated, but assumed. That part of the conditional is made up of what may very well consist of an infinite chain of assumptions, arguments, and concepts, some of which are perhaps very tenuously linked to one another, that the inquirer must use to allow his or her question or problem to be addressed at all. To be sure, any one of those assumptions may in fact turn out to be false; any one of the arguments may turn out to be invalid or unsound; or the concepts, definitions, and distinctions may turn out to be confused or inadequate.

In order for inquiry to proceed at all, however, we must start somewhere. But it is the wise and careful inquirer or scientist who recognizes that there is always a risk involved. There is a risk that we may find, sooner or later, that one or more of those prior assumptions lurking in the background may be false. Or we may find that we are working with a very confused concept or inadequate definition; or that we may be committing some subtle fallacy not yet recognized. Worse, we may go our merry way without discovering any of this *because we never asked the questions we should have asked in the first place.* We have not recognized that there is a problem.

This is the crucial difference between *rational* inquiry and that pseudo-inquiry sometimes posed by ideologues. *Ideologies,* and those who defend them, generally demand that some or all prior definitions, concepts, assumptions, and methods never be questioned. They demand that some or all of these be *uncritically accepted without question* as a basis on which to proceed with anything at all. This is a demand for reason to step aside and be replaced by obedience to authority, even if it is only the authority of some written script somewhere whose author is unknown or who is unavailable to answer questions. It is obedience to authority whose reasons, evidence, and even credentials remain beyond your reasonable assessment and critical scrutiny. Ideology demands blind acceptance and obedience to unquestioned authority. Rational inquiry demands critical assessment of anything and anyone in terms of reason and evidence.

The "*. . .then. . .*" part of the above conditional sentence is generally followed by a question or problem. The problem should always consist of an explicitly formed question, not formed as a hidden assertion or a declaration that we already believe is true. The question should not be rhetorical since these always contain one or more hidden assumptions that are implicitly claimed to be true. An example of a rhetorical question is the famous or infamous "When did you stop beating your wife?" The question contains the not-so-well-hidden and perhaps unproven assumption that the one being questioned is guilty of beating his wife.

In essence, we must make it very clear that we are asking a genuine question and not presumptively claiming something to be true. An example is the following: "If the laws of mathematics are true, then can elements of the real number ***R*** and complex number ***C*** fields be used to describe and explain intelligence?" That is an example of a rationally formed problem, a question, posed within a universe of discourse.

But when we become aware of a problem within a given subject matter and seek to pose a question to solve it, we are already beginning to demarcate what is called a "problem space" within that larger universe. We are beginning to "carve up" the problem space *by our very conceptualization of the question we seek to answer*. Our question has packed within it many assumptions and concepts that may have been uncritically accepted in our minds without proper questioning and close scrutiny.

2.1.2 Careless Carving

This is the point at which any rational inquirer must be very careful indeed. For if the problem space in that universe is carved carelessly or if the inquirer cuts the problem space on a bias, that cutting will henceforth doom the inquiry to a skewed, fanciful, partial or completely false answer. As one scientist put it, "We should not be surprised that when we put crooked questions to Nature that she then returns crooked answers" (Geach 1971).

Worse yet, that careless carving may create harm that is not even seen. It may not be seen because it becomes the accepted way of looking at the problem. Worse, the "crooked answers" may become the accepted solution. Without a careful examination of at least the core necessary prior assumptions, concepts, definitions, and distinctions that make the inquiry possible, followed by that very careful posing of the question, the inquirer can create harm that may take generations if not centuries to correct.

Unlike the hypothetical nature of the rational method of inquiry, the harm that may follow may not be hypothetical at all. In the past, crooked questions leading to crooked answers have led to much suffering and the deaths of many innocent lives, sometimes over many years or even many centuries.

The subject matter here is intelligence, which falls within the broader domain of behavior generally. Speaking in the language of set theory, intelligence is a set of elements, or subsets that may themselves be sets of elements. And the set of intelligence itself, with its subsets, is another set within a larger set of elements. We will want to consider the subsets of the set of intelligence; and we will want to consider the larger set of which intelligence is a part.

Intelligence in general is often taken to be a subject of study by psychology, and that will be the approach we will initially take. Lately, however, it has become the focus of the neurosciences and has for some time been the focus of computer sciences and engineering. In earlier centuries, it was the subject matter of philosophy. Philosophers tended to analyze the human mind or psyche in terms of a universe of possibilities ("what is it possible to know?") as well as in terms of the power of human reason to achieve knowledge of that universe. They asked fundamental questions about the nature, extent and power of human reason of certain kinds.

2.2 Classical Origins and Fabric of Intelligence Theory: Cut on Biases

For about 2500 years philosophers and scientists have written much about the nature of intelligence, its categories and kinds. They have also written much about who and what has what kind and who and what does not.

The nature of intelligence was often couched by philosophers within theories of the mind, soul or psyche. It is at the core of fundamental questions about the nature of humanity itself. It is also at the core of fundamental questions about the relation between humanity and the rest of the world, if not the rest of the universe. It is central in fundamental questions about the relation between body and mind and issues about human freedom. Indeed, for some philosophers intelligence was conceived as that part of us that we have in common with the gods.

Moreover, it is also at the core of issues related to what knowledge and knowing are and what they are not; what can be known as well as what can only be believed. It is also about who, by virtue of intelligence, can attain the highest good and who cannot. It is about who and what rules and who and what gets ruled in societies and in nature.

Alongside religious theories of relations between gods and humans, intelligence ranks high among those things that have been held by philosophers and some scientists to justify power and privilege rankings among people in societies and governments.

2.2.1 Plato and Aristotle's Conflicting Theoretical Stage

Setting a theoretical stage to be followed by many scientists and philosophers over the next almost 2500 years, both Plato (427–347 B.C.) and Aristotle (384-323 B.C.), in their own ways, set forth theories of intelligence that one way or another have been used to both elevate some groups to rule others and condemn still other groups to be ruled.

Between the two, however, it is Plato who shows the most egalitarianism. It is first in Plato that we find a theory of intelligence that emphasizes the role of reason, and a theory that *all* are capable of it. For Plato, pure reason and the senses are polar opposites. The role of reason in human life is to unveil the ideal eternal forms behind those sensible appearances that we experience with our bodily senses and that always mislead us. It is only the objects of intelligence, pure reason, that are perfectly knowable because only they are the perfectly real. The objects of the senses, on the other hand, are fleeting appearances, mere illusions of our bodily senses.

Plato's theory of intelligence was part of his theory of the psyche or soul. These fit into his much larger theory of the ideal state, the Republic. He held that the soul, our psyche, was divided into three capabilities: reason (the philosophic capabilities); will (or spirit); and appetite. One can identify three parts of the soul, he claimed, because they are sometimes in conflict with each other. A person may have an appetite for something, yet have the willpower or spirit to resist. A correctly operating soul, just as a correctly operating state, requires the highest part, reason, the philosophic capabilities, to control the lowest part, appetite, with assistance from the will or spirited capabilities.

The appetitive capabilities are usually associated with the body or sensuality, and with material gain. But these can be ruled by reason. The spirited capabilities, the will, can give rise to self-consciousness and self-assertiveness, a sense of self or pride. These too can be ruled by reason. The will can give rise to courage, when ruled by reason; but it can also give rise to brutality, ambition, contentiousness, indignation, as well as temper when it is not. It is only the philosophic capabilities, reason, which can develop into wisdom.

Plato's Dichotomy of Mind and Body

It is clear that Plato's view of intelligence was also based on a larger complex view of the human. As a metaphysical dualist, he regarded the body and soul as separate entities, positing an "unreal" world of the senses and physical processes as well as a "real" world of ideal forms. He held that only the soul or reason could perceive the ideal forms, exemplified in truths of mathematics. When the body and the soul combine, the body obstructs the soul's ability to recall the ideal forms. "Knowledge is not given by the senses but acquired through them as reason organizes and makes sense out of that which is perceived" (Zusne 1957; Plucker 2003).

Plato's theory of the human, *all* humans, was the first attempt to set forth a taxonomy, a systematic classification, of human being or behavior. He set forth that classification in terms of capabilities. In the ideal state, those persons whose capabilities for reason are strongest are to rule the rest. They may be male or female, slave or free. And the good life is the life of proper development of the psyche or soul, in which reason rules the appetites with the help of the will.

Later, however, Plato's student Aristotle (384-323 B.C.E.) held that the highest levels of reason are found only in certain persons. The highest levels of reason, said Aristotle, are found in men. Some will rule and others will be ruled *by birth*. ". . . from the hour of their birth, some are marked out for subjection, others for rule" (McKeon 1941).

In *Politica*, containing his writings on intelligence, Aristotle claimed that man rightly takes charge over woman because he has superior intelligence. Moreover, he compared this to the relationship between human beings and tame animals.

"It is the best for all tame animals to be ruled by human beings. For this is how they are kept alive. In the same way, the relationship between the male and the female is by nature such that the male is higher, the female lower, that the male rules and the female is ruled" (McKeon 1941).

For these same reasons, Aristotle continues, some people are *by nature* destined to be slaves:

"That person is by nature a slave who can belong to another person and who only takes part in thinking by recognizing it, but not by possessing it. Other living beings (animals) cannot recognize thinking; they just obey feelings. However, there is little difference between using slaves and using tame animals: both provide bodily help to do necessary things."

Invoking an *a priori* theory of natural kinds, instrumental value, and limitation of social and civil rights applied to those of purported inferior intelligence, he claimed that a slave is no more than a tool of his master.

"Together with the wife and the ox, a male or female slave is a householder's indispensable beast of burden. . .But slaves have no right to leisure or free time. They own nothing and can take no decisions. They have no part in enjoyment and happiness, and are not members of the community" (McKeon 1941).

Aristotelian Dictum: Anatomy and Intelligence are Destiny

In part, Aristotle held that the development of intelligence follows biological development. Stating a principle that continues to be repeated today, he held that those that are larger and stronger are more intelligent. Female animals, including human females, are smaller and weaker than male animals. From this fact it followed, according to Aristotle, that women are not only smaller and weaker than men but rationally inferior as well.

For these same reasons, he argued that it is right even to go to war against those who are inferior as are slaves and women. They are inferior in intelligence. They are therefore inferior by *nature*. Thus it follows, according to Aristotle that it is only right that they should be conquered, ruled, and taken as slaves.

It is first in Aristotle that the Western world finds a theory of the superiority and inferiority of certain people tied directly to a theory of their biology which in turn is directly tied to a theory of their intelligence. This interconnected theory of intelligence in whole or in part we continue to find today.

Moreover, contrary to his teacher Plato, Aristotle held that the body and the mind exist as facets of the same material being. The mind is simply one of the body's functions. The mind is the intellect which consists of two parts, the passive and the active. The intellect is not a set of capabilities as found in Plato's writing, but is:

"in its essential nature activity. . .When intellect is set free from its present conditions, it appears as just what it is and nothing more: it alone is immortal and eternal . . . and without it nothing thinks."

The psyche or soul is the actualization of the capacities of the human: nutritive, sensitive, and rational. The capacities are of the body hence the soul cannot exist without the body. He described the psyche as the "form" of the intellect, a substance able to receive knowledge.

Knowledge is obtained through the psyche's capability of intelligence, and the five senses are also necessary to obtain it. Foreshadowing later philosophers such as Hume, as well as some naturalists today, Artistotle also held that the senses are stimulated by phenomenon in the environment. Memory is merely the persistence of sense impressions. Mental activities are all primarily biological. He also held that the body and the psyche form a unity and that thinking requires the use of images.

While some animals can imagine, only man thinks, according to Aristotle. Knowing (*nous*) differs from thinking in that it is an active, creative process leading to the recognition of universals. Universals are the essences of particular existing things. Induction is the primary way of *knowing*. It is a direct insight into those essences in particulars. Particulars are individual, non-repeatable, unique. Everything that exists is particulars. Knowing is akin to intuition; it does not cause movement, and it is independent of the other functions of the psyche (Zusne 1957).

Early Differences Between Theory and Practice

Aristotle sorted *theoria* (speculating or contemplating) from *praxis* (acting). This was the original division between theory and practice, a division that would develop and be reinterpreted through the centuries fundamentally splitting body and mind.

These two fundamental distinctions between body and mind, and between *theoria* and *praxis*, came to have the most profound effect upon our present day concept of intelligence. After Aristotle and culminating in the works of Descartes with full sanction of powerful religious authorities, these distinctions were later conceived as unbreachable dichotomies.

For Aristotle, however, *theoria* consists of rational activities related to knowledge of universals, and *praxis* consists of rational activities related to moral activity. But in our present day notion of intelligence, and as this distinction has been reinterpreted over many centuries, rational activities related to theory consists of language representations of *knowledge that*. It is this that now largely defines intelligence while rational activities related to any practice or acting— *knowing how* –related to a broader concept of *praxis*—are left out of the intelligence picture altogether. Today, *theoria* is intelligence while *praxis* is merely doing.

2.2.2 Anthropocentrism, Language, Gender, Race, Size, Wealth, and Place

It is interesting to note that before Aristotle, Plato had a far more complex view of women and slaves. He did not regard either as inferior in intelligence or as inferior by nature. In principle, they could attain the highest levels of human achievement in intelligence and in all fields of endeavor. ". . . all the pursuits of men are the pursuits of women also. . ." he said (Gould and Wartofsky 1976).

But later religious and most philosophic tradition followed Aristotle on the intellectual and "natural" superiority of men. This view was reinforced with the rise of Christianity, especially the Catholic Church, and the influence of Aristotle. Texts of the Old Testament were translated in terms of Aristotelean theories.

It was St. Augustine (A.D. 354-430) who later attempted to reconcile apparent contradictions in Biblical passages that imply that woman as well as man reflects the divine image, while St. Paul claimed women were inferior. St. Paul called *man* the glory of God, while woman is the glory of man. He also prescribed different rules of conduct for each, greatly restricting the range of women's freedom and liberty. Augustine concluded that it is in man alone, as well as man together with woman, that the divine image is reflected. By herself, however, woman, created as the helpmate of man, does *not* reflect the divine image.

Centuries later, with the rise in power of the Catholic Church, Thomas Aquinas (ca. 1225–1274) based many of his own ideas on those of Aristotle, metaphysically interpreting them to fit his Christian theological framework. He claimed in the *Summa Theologiae* and elsewhere that the inferiority of women lies not just in bodily strength *but in force of intellect.* Nonetheless, he says, such inferior intelligence contributes to the "order and beauty of the universe." Arguing a principle that is also repeated today, woman's inferiority is just part of the nature of things.

Thus just as Augustine and Aristotle before him, Aquinas reduced the value of women to instrumental worth because of her weaker physical strength and because of her inferior intelligence. Unlike men, she has no intrinsic value of her own. Worse, women's supposed inferiority was part of God's plan. It was in the *nature of things* for women to be inferior. And any presupposition to the contrary was "unnatural" and therefore against the will of God.

The Intrinsic and Instrumental Intelligence Difference

Just as others had done so before, Aquinas invoked Aristotle's *a priori* theory of instrumental value to rationalize claims about the intellectual inferiority of women and their use by men (George 1999). That theory is found implied in biblical passages such as 1 Corinthians 11:10, in which St. Paul says that "man was not created for the sake of woman, but woman was created for the sake of man." This passage echoes Genesis 2:18-19: "It is not good that the man should be alone. I will give him a helpmate."

In the theory of value, when it is claimed that one thing exists for another, then that thing is inferior to the other for which it exists. The inferior exists as an instrument to be used by the superior. For Aquinas, biblical sources were absolute authority and not to be questioned, not even with his rather profound intelligence. The Christian bible claimed that woman was created for the sake of man, so it had to be.

It followed that if women were inferior by nature in their intelligence, then they were likewise inferior also in society and in the home. He took other passages as indicating that intelligence is the core of woman's "divinely ordained" inferiority. In 1 Corinthians 11:3 St. Paul says "man is the head of woman," and in Ephesians 5:22 "a husband is the head of his wife." Aquinas takes it as evident that men are meant to rule women by virtue of intellectual superiority because they are by nature superior to them.

The views of Aristotle and Aquinas are reflected in varying degrees throughout Western philosophy, even today. Their basic arguments are still heard especially, though not exclusively, in religious circles. Theories of intelligence coupled with theories of the brain and behavior are found merged with underlying theories of the moral or instrumental worth of whole groups in larger political contexts (Gottfredson 1998).

The Intelligence Center of the Universe

In most philosophic theories during the classical and medieval periods, mankind—but more specifically the human male—was regarded as the center of the universe. The highest levels of intelligence, found only in human males, according to Aristotelian doctrine, adopted as dogma by the Church, made human males God's centerpiece.

As such, the human male became regarded as the highest point in the hierarchy of all being in the entire universe. The human male was regarded as a reflection of God, having been made in His image by virtue of his superior intelligence. His was a superior life, based upon this reflected divinity, based in part on his assumed superior intelligence.

As earlier noted, such theories were used to rationalize slavery and war against those considered inferior. Before the period known as the Enlightenment, monarchies headed by kings together with religious leaders used the same theories embedded within so-called sacred texts and legal systems to make women and others legally dependent upon and in some cases owned by husbands and masters. Though the state may own all men not of royal birth, it was men who owned women.

Under various such legal systems as the Napoleonic Code, for example, women were made totally dependent upon their husbands. It was the husband who chose where she lived, controlled her property, if she had any, and had absolute authority over her and her children. If she was caught committing adultery, the husband could order her imprisoned, divorce her, and even kill her.

Theories of intelligence are not mere academic exercises. Such theories can and do have far-reaching, sometimes unimaginably horrible consequences to the value, quality, and length of life itself.

2.2.3 The Fabric of Concepts Defining Intelligence Since Darwin

It is fitting that our review of the Western origins of theories of intelligence began with Classical theories of Plato and Aristotle since it is Aristotle who is later regarded as the father of psychology.[2] Until around the mid-nineteenth century, writings about human intelligence and mental ability generally were dominated by both Platonic philosophy and Christian theology interpreted with Aristotelian principles, especially as found in Aquinas.

It is with the publication of Darwin's *Origin of the Species* in 1859, however, that human and animal behavior began to be seen as products of mechanical evolution in natural selection. Heredity became a focal point of the science of intelligence, yet still within the context of theories of the psyche, mind, or soul left over from 2,500 years of philosophy mixed with religion.

Since the late nineteenth century, psychologists have analyzed human behavior in terms of what are called theoretical "constructs" and causal chains of certain kinds. In some cases, American psychologists took a theoretical construct meant to explain a small segment of human behavior and expanded it to explain everything in human behavior.

---

[2] Aristotle's *De Anima* (On the Soul) is sometimes referred to as the first book of psychology.

The concept of Intelligence Quotient (IQ) is just such a construct. Wrested from the context of empirical findings and warnings of its originator, Alfred Binet, the concept of IQ was redefined and born on the crest of a wave of political conservative forces in the United States. It was taken to explain everything, including deplorable social conditions that accompanied huge poverty-stricken immigrant populations pouring into the United States in the 19th century to fulfill the needs of industrial development.

Early intelligence tests were used at Ellis Island to deny the "feebleminded" entrance to the country; they were also used by the U.S. Army to label entire groups of men as morons who nonetheless could be trained to fire a rifle; and they were used to inspire a eugenics movement resulting in forced sterilization of minorities, among other often heinous and cruel abuses (McLaren 2005).

The general concept of intelligence we currently find in the U.S. is an outgrowth of this history. It is related to a number of other concepts, some of which are inherited from earlier Classical and medieval theories. Certain of these are central to an understanding of what intelligence is taken to be at its core, while others are peripheral though they may nonetheless play a significant role in our scientific understanding of it.

What the central concepts are and how they have been claimed to be related to intelligence is something we should review. This is so because the meanings of these concepts *determine the very scope as well as meaning of the universe of intelligence* as it is understood today. These in turn can direct intelligence research efforts in a certain direction, and they can be used to effectively quash efforts in other directions.

Though each discipline may "carve up" the problem space of an object of inquiry in its own way, they each contribute to a broader understanding of that object, sometimes sharing tools such as methodologies, experimental and clinical strategies, concepts, distinctions, definitions, and assumptions to do so. Clearly, intelligence generally, and natural intelligence in particular, are objects of both multi- and interdisciplinary inquiry.

The concepts related to intelligence, as they are currently largely understood, determine the scope as well as the content, the elements, of the set or universe of intelligence. Taken together, they form an interrelated network that is so tightly woven together that the larger integrated picture must be seen before we can assess whether or not what we are looking at is distorted.

In other words, we must get clear on how the problem space has already been carved to determine any biases to the cut. Though the above historical review of Classical and some medieval theories already demonstrates many biases as to the *distribution* of intelligence among persons, and resulting inequalities and social abuses following that distribution, we did not look at the nature of the concept of intelligence itself for biases. That is what we will now do.

Reason, Logic and Language

Following the ancient theoretical stage set by Plato and Aristotle, "intelligence" has been defined almost exclusively in terms of the ability to *reason*. And reason, divided earlier by Aristotle into *theoria* and *praxis*, with clear preference given to *theoria* as that kind humans have most in common with the gods, emphasizes the use of rational tools by the mind. That is, it emphasizes the use of logic, also originally conceived and formulated by Aristotle in the form of deductive syllogisms, though logic has since developed far beyond those.

Minimally, reason includes the ability to plan and solve problems. It is obviously a core concept of intelligence. Following Aristotle, the concept "reason" eventually came to be defined almost exclusively in terms of concepts and rules of *logic*,[3] concepts and rules of evidence (which are sometimes taken to be reducible to rules of inductive logic), and the use of number. These rules are necessary to plan and solve problems. Thus logic also became another core concept of intelligence because of its integral relation with reason: reason is understood to be wholly dependent upon logic.

Even more narrowly, however, logic is itself sometimes taken to include only those concepts and rules that are publicly expressible by means of certain elements (not all) of public formal *language*[4] *or symbolizations*. Performing deductions and inductions expressed as propositions[5] in declarative sentence form or their tokens was (and is still)

[3] The discipline of logic eventually expanded far beyond Aristotle's syllogistic (deductive) forms that dealt solely with categories. Syllogistic logic is limited in part because it leaves out all of relational logic, does not utilize operators and connectives, among other things. Additionally, Aristotle and later logicians recognized the power of generality that can be gotten by substituting special symbols for terms of arguments, as well as limiting the ambiguities involved in natural languages. This eventually led to the development of highly technical symbolic notation to permit the expression of complex ideas.

[4] A formal language is one completely defined in terms of its syntactics (rules of form); semantics (rules of content); and pragmatics (rules of function). It is an artificial language with rules laid out from the start which completely, with the exception of primitive terms, defines it. Formal languages are in contrast to natural languages which are historically given, are described as having "semantic thickness" or meaning, and which evolve with use.

[5] A proposition is a complete unit of thought that can be asserted in a declarative sentence. It is that unit in language that is considered either true or false. Such complete thoughts as "The chair is red" or "Steve is late" are propositions. They are either true or false,

thought by some to be both the necessary and sufficient conditions for reasoning about anything. Thus logic in turn is understood to be wholly dependent upon the grammar and elements of language. To the extent that the relation between logic and formal languages is not held to be as strict as this interpretation suggests, nonetheless, the rules of logic are held to be necessary and sufficient conditions to reason.

The network of concepts and definitions so far says that the meaning of intelligence is that it is entirely tied to reason, which in turn is defined entirely in terms of logic, taken to mean performing deductions and inductions expressed in declarative sentences or their tokens in language. All the network of concepts related to intelligence in general, including natural intelligence, is within the broader domain of language ability, specifically the use of public language, including its elements and grammar.

Number

It should be stressed that this understanding of intelligence also applies to the use of number, indeed to the entire scope of numerosity. A highly arguable and *unproven* if not outright false assumption is that the ability to develop and use numerical concepts depends upon the ability to use language. It is assumed that all of mathematics is merely a subset of the larger set of formal languages and that the ability to use numbers or other concepts of mathematics is entirely derivable from the ability to use language (Gelman and Butterworth 2005). I will raise significant arguments and empirical evidence against this assumption later. Needless to say this theory of numerosity is based upon the same network of questionable and problematic concepts that define the larger theory of intelligence of which it is a part.

In effect this integrated fabric or network of concepts and definitions so far paints a picture of intelligence as a subset of the domain of public alphanumeric language ability, including both formal and natural languages. This point should be stressed about this development of the concept of intelligence: there is no part of the entire domain of intelligence that is not included within the set of public language ability. Intelligence is understood as subsumable within speaking, reading, and writing natural or formal languages. It should also be stressed that this view of intelligence as within the domain of public language ability functions solely according to the rules and principles of logic.

---

while the directive "Close that door," is neither true nor false. Though philosophers often talk of propositions as abstract entities that may or may not get expressed in declarative sentences, for all practical purposes we can think of them as such.

However, it is not clear at this point that the converse holds. That is, it is not clear that the entire domain of public language ability is included within the domain of intelligence. In fact, we will see that this converse does not hold. Indeed, this view of intelligence omits certain forms and functions of public natural language itself.

This network of concepts and definitions reviewed so far demarcating ("carving") the problem space of intelligence has carried over to *natural intelligence* where that includes the intelligence of lower animals. Natural intelligence has been taken to be a subdomain, a subset, of the larger category of intelligence.[6] Until recently, natural intelligence, indeed intelligence in general, was largely thought to be found solely within human beings. Only human beings were thought capable of performing logical or number operations with their minds, so the very notion of lower animal intelligence was left out of the picture of intelligence altogether. Besides that, only humans had language abilities. Or so it was thought.

Knowledge

Moreover, continuing on the stage set by the Classical and some medieval philosophers, intelligence was tied to the concept "*knowledge*". The aim of our ability to reason was to give us knowledge of reality. Recall that for Plato only the perfectly real is the perfectly *knowable*; these are the Forms. It is the philosophic capability of pure reason that permits human beings to know anything at all. Aristotle disagreed, as did many later philosophers, but they all tied the notion of mind and reason to some kind or kinds of knowledge. This in turn affected their theories of how one learns or comes to know anything.

Our contemporary notion of "knowledge" is that it is defined in terms of certain elements of language, called "propositions."[7] These are elementary units of thought captured or represented in language or symbolizations that are at least in principle capable of being shown to be true or false. In general, the philosophic community still does not recognize a broader concept of *knowing* that extends beyond language propositions. As currently conceived reason deals *only* with propositions

[6] Again, readers should keep in mind that I use the terms "intelligence" and "natural intelligence" interchangeably, in contrast to artificial intelligence.

[7] A proposition is a complete unit of thought that can be asserted in a declarative sentence. It is that unit in language that is considered either true or false. Such complete thoughts as "The chair is red" or "Steve is late" are propositions. They are either true or false, while the directive "Close that door," is neither true nor false. Though philosophers often talk of propositions as abstract entities that may or may not get expressed in declarative sentences, for all practical purposes we can think of them as such.

that are asserted in declarative sentences. Only those propositions shown to be true can be put together to comprise knowledge of any kind. Thus intelligence is further defined and delimited in terms of this concept of knowledge.

Moreover, this tapestry of concepts defining intelligence also demarcates those who "have" it from those who do not. According to some contemporary intelligence researchers, even if lower animals have some primitive form of language, they cannot be said to *know* anything. This is so in part because their languages (or any other means of communication), if they have any, do not operate according to recursive rules permitting an infinite number of possible combinations (Pinker and Bloom 1990; Pinker 1994; Dennett 1994).

Contradicting increasing evidence to the contrary (Pot 1997; Fagot et al. 2001), only human language can do that, they say. Moreover, animals do not perform logical operations because it is held that they do not have minds, though they have brains. And they clearly have no concept of truth, a defining concept of "knowledge." Animals behave solely in terms of "hard wired" instincts, primitive senses, and motor responses, it is argued. And instincts, primitive senses, and motor responses have nothing to do with intelligence, so it is still claimed. Though some may speak of the senses and instincts in some metaphorical sense as "intelligent," they really are not, so the argument goes.

Thus far, this family of related concepts seems to paint a rather concise and decisive picture of what intelligence is and what living things "have" it and which living things do not. Human beings have it; other living things do not.

To summarize briefly, intelligence in general came to be defined in terms of reason which is defined in terms of rules of logic, which in turn *define* the rules of reason; and in terms of knowledge, consisting of truthful assertions that are in turn expressed in terms of certain elements and the universal grammar of public language.

This notion of intelligence was later expanded by intelligence test designers to include reasoning about designs and patterned blocks, with the capacity to put them in some logical order, which means according to a logical rule describing a pattern. This latter expansion was about as far as any notion of practical reason, what Aristotle called "rational activities related to activity" or *praxis*, was allowed to go.

The Continuing Cartesian "Split": Body and Mind

Until about the early 1980's, the intelligence universe of discourse was largely carved just this way by researchers and test designers. The larger universal set was *human behavior* divided into three domains. The domain or set of intelligence within that universe was demarcated in terms of the intersection of reason (rules of logic), public language, and knowledge. These included within them the rules of grammar and the application of rules of logic. Problem solving was viewed as the application of rules already known, such as rules of logic and numerosity. This intersection was labeled by psychologists as the *cognitive domain*.

The other two domains of human behavior include the *affective* (or emotive) *domain* and the *psychomotor domain*, but only the cognitive domain is taken to include intelligence. These three major domains were held, and are still largely held, to be mutually exclusive, and to exhaust all kinds of human behavior.

Some recent research has looked at areas of possible overlap between the domains, particularly emotion and reason. Most if not all of that research focuses upon how emotions *contribute* to reasoning effectively about our daily or professional lives. Additionally, some research also focuses upon how to reason about one's and others' emotions. These studies sometimes refer to Emotional Intelligence as though they are referring to a separate kind of intelligence (Grewal and Salovey 2005). However, as far as I can determine, none of the research addresses fundamental mechanisms of affect *within the structures of intelligence or reason itself*.

In other words, emotion and reason may be related in that one may act on the other but they are still fundamentally mutually exclusive categories. Reflecting more than one dichotomy going back to Aristotle, reason is cognitive; emotion is still not cognitive but may have *contributory or instrumental value* to successfully reason (Grewal and Salovey 2005; Damasio 1994; Ekman 1980; Feldman et al. 2001).

The view of many in the behavioral science community was and still is that if there are kinds of human behavior that do not fit into one or the other of the three domains, then it should be put into a "junk" or "other" category. The junk category has included such behavior as improving or improvising, innovative, and even merely "different" ways of approaching a practice or accepted way of doing something.

In the history of psychology, for example, the junk category often included kinds of behavior or thinking "outside the box" that cannot be described or explained within the three-domain framework as kinds of intelligence or as emotive or as psychomotor behavior.

Underlying this network of related concepts, assumptions, and definitions is the notion that all intelligence is a *single* thing. Intelligence is largely conceived to be a single human ability or aptitude exhibited in public language ability governed by grammar and logic used to assert or record knowledge or solve problems with rules already known. The presumption was and still is that intelligence is found only in the cognitive domain, not in either the emotive or psychomotor. And of course, the cognitive domain was and is still defined in terms of language and logic, which in turn defines intelligence.

Language containing the clearest expressions of intelligence was entirely stripped of everything except the formal rules of logic and propositions or their symbolic tokens. It was stripped of all emotive forms and functions, exclamations, and all idiosyncratic, figurative, metaphorical, indexical and personalized expressions. Certainly, it was stripped of spontaneous and commonly used three-dimensional manual gestures human beings often make while speaking.

Likewise, it was stripped of all sentences referring to the senses or any other internal states unless objective instruments could independently and publicly verify them. Truth, a defining property of knowledge, was operationally defined; it was placed solely within the domain of public syntactical rules of language and logic.

Additionally, that single thing, intelligence, was identified with only one part of the human body, the brain. But it was not found in the *total* brain, only one part (or at most two parts) of it. It was only found in the language centers of the cerebral cortex. It was definitely not found expressed or exhibited by activities of the rest of the human body unless *descriptions* of what the body did were reducible to *descriptions* of what the brain told the body to do. Then it was reducible to propositions, to declarative sentences of the form, "The brain tells (causes) the body or central nervous system, CNS, to do *x, y, z* (effects)." And those sentences were operationally defined, hence verifiable. They could be shown to be true or false relative to a set of rules defining verification, truth and falsity. They were reducible to public knowledge hence they were expressions of intelligence. If they were not so reducible, then they were not expressions of any intelligence.

By virtue of limiting intelligence to language asserting or recording knowledge, anything that a human being *knows how* to do with their body was held to be *motor*, not cognitive, behavior. Anything that a human being *knows how* to do not reducible to language sentences about what they know how to do was not considered a part of intelligence.

In case it may be missed, it is important to stress the underlying Cartesian split between body and mind, as well as the split between theory and practice inherent in this carving of intelligence space. On the Cartesian mind- and proposition-centered view of human intelligence, the body is merely a reactive machine doing what the mind tells it to do.

In the Cartesian world-view, if a person knows how to do something, say, play a viola, what they know how to do must not only be reducible to a set of sentences, propositions, *describing* what they know how to do, and *prescriptions* telling them how to do it. The person must first read and understand the prescriptions telling them how to do it and then follow those prescriptions to put into practice what the prescriptions tell them to do. Only then can a person actually successfully and intelligently do anything.

On the Cartesian view, the location of intelligence is in the mind which though found in the brain is not reducible to it. This is so because the brain is a material substance and is therefore a machine. To some extent following the Platonic model, the mind controlled by pure reason is that part of the human we have in common with God. The body is mere material substance that must be brought under our control, the control of our proposition-embedded intelligence.

The Cartesian bifurcation of body and mind/intelligence is a view well-established in the Western world. In some materialistically modified ways, it is also firmly established in the usually accepted taxonomy of human behavior in psychology and more recent theories of human behavior and intelligence.

More to the point, on the Cartesian view, intelligence is not found in the rest of the body or what the body does. It is certainly not found in the senses or in any motor behavior. It is given that the mind controls the body, so the body gets its instructions for doing whatever it does from the mind. It is worth repeating: knowing how to do something that involves one's body, like playing the viola or playing tennis, or knowing how to throw a curve ball, therefore, must be reducible to sentences *about* knowing how to do something. Knowing how to do anything must be reducible to sentences and instructions in order for it to be included in the domain of intelligence because that's what intelligence is.

In outline, this is the fundamental story approved by Church authorities when Descartes first put it forth in 1637, a *post hoc* justification by a famous philosopher for many abusive Church practices. It was directly taught or assumed by thousands of priests and teachers and philosophers over the next more than 350 years.

However, in the 1940's, setting a new stage to be followed over 60 years later by philosophers and neuroscientists alike, Gilbert Ryle in the *Concept of Mind* nailed his theses to the door of Cartesianism, overthrowing one of its most fundamental principle of faith: The reducibility of *knowing how* to *knowledge that.* Though barely noticed at the time, and still largely ignored, that reducibility was shown to be logically impossible. The sacred mind-body dichotomy used to shore up so many other false dichotomies and myths was itself shown to be a myth, to encompass legends that served merely to prop up other arguments that turned out to be just so much puffery.

Logically and empirically, *knowing how* to do something is not derivable from a description or prescription *about* knowing how. Knowing how is found *in the person*, not in language. It is found in the person actually doing something whether it is performing some physical task with the body or performing mathematical proofs with the mind.

It turns out, Ryle showed, that *knowing how* is a different kind of intelligence. It is altogether different from the traditional Cartesian- and Church-approved *knowledge that*. Knowing how is a kind of intelligence. It is not a set of "mindless" motor responses.

With Ryle's overwhelmingly tight arguments shaking the foundations of Cartesianism, those who held to the faith were left scrambling for a counterargument. They *had* to show him wrong in order to shore up the long-held mind-body dichotomy and its principles. They had to shore up the dichotomy because so many other things were built upon it. It was that basic dichotomy, after all that allowed the Church, kings and princes to keep whole groups in virtual subservience as instruments, purportedly "in the service of God." It was that dichotomy that formed the foundation of rationales for keeping whole groups such as women, serfs, and slaves from learning how to read and write and enter certain professions. It was that dichotomy that served to support the very *status quo* and foundation of Church-approved society itself.

In spite of overwhelmingly powerful arguments, the tight hold of the mind-body dichotomy, declarative sentences, propositions as the pinnacle of intelligent expression, would not die. It would not die because of the mind-in-the-brain-centered, linguistic, verbal notion of human intelligence.

In effect, the dogma continues today shoring up the intelligence testing industry. Knowing how to do anything unrelated to language speaking and writing ability is not really knowing anything at all. The problem space was carved such that professionals in the fields studying intelligence could not and still largely do not acknowledge *nonpropositional* intelligence.

Later, when sciences showed those portions of the brain where language and cognitive activity (so defined) occurred, it was claimed that the Western world had finally located the "seat" of intelligence in the brain. Language centers in the cerebral cortex of the brain, conceived as the "controller" of all intelligent activity, are held to be the center of that intelligence, excluding other parts of the brain and the rest of the body.

Making the Natural Artificial

Intelligence was stripped of the body, except for the brain, or certain parts of it; stripped of the senses; stripped of emotion; stripped of things the body can do that are not dependent upon the use of language, those sensorimotor behaviors. In an interesting twist, the concept of natural intelligence was largely made *artificial*. In other words, this single thing, this natural intelligence found in humans, was stripped of everything that made it "natural" in the first place.

Indeed, tests of *natural* intelligence, the intelligence found in living things (specifically humans), that have gained much attention, actually pit humans against machines. Human intelligence is pitted against computers, archetypes of artificial intelligence. It is the *artificiality* of artificial intelligence that has, in some professions, become the standard for *all* intelligence.

It is certainly not the *naturalness* of intelligence found in living things that is the standard for all intelligence. This is so, first of all, because we do not know what that naturalness is; and secondly, because our Western concept of intelligence has so evolved that we have stripped everything from the concept that might have led us to a greater understanding of it. Patterned after deterministic causal machines, intelligence is held to be determined by the mechanical application of rules of logic, number and grammar.

The Western world's notion of intelligence is a narrowly algorithmic, rule-*governed* notion, leaving out any notion of intelligent rule-*bound* activity that includes knowing how to do anything at all beyond engaging in verbal behavior.[8]

The Intelligence of the Large and Small

In one way or another, size has almost always been correlated with intelligence. For some time it was generally supposed that because human males tend to be physically larger than human females, they must therefore be more intelligent. This was a belief firmly held by Aristotle, who also held, contrary to his contemporary Galen, that the location of thought was in the heart, not the brain.[9] When mental activity was later shown to be related to the brain, it was then held that humans with larger, heavier brains are more intelligent than those with smaller ones. Males have larger and heavier brains than females, hence (so the argument goes) they are more intelligent.

[8] Needless to say, it also excludes creative behavior. One psychologist at a major state university in the state of Texas once stated to me that "if a student wants to study creative behavior, we send him or her to the religion department."

[9] Like many Classical Greeks, Aristotle held that the heart and not the brain was the center for human thought. According to the Classical Greek view, all living things obtained their vitality or life from *pneuma* (air). Growth, movement and thought were made possible by alterations of *pneuma*. Disagreement over the role of the brain in mental activity persisted until at least the 18th century.

Human males are thought by some even today as more intelligent than human females in part due to their (on average) larger, heavier brains. However, the brain of the sperm whale on average weighs 7,800g, the elephant's weighs 7,500g, and the typical human male brain weighs 1,500g. By parity of reasoning, if size and weight of the brain are used to determine intelligence, then the sperm whale and the elephant are about five times as intelligent as the human male.

It is also believed today by some that the relationship between brain weight and body size is a criterion for comparing intelligence. In that case, if we ranked humans and the above animals along with mice in order of how large their brains are in relation to their body weight mice would rank ahead of them all with their brain comprising 3.2% of its body weight.

In fact, however, we know that neither absolute brain weight nor the relationship between brain weight and body size provide us with rational standards for comparing the intelligence of different species.

Nonetheless, many who study intelligence apparently prefer to find it only in creatures that are about our size and who speak a language we understand. It is a lot easier to study ourselves and others like us than those creatures who may be a lot smaller or a lot larger and who may not speak at all.

Until relatively recently, larger animals such as elephants, whales, and giant octopi have been excluded from intelligence research. Some of the exclusion may be attributable to the difficulties with extending research into natural environments or setting up research laboratories and other conditions to permit studying them.

Brainless Intelligence and Intentionality?

Even on the other end of the size spectrum, amoebae are still excluded, not just due to their size but because they have no brain. Usually seen under a microscope, with no central nervous system or brain, they are automatically excluded as intelligent creatures. This is so in spite of the strong reservations of some scientists (Stewart 1998).

At a minimum, we require that intelligent beings, no matter what size they are, must have a brain. This is a fundamental unproven assumption tied to yet another fundamental unproven assumption. In observing behavior, we tend to attach intelligence to behavior that is intentional, purposeful. An underlying assumption is that only creatures with brains can act with intention and purpose. Therefore, the argument goes, only creatures with brains can have cognitive abilities, can have intelligence.

Even more narrowly, some argue that only human brains can act with intention and purpose because only human brains have language centers with universal grammar. Yet examined closely, though we are quick to attribute such intentionality, cognition, and intelligence to ourselves and others like us, we demur attributing these to other creatures, based largely on some of the same unproven assumptions.

In the case of both ourselves as well as the *amoebae* we are not even certain what constitutes *intentional*, cognitive behavior. Though the *amoeba* may look as though it is acting intentionally and with purpose, we claim that it really is not. After all, how could it, given that it has no brain? However, just what evidence do you have that even though your distant neighbor, with whom you have never spoken, and who acts as though he or she has intentions and purposes, he or she *actually does have* intentions and purposes?

Aside from observing your neighbor's obvious behavior, just as you examine the *amoeba's* behavior under a microscope, you have no way of knowing. Yet because your neighbor looks like you in relevant respects—and you are pretty certain that your neighbor has a central nervous system (CNS) including a brain while the *amoeba* does not, you freely attribute intentionality and purposiveness, hence intelligence, to your neighbor that you withhold from the *amoeba*. Do we have a rational basis for withholding intentionality, purposiveness, and intelligence from amoebae?

Even focusing solely upon higher primates, we do not yet know enough about natural intelligence even among the higher primates other than ourselves to claim that they do not "have" intelligence or that they do not possess intelligence at higher levels.

We are also not even certain what "higher" means here. Given the tendency to establish a hierarchy when carving the space of such an important concept as intelligence, we must view with some suspicion those hierarchies already noted that were established within a network of biases formed over many centuries sometimes solely on the basis of the approval of unquestioned religious authorities.

Historically the concept of "higher" intelligence has been correlated with the use of certain abstract logical and mathematical concepts in problem solving. Where this has fallen short of good predictions of intelligent behavior, then it has been correlated with the *speed* with which such concepts are used (Jensen 1998).

But these claimed indicators of "higher" intelligence are still tightly woven within the above historical network of concepts that have been taken to define intelligence in the first place. Given the dubious assumptions interwoven within that network, turning to fast performances on cognitive tests appears to be yet another form *of begging the very question(s) at issue*. It assumes what it seeks to prove.

It is that network of concepts and underlying questionable, unproven assumptions that are subject to challenge. Among other things, evidence we will examine later shows that human ancestors were certainly endowed with some forms of intelligence long before they became language speakers and long before they acquired an understanding of explicit concepts of logic and mathematics. So among other assumptions we have already examined, the assumption that intelligence requires language will not hold either.

Those who claim that we *do* know enough about intelligence to establish this hierarchy are basing their claims upon the above questionable network of concepts, definitions, and assumptions that largely rest on the ability to use language and logic. Their claims also rest on a network of assumptions related to size, gender, and outright anthropocentrism that appears to have been largely left over from pre-Enlightenment centuries of adherence to religious dogma and magical belief. Taken collectively, these concepts are largely little more than a form of begging the very question at issue.

2.3 Today's IQ Tests: Circularity, Bias, and American Eugenics

Even assuming what we already claim to know about intelligence, we do not know enough to assume or claim that only one gender, race, or ethnic group tends to "have" it at its purportedly higher levels (whatever we take that to mean) than others. Though some researchers have set forth highly elaborate arguments to show that such differences are due either to genetics or "cultural pathology" of those groups scoring low on standard IQ tests, their arguments have been shown to fail to meet scientific evidentiary and logical standards.

Some of the problems are related to the way otherwise respectable scientific techniques are used to make the argument. For example, the use of principal components analysis combined with factor analysis to establish the existence of *g* is fraught with errors of reification and other fallacies. Moreover, extending such arguments way beyond existing data to appeal to over-arching genetic or "cultural pathology" explanations is not warranted either by genetics or test data. Unfortunately, these arguments sometimes appear to support the authors' biases more than provide explanations of statistical differences in IQ test performance, even given some of the dubious assumptions those tests are based upon.

Contrary to some researchers, there are alternative explanations to statistical differences in IQ test performance between genders, races, ethnic, and socio-economic groups. Moreover, IQ tests are based not only upon the same set of assumptions, concepts, and definitions we are challenging here, but on other unproven and dubious assumptions as well.

2.3.1 The Economic Argument

One of the most insidious assumptions underlying the arguments of those who hold that current IQ tests are a valid measure of general intelligence or *g* is the assumption that Darwinian survivability in a U.S. type *laissez faire* economy is the best broader test of general intelligence. This assumption is held by those researchers advancing a theory of genetic superiority/inferiority and cultural pathology, such as criminal behavior, to explain racial differences in IQ test performance.[10] Their argument goes something like this:

1. Those who score well on current general intelligence tests tend to do well economically in the United States.
2. Those who do not score well on those same tests do not do well economically in the United States.
3. Those same populations who do not score well on those same tests and who do not do well economically in the United States are statistically often the same populations who end up on welfare rolls and in trouble with the criminal justice system.
4. Thus, those groups not scoring well on IQ tests and not doing well economically in the U.S. are genetically or culturally *determined* to have inferior intelligence.

Among other problems, the premises with evidence missing are those tying low IQ scores, low economic performance in the U.S., and increased numbers on welfare and in trouble with the criminal justice system, to genetic profiles of whole racial groups.

Setting the genetic determinist argument aside for the time being, however, the fact that these same populations in other cultures with more humane economic arrangements tend to score on a par with more "successful" groups appears to elude these researchers.

[10] This view is shared by many in the single capacity intelligence research community; it can be found spelled out in some detail in many sources. I recommend Gottfredson 1998 as a start.

Moreover, those same populations in other cultures with more humane economic arrangements do not tend to end up on welfare or in trouble with the law in any significantly greater numbers than do other racial groups. The mountain of prior hidden, uncritical and biased assumptions, concepts, definitions, and methods underlying this view literally begs for critical scrutiny and question. We will return to specific issues surrounding genetic and biological determinism later.

More to the point, however, what good reasons are there to suppose that intelligence is a *single* ability or capability that can be measured by standard IQ tests or by Darwinian survivability in the U.S. *laissez faire* economy? *What does it mean to claim that such survivability in a specific cultural economic context is a valid measure of intelligence anyway?*

The Issue of Test Validity

The validity of a test is that it actually measures what it *says* it measures.[11] But the validity of a test must be subject to measures that are *outside the scope* of the test itself. Otherwise, what one gets is logical circularity, a question-begging kind of "self-fulfilling" condition that may not actually measure anything.

What do standard IQ tests claim to measure? Intelligence, as defined in terms of "*g-theory*," a single group of related cognitive abilities taken to define intelligence in the first place. How does the validity of standard IQ tests get measured? The validity of these tests is measured by the fact that they claim to measure what they measure. That is, the tests are designed to define, measure, as well as assess their own validity to measure the very things the tests say are the things they measure. In sum, it is the tests that define and measure what they say is intelligence and it is the tests that define and assess what they say is their own validity.

The circularity here should be apparent. This is not an innocent or benign circularity. It is vicious. Moreover, going beyond this to claim as well that the best general test of intelligence is the U.S. *laissez faire* economic system as a whole, with virtually uncountable and certainly uncontrollable variables, is worse than merely vicious. It becomes a brutal disparagement of poverty-stricken groups, let alone a scientifically invalid argument.

How did the subject of intelligence end up with this kind of politically charged as well as logical circularity? How did it end up with the extrapolation of test scores far beyond any justifiable purpose?

---

[11] There are different kinds of validity in test design, content, and construction. I am setting those other concepts aside.

When Alfred Binet founded the French School and devised intelligence tests, he regarded those as just practical ways of separating fast learners from slower ones. An intelligence quotient (IQ) was just an average of a number of very different abilities that he used so that he could determine how to help slower students. He did not regard IQ as a *real* existing thing found within the brain with properties that could be studied.

2.3.2 Reification and the Eugenics Argument

On the other hand, Sir Francis Galton (1822-1911), widely recognized as the father of behavioral genetics, promulgated the theory that intelligence was very much a real thing with a basis in biological, *e.g.* eugenics fact. Galton believed that this real thing, intelligence, could be studied using reaction times on cognitive tasks.

Galton is credited with having founded the English School of Intelligence that had more influence on the development of the concept of intelligence in the U.S. than did Alfred Binet's French School.

Influenced by his cousin Charles Darwin, Galton began to study heredity in 1865. He became convinced that success in life was due to superior mental faculties inherited by offspring through heredity. He later published a series of studies and books advocating the principle that "human mental abilities and personality traits, no less than the plant and animal traits described by Darwin, were essentially inherited" (Seligman 2002).

His studies and publications along with the growing popularity of his ideas, gave rise to the eugenics movement calling for methods of improving the biological make-up of the human species through selective parenthood. At one point, he proposed human breeding restrictions to curtail the birth of "feeble-minded" babies (Irvine 1986).

Though eugenics later became very popular in the United States as well as in Europe, the rise of the Nazi movement and WWII brought an end to that popularity. The Holocaust, based in part upon principles of eugenics, brought disrepute to the movement.

In 1925, however, Galton's theories were greatly advanced by Lewis Terman (1877–1956) who directly tied them to scores on the Stanford-Binet intelligence test. With this, "Galton's belief in the adaptive value of natural ability became thereby translated into widespread conviction that general intelligence provides the single most critical psychological factor underlying success in life"(Sternberg 2003). Among some racist groups, it also became one of the most critical factors underlying the worth of some human lives.

When Charles Spearman (1863–1945) invented what is called "factor analysis," Galton's English School of Intelligence gained favor among scientists. Using Galton's concept of correlation and his own concept of factor analysis, Spearman found that all tests of intelligence are positively correlated. In a matrix of positive correlation coefficients, he calculated a first principal component that he called *"g"* or general intelligence.

Confusing multiple correlations with *causal* explanation, he also fallaciously reified and interpreted *g* as a real entity. As Gould notes, Spearman "imagined that he had identified a unitary quality underlying all cognitive mental activity—a quality that could be expressed as a single number and used to rank people on a unilinear scale of intellectual worth" (Gould 1981, p. 251).

For the reader somewhat unfamiliar with correlation matrices, we should explain the rationale underlying the argument for *g*. The degree to which any two or more tests measure something in common is indexed by their correlation, labeled *r*, ranging from –1 to +1. A positive *r* means that individuals who score high on a given test also tend to score high on the other. On the other hand, a negative *r* means that high scores on one test go with low scores on the other.

When a number of different tests are administered to the same group of individuals, an *r* can be computed for each pair of tests considered separately. The results of those computations are what are called a "correlation matrix". On intelligence tests, the correlation matrix tends to consist of all positive *r*, but below 1.00.

When Spearman showed the first formal factor analyses of such correlation matrices, he concluded that a single factor, *g*, accounted for the positive correlations among tests. This is a notion still accepted by many psychometricians, but it is a notion fraught with serious theoretical and logical risks. Those risks usually involve the fallacy of reification, of taking a first principal component calculated for a matrix of correlation coefficients as a "real" entity, when it may not be.

Concluding that *g* is a real entity is not warranted unless there is independent evidence of *g* beyond the fact of correlation itself. The mere fact of a number of positive correlations of mental tests does not warrant the inference to the existence of *g*.

Many psychometricians consider the intelligence test that best measures *g* is the one that has the highest correlations with all other tests of intelligence. These are called "*g*-loaded" tests. Not only did Spearman regard *g* as a real existing entity, *general intelligence*. Some of his adherents have argued that it is genetically determined.

Later, however, L.L. Thurstone (1887–1955) rotated the factors in the analysis and found *several* primary mental abilities instead of only one found by Spearman. Thurstone thus disputed the prominence assigned to general intelligence, the *g* factor. Along with the spurious confusion of multiple correlations with causation, and the lack of any evidence of *g* independent of the correlations themselves, Thurstone's early findings, followed by many others, provide good reasons to seriously question or reject the single-capacity or "*g*" theory of intelligence altogether.

In the current field of psychological theories about intelligence, however, it appears that there are two opposing theoretical approaches. Some psychologists who follow the Galton-Spearman hereditarian argument hold there is one unilinear construct of general intelligence, *g*, ranking above and controlling all lower abilities and aptitudes. This construct is modeled upon a hierarchical grouping of capabilities with one general capability, *g*, held to be the controlling factor for all lower ranking capabilities.

However, increasing dissatisfaction with standard IQ tests and theoretical and evidentiary reasons have resulted in a growing consensus among other intelligence specialists from a variety of related professions that there are *many* different intelligences. Though this is not the prevalent view, the major disputes between the two camps indicate that a more fundamental, unified, logically consistent and exhaustive approach to intelligence is both warranted and necessary.

2.3.3 A Static Hierarchy: *g* the Controller

For more than one hundred years psychologists have addressed the nature of intelligence. Their efforts have produced a hierarchically arranged stack of conceptual "blocks" or aptitudes of behavior with general intelligence, *g*, as the top block on the stack. That top block is argued to be thoroughly measured by standard IQ tests, which provide a single quantity based upon a correlation coefficient of a series of declarative language, paper and pencil tests.

The stack of blocks is in essence a top-down, linear, functional-block oriented, serial processing model. In many ways it mirrors the Classical and medieval religious as well as Cartesian anthropocentric view, with "*g*" placed at the top of the hierarchy directing everything else falling below it.

General intelligence, *g*, is the top block on the stacked pile of mental attributes or abilities because it is conceived to be the high-level "controller" of all other blocks below it, the aptitudes "arrayed at successively lower levels" (Gottfredson 1998). Overall, the view of intelligence is fundamentally conceived as a rigid, static arrangement. Intelligence is not conceived as a *dynamic*, self-organizing, and adaptive *system integrated throughout* all other kinds of behavior. It is not self-acting or self-directive, but largely reactive.

As the concept "*g*" is explicated by its proponents to be general intelligence, the top block on the pile, while ranking categories of specific aptitudes and abilities below it, we need to take a very hard look at what these theoreticians specifically take *g* to be and how it relates to those other aptitudes and abilities that it controls. Just as importantly, we need to take a hard look at what they take to *not* be intelligence.

One of the proponents of *g* states:

"The *g* factor is especially important in just the kind of behaviors that people usually associate with "smarts": reasoning, problem solving, abstract thinking, quick learning. . ." (Gottfredson 1998, pp. 24-29).

It should quickly be noted that the kinds of behaviors she mentions are logico-linguistic, emphasizing verbal and written language, as well as the speed of their processing (*e.g.* "quick learning"). She continues by explaining the historical significance of these, given the findings of psychometrics.

"Several decades of factor-analytic research on mental tests have confirmed a hierarchical model of mental abilities. The evidence. . .puts *g* at the apex in this model, with more specific aptitudes arrayed at successively lower levels. . .verbal ability, mathematical reasoning, spatial visualization and memory, are just below g, and below these are skills that are more dependent on knowledge or experience" (Gottfredson 1998).

Based upon this particular researcher's descriptions, *g* is a mental aptitude that is found in "reasoning, problem solving, abstract thinking, quick learning. . ." And just what might comprise that aptitude expressed in reasoning, problem solving, abstract thinking, and quick learning? Clearly, the researcher is referencing kinds of rule-governed behavior where the rules are primarily those of logic.

The mental aptitude comprising *g* appears to be logic rules, along with rules governing number. This becomes more evident in the following description of tests and test items:

"Some tests and test items are known to correlate better with *g* than others do. In these items the 'active ingredient' that demands the exercise of *g* seems to be complexity. More complex tasks require more mental manipulation, and this manipulation of information—discerning similarities and inconsistencies, drawing inferences, grasping new concepts and so on– constitutes intelligence in action.

Indeed, intelligence can best be described as the ability to deal with cognitive complexity. . ." (Gottfredson 1998).

The concept of "mental manipulation" of information, unpacked to mean "discerning similarities and inconsistencies, drawing inferences, grasping new concepts," is a description of the application of rules of logic. For this researcher, applying rules of logic *is* "intelligence in action."

Moreover, the "cognitive complexity" referenced in the above paragraph turns out to be dealing with instances of logic problems "discerning similarities and inconsistencies, drawing inferences, grasping new concepts." So we know that *g*, this ability to deal with cognitive complexity, is expressed in the ability to use the rules of logic.

The scope of *g*, as explained by its proponents, is overlaid with elements of language. This is easily verified by examining the description the researcher gives of test conditions for determining individual differences in *g*.

"The general factor explains most differences among individuals in performance on diverse mental tests. This is true regardless of what specific ability a test is meant to assess, regardless of the test's manifest content (whether words, numbers or figures) and regardless of the way the test is administered (in written or oral form, to an individual or to a group). . ." (Gottfredson 1998).

These diverse tests are, in one way or another, tests of a person's language ability. They are tests of *knowledge that*. They are not tests of *knowing how* nor are they tests of visual and any sensorimotor intelligence.

Missing From g: Experience

Perhaps the most astonishing cut in this widely accepted carving of intelligence space is the relation between intelligence, knowledge, actual experience, and *knowing how*. This relation is at best muddled. At worst, there is no relation. This is evident in the hierarchical model as noted in the above quotation, partially repeated here, with my emphasis:

"The evidence. . .puts *g* at the apex in this model, with more specific aptitudes arrayed at successively lower levels… and below these are skills that are more dependent on knowledge *or experience, such as the principles and practices of a particular job or profession. . .*"

The reader should note where the researcher places actual experience. So pervasive is the "*knowledge that*" propositional bias in intelligence studies, that the prevailing view is that actual experience working with day-to-day problems or even knowing how to do things in one's job or profession, rank lower than *g*.

Underlying this assumption is that all the products of civilization, from actually setting out complex mathematical proofs, building supersonic jets, expansive bridges spanning our widest bodies of water, to the development and testing of medicines, super technology such as computers and even language itself, are all reducible to written descriptions and "how to" manuals to build them all. Knowing how to make anything based upon the awareness that comes with experience is left out of the intelligence picture altogether.

The powerful, far-ranging and demonstrably *false* assumption underlying this view is that *knowing how* to build any of these or operate with any of the machinery, including language itself—is reducible to sets of descriptions and prescriptions, our *knowledge that.* Yet of experience and *knowing how*, or "practical intelligence," the author states:

"Practical intelligence like 'street smarts,' for example, seems to consist of the localized knowledge and know-how developed with untutored experience in particular everyday settings and activities—the so-called school of hard knocks" (Gottredson 1998).

However, *knowing how* names a *different kind of intelligence* than the one purportedly found in *g*. It is one that is not measured by standard IQ tests. The usual interpretations of the *g*-theory dismiss *knowing how* as a kind of "street smarts" acquired by experience. It is astonishing that experience itself is given such short shrift in any theory of intelligence since it is by all other accounts the larger domain of which intelligence is a part.

Given the constraints the *g*-theory imposes on both facts about the objective reality of natural intelligence, as well as the means by which to measure it, it cannot give any account whatsoever of this kind of intelligence found pervasively underlying the most magnificent achievements of human kind.

More to the point, the researcher makes clear that knowledge and experience are not identical with *g*. Though she earlier made clear that one's accumulated knowledge often correlates with *g*, it is not *g*. Experience, or "street smarts" as she refers to it, "seems to consist of the *localized* knowledge and know-how developed with untutored experience. . .In contrast, general intelligence is not a form of achievement. . ."

As to other forms of intelligence, however, the author states the following:

"Other forms of intelligence have been proposed; among them, emotional intelligence and practical intelligence are perhaps the best known. They are probably amalgams either of intellect and personality or of intellect and informal experience in specific job or life settings, respectively" (Gottfredson 1998).

Why the researcher would cite emotional and practical intelligence as opposed to considering the seven or eight categories of kinds of intelligence proposed by Gardner (1993), is puzzling. With respect to practical intelligence, however, the researcher's view is certainly not in accord with research and arguments leading back in time as far as Aristotle's much earlier theory of *praxis*. Recall that he specifically defined this kind of intelligence as reason related to moral activities. It was not conceived by him to be idiosyncratic in any way, tied to personality or "informal experience."

With regard to emotional intelligence, the literature is not at all clear on this. As pointed out earlier, the current research appears not to address itself to specific affective mechanisms within reason itself, but to look for the *contributory or instrumental* value of emotions to effectively reason in one's everyday life and work. As far as I can determine, this is not another "form of intelligence."

Arguing that *g* is not affected by social context, the author goes further, offering conclusions based upon the above network of dubious concepts and even more dubious and some outright false assumptions. She states:

"the fact that *g* is not specific to any particular domain of knowledge or mental skill suggests that *g* is independent of cultural content . . .tests of different social groups reveal the same continuum of general intelligence. This observation suggests either that cultures do not construct *g* or that they construct the same *g*. Both conclusions undercut the social artifact theory of intelligence" (Gottfredson 1998).

Yet close analysis of what general intelligence is taken to be shows that *g is* specific to a particular domain of knowledge and mental skill. It is specific to the body of knowledge and skill called "logic" and it is specific to language skills, including the use of number. The rules of logic and even natural language grammars have universal rules and attributes which can be found cross-culturally, accounting for the claim "different social groups reveal the same continuum of general intelligence. ."

Of course, ultimately the conclusion that the author argues for is that intelligence is based in genetics and is, therefore, largely unmodifiable at birth. This in turn is based upon the network of interrelated dubious concepts and arguments defining intelligence as *g* in the first place, and upon dubious and outright false assumptions that largely beg the very questions at issue.

2.3.4 Biological Determinism Revisited

The *g*-theory architects and its advocates make no pretense at being anything other than genetic determinists when it comes to intelligence. The primary theses of *The Bell Curve* (TBC) include the claim that intelligence is predominantly due to genetic factors that are largely unmodifiable by education or experience.

Moreover, according to the authors, it largely determines life outcomes and involvement in poverty and crime. Those scoring low on IQ tests tend to have poorer life outcomes and show a higher incidence of involvement in crime. Nurture plays little or no part at all in life outcomes of individuals; it is only genetically determined intelligence that counts.

In spite of numerous research studies contradicting these and other claims, one of its authors recently asserted:

"As for the special case of the heritability of IQ, we can all sit back and relax. The answer is on the way, not from psychometrics but from genetics" (Murray 2000).

It should be noted at the outset that the Human Genome project was not complete when Murray and Herrnstein were writing *The Bell Curve*. Empirical data that might have been used to support such arguments was not available. Nor was it available when Arthur Jensen stated:

"differences in allele frequencies between populations exist for all heritable characteristics, physical or behavioral, in which we find individual differences within populations" (Jensen 1998).

But even if the Human Genome project had been complete at the time, it would not have provided evidence to support these and other genetic determinist claims.

Neo-Darwinism and the Heritability Argument

These authors' point of view on intelligence and later ones to be noted, reflect what is known as a Neo-Darwinist view of genetics. Not only is it a distortion of the science of genetics, but when used as a basis for determinist arguments about intelligence, it greatly undermines any rational, scientific understanding of the subject. A quick review of some basics of genetics, and behavioral genetics in particular, easily shows the problems with such arguments.

Some clarification of terms is in order. The term "allele" refers to a different form that a gene can take. Alleles determine the characters (the characteristics) of an organism. The word "character" is geneticists' term for any definable feature of form, pattern, or behavior of an organism. For example, the gene for seed color has at least two alleles, yellow and green.

For Neo-Darwinists, only genes and alleles matter. Organisms such as human beings are only a kind of secondary by-product of genes. As Stewart notes: ". . .the source of all important action [for the Neo-Darwinist] is seen as the molecule DNA" (Stewart 1998, p. 102).

Since its origin with Galton, the science of behavioral genetics generally has sought to understand the genetic and environmental contributions to individual variations in human behavior. This is enormously difficult if for no other reason than the fact that it is often difficult to define the behavior in question. We have seen that issue with the concept of intelligence itself.

Having defined "intelligence" for scientific purposes, however, as *g*-theorists claim, it is also necessary to measure the behavior with acceptable degrees of validity and reliability. As one noted genetic scientist observed:

"Sometimes there is an interesting conflation of definition and measurement, as in the case of IQ tests, where the test score itself has come to define the trait it [purportedly] measures. This is a bit like using batting averages to define hitting prowess in baseball. A high average may indicate ability, but it does not define the essence of the trait" (McInerney 2004).

This conflation of definition and measurement throws all claims to validity and reliability into question because arguments supporting it are fallacious. As noted earlier, they are instances of circularity or begging the very question at issue. More egregiously, any arguments and measurements based upon these are themselves in turn fallacious as well. Entire research programs based upon such fallacious claims and arguments cannot be valid and neither can be any conclusions drawn from them.

Adding to the complexity for such intelligence research are the enormous scientific obstacles to correlating *genotype* (an individual's genetic endowment) and behavior. Not only is there the above problem of defining the behavior at issue, but there is another problem in identifying, defining, and *excluding* other possible causes, thus allowing a scientific evaluation of any supposed correlation. This is made almost impossible for correlations between genotypes and such complex behaviors taken to define intelligence.

Added to this is the fact that individuals sharing a given genotype may end up with very different behaviors. Twin studies showing schizophrenics whose identical twins are unaffected by the disease are a case in point. Such problems are even greater for studies of intelligence, even given an acceptable definition of intelligence, which we do not yet have.

Like all complex traits, intelligence involves *multiple* genes. Hence the genetic and psychological research task is compounded many times over for multiple genetic contributions to highly complex behaviors. In general, as one scientist noted, "it is easier to discern the relationship between biology and behavior for chromosomal and single-gene disorders than for common, complex behaviors" (McInerney 1999). *There has been no such productive research to date with complex genetic contributions to intelligence.*

Nonetheless, the authors of *The Bell Curve* tied their conclusions to "substantial heritability" of intelligence. In something of an irony, this is in fact another obstacle to their genetic determinist arguments. Heritability is a statistical construct that estimates the amount of variation in a population attributable to genetic factors. But the explanatory power of heritability arguments is severely limited. Heritability figures *apply only to the population studied and only to the environment in place at the time the study was conducted.*

"If the population or the environment changes, the heritability most likely will change as well. Most important, heritability statements provide no basis for predictions about the expression of the trait in question in any given individual" (McInerney 1999).

The Neo-Darwinist view underlying much of the *g*-theory of intelligence reflects much misunderstanding of genetics. In the life of an organism, the role of genetics is not to determine everything about it. Genes orchestrate many physical and chemical processes carried out in an environment. They are not a blueprint for the fully developed human. Environments always have a determinant effect on what one's genetic endowment does. From the level of activity of individual cells, including cell division, to the level of behavior of entire colonies of organisms, we do not find a world "where everything obeys instructions coded in its DNA and nothing else matters" (Stewart 1998).

Indeed, as Stewart goes on to explain, DNA seems to exploit mathematical principles of growth and form found in the natural world surrounding it. DNA mechanics must be understood in conjunction with universal mathematical principles found in the rest of nature. Natural intelligence is more properly seen to be integrated complex patterns emerging *in part* from the acting, interacting, reacting, and transacting operations of the intentional unique person, their cultural and social context, their DNA, and universal dynamic mathematical principles of the natural world (Stewart 1998).

Contrary to the Neo-Darwinist arguments of *g*-theory proponents, there is no rational basis for ascribing every aspect of an organism to information in its genes. Indeed, once an organism gets going, genes appear to take on a *minimal* role in the growth and form of its intelligence.

The faulty Neo-Darwinist genetic determinist argument, however, concludes that intelligence is largely if not entirely due to genetic factors and it is largely if not entirely *unmodifiable* for that reason. Moreover, when public social policy is then based upon such arguments, institutions such as public education are seen to be largely an expensive waste of time. This point of view is rather succinctly stated by the following adherent of the *g*-theory:

"People are in fact unequal in intellectual potential. . . Although subsequent experience shapes this potential, no amount of social engineering can make individuals with widely divergent mental aptitudes into intellectual equal" (Gottfredson 1998).

Setting aside the red herring claims using terms such as "unequal" and "equals", not only is this argument based upon faulty genetic theory, it is contradicted by numerous empirical studies on the efficacy of public educational intervention alone to boost cognitive abilities. The following are just a few such studies cited in a five-year follow-up of research conducted after the publication of *The Bell Curve*:

"There appears to be a broad consensus that preschool Head Start-type programs .do not produce lasting improvements in cognitive ability, unless they are extremely intensive. One such intensive program, however, is the Abecedarian Project, the results of which showed a five-point IQ difference in favor of the treatment group at age 15" (Reifman 2000).

The author notes as well that even Jensen concurred that these results indicated a probable rise in the level of *g*. Additional research includes the following:

"These investigators found. . .black students' scores on the Armed Forces Qualification Test (the same cognitive ability test used in TBC [*The Bell Curve*]). . *rose sharply over the four years of college*, in comparison to a relatively flat line for whites. . .This shows the apparent error in Herrnstein and Murray's characterization of cognitive ability as a relatively static entity" (Reifman 2000, emphasis mine).

Additionally, (Reifman 2000) evidence was cited in the *Cognitive Acceleration through Science Education* (CASE) program, conducted with 11–12 year-olds, in which some program effects were present at a two-year follow-up, with transfer even to subjects such as English. Contrary to fallacious arguments for the genetic bases in determining life outcomes and involvement in poverty and crime, other studies were also cited that cast serious doubt on the significance Murray and Herrnstein attributed to IQ. With respect to involvement in crime, other researchers noted the following:

"Herrnstein and Murray would have first identified the known predictors of crime and then sought to demonstrate that IQ could explain variation above and beyond these criminogenic risk factors. . .By limiting their analysis primarily to three factors —IQ, SES [social economic status], and age — they risk misspecifying their model and inflating the effects of IQ" (Cullen et al. 1997).

The author rightly concludes that the available evidence contradicts the major genetic determinist theses of *The Bell Curve*.

A Short History of Rising IQ Scores

But the genetic determinist arguments of *The Bell Curve* are also contradicted by the history of rising scores on IQ tests themselves. Average scores on these tests have been rising substantially and consistently all over the world since the tests were invented, at a rate of about three IQ points per decade (Neisser 1997). Though the cause is not entirely known, Neisser hypothesizes that such environmental nurturing factors as better nutrition, schooling, better child-support systems and advances in technology may be part of the explanation.

In the United States, the most widely used and best known IQ tests are the Stanford-Binet and various Wechsler scales. These include five verbal subtests such as information, comprehension, arithmetic, vocabulary and explaining similarities. They also include five performance subtests in which a child must *copy* designs using patterned blocks, put related pictures *in a proper order*, and so on.[12] Scores on these subtests are added

[12] The reader should take note that these performances must be done in accordance with a test-approved rule. Independent individual problem solving and critical thought, especially thinking "outside the box," are not included in this scale of intelligence.

up and converted to an IQ by noting where it falls in the established distribution of WISC [Wechsler Intelligence Scale for Children] scores for the appropriate age.

To those unfamiliar with how IQ scores are determined, the *distribution* is the crucial reference for assigning a score. Conventionally, the mean of each age group in the standardization sample taken when the test was initially standardized defines an IQ score of 100. The standard deviation of the sample defines 15 IQ points. Given appropriate sampling and a normal ("bell curve") distribution, this means that about two-thirds of the population in any age group will have IQs between 85 and 115 (Kerlinger 1973).[13]

Thus IQ actually reflects *relative standing* in an age group. It does not reflect any *absolute* achievement of any kind. In spite of this, *g*-theory advocates nonetheless take IQ scores to define *g*, which they in turn take to be a real existing thing with properties found inside the individual's brain that is genetically determined. However, one child getting a raw score on the WISC that is higher than another may mean they both have the same IQ. And though raw scores may rise throughout their school years, IQ scores themselves "rarely change much after age 5 or 6" (Neisser 1997).

But IQ tests themselves *are* often changed. Revised versions of these tests are standardized on new samples and scored relative to those samples alone. The only way to compare difficulties of two versions of a test is by conducting a separate study in which the same subjects take both tests. Such studies have been carried out by Flynn (Flynn 1984) with results showing that in virtually every case the subject achieved higher scores on older versions of a test.

On versions of the Wechsler Adult Intelligence Scale-Revised [WAIS-R], a group of subjects who averaged 103.8 on the new WISC-R had a mean of 111.3 on the older WAIS. According to Neisser (1997):

> "This implies that the actual IQ-test performance of adults rose by 7.5 points between 1953 . . .and 1978."

Flynn's research showed increasing raw scores in every age range on every major test in every modern industrialized country. It has been "continuous and roughly linear from the earliest days of testing to the present" (Flynn 1987).

As to the purported genetically driven differences in mean test scores between African-Americans and the general white population of the

---

[13] See especially Kerlinger's section on factor analysis.

United States, which is about 15 points or one standard deviation of the IQ distribution, Americans have gained about 3 IQ points per decade, or 15 points over a 50-year period (Neisser 1997). And surprisingly, the largest "Flynn effects" [increasing raw scores] appear on highly "*g*-loaded" tests such as Raven's Progressive Matrices, a test very popular in Europe.

As Neisser notes, such increases in raw scores are entirely too rapid to result from genetic changes. ". . .*the sheer size of the gains undermines the very concept of 'genetic intelligence'*" (Neisser 1997, emphasis mine).

Moreover, alternative explanations for the Flynn effects, such as increasing levels of test-taking sophistication, do not work.

"An ongoing rise of 0.3 IQ points per year means, however, that if a representative sample of the American children of 1997 were to take that 1932 [Stanford-Binet] test, their average IQ would come out to be around 120. This would put about one-quarter of the 1997 population above the 130 cutoff for 'very superior'—10 times as many as in 1932. Does that seem plausible?" (Neisser 1997)

Additionally, doing a kind of "backwards" assessment:

"Judging the American children of 1932 by today's standards. . .we find that their average IQ would have been only about 80! . . Either America is now a nation of shining intellects, or it was then a nation of dolts" (Neisser 1997)

The enormous rise in IQ test scores, particularly on the most *g*-loaded tests, casts more than just serious doubt on the genetic determinist thesis of the authors of *The Bell Curve* and their followers. Indeed, it casts serious doubt on the legitimacy of the tests to measure intelligence in the first place.

Suspect Racial Sorting

Moreover, it should also be mentioned here that the authors of *The Bell Curve* sorted African Americans from other groups in their assessment of longitudinal, socioeconomic, and IQ tests. However, such a sorting is itself suspect. This is so even if it reflects self-sorting by the subjects themselves.

Research and knowledge obtained from the Human Genome Project (Royal and Dunston 2004) has challenged the applicability of the term "race" to human population groups at all. Human population groups are more appropriately described in terms of *genome variation*. Sorting human groups according to race in assessments of IQ test scores casts doubt or suspicion on any conclusions based upon any such racial "sorting."

But the evidence has not deterred followers of the *g*-theory who have turned to still other sources and methods to buttress faulty genetic determinist arguments. The emphasis upon DNA as the source of all things intelligent has been abetted with an increased focus upon the precision of mathematics in measurements of *g*. The study of intelligence has become highly mathematical. Mathematical methods hold the promise of objectivity precisely because of their precision.

But as noted elsewhere and clearly demonstrated even with their own psychometric instruments, especially as demonstrated in trends shown by the Flynn effects, precision is not the same as *accuracy*. One can very precisely measure the wrong thing, or make faulty inferences based upon very precise measurements.

Precisely measuring physical stature, brain size, speed of nerve conduction, and brain wave correlations under the pretense of measuring intelligence are examples of building additional faulty arguments on top of already thoroughly fallacious ones which in turn are based in part upon outright false assumptions.[14]

2.4 Summary

The history of carving intelligence space is that it has been cut on too many biases. Many of those are held over from centuries of unquestioned obedience to authority combined with a lack of critical thought.

It has been cut on the human bias because until recently lower animals have been largely excluded for a variety of reasons. Descartes held that animals had no intelligence because he held they had no souls; more recently, they have been excluded because it is claimed they do not have recursive language capabilities.

Intelligence space has also been cut on nothing more than a convenience and comfort bias, ruling out some animals as too large or too inconvenient to study.

Intelligence space has also over many centuries been cut on the language, knowledge *cum* propositional bias since it is only there, some argue, that we can have true sentences, the gold standard of intelligence.

Moreover, intelligence space has been cut on the public and verificationist bias ruling out references to internal states such as kinds of sensation, imagery, and sensorimotor awareness.

---

[14] This is amply illustrated by Gottfredson. See the reference above, in which she states: "brain size as determined by magnetic resonance imaging is moderately correlated with IQ (about 0.4 on a scale of 0 to 1). So is the speed of nerve conduction. The brains of bright people also use less energy during problem solving than do those of their less able peers."

It has been cut on the test bias, excluding those parts of intentional, purposeful behavior that are not amenable to existing data collection instruments. It has been cut on the brain bias excluding organic life that has no brain or spinal cord. More specifically, it has been cut on the language centers in the cerebral cortex bias, excluding other parts of the brain that we have known for some time play a large part in otherwise intelligent activity and practices.

Moreover, intelligence space has also for centuries been cut on gender and racial biases that still hold sway in spite of all evidence to the contrary. These biases are often based on spurious economic and other arguments by many professionals who ought to know better.

The domain of intelligence generally, but natural intelligence in particular, has been demarcated far too narrowly and too many of the cuts have been crooked. Of course not all of these biased cuts of intelligence space are accepted or implicitly assumed by all intelligence researchers. Nonetheless, some or all of the cuts are found liberally peppered throughout the research literature and current research strategies of many intelligence researchers.

3 The Genesis of Intelligence: Innate and Emergence Arguments

One of the most crucial issues in human intelligence research today is determining how the mind or brain makes sense of the world. Generally, this issue revolves around explaining how the brain forms concepts (also known as universals) or categories utilizing the extraordinary variation it experiences by way of sensory receptors processing multiple perceptual signals from the environment and within ourselves. This issue in turn is directly related to our understanding of major mechanisms by which the brain and body deal with information and form knowledge of anything at all.

Of course, the very framing of the issue is often in terms of language. Categories and concepts are often questionably taken to be solely linguistic or verbal elements. Unlike classes and representations, however, categories and concepts cannot be easily explained as language products. Moreover, to refer to processes or operations such as "abstraction" in order to explain how concepts and categories are formed is often to muddle the issue even more. The issue is then usually redirected to whether or not the mechanisms that generate concepts and categories, as well as classes or representations are "innate" or acquired.

There is a great deal of confusion attached to these terms themselves, hence we should get clear on those first then look at the issue of innateness and whether or not the entire problem should be framed the way it is.

3.1. Categorization, Classification, Concepts and Representation

To categorize a thing is to classify it as a kind. In effect, categories and classes are logically equivalent concepts. Classes and kinds are in turn defined in terms of groups of entities that share certain properties in common. Those properties are taken to define membership in that kind or class. In effect, our concept of the class in question is defined in terms of those properties shared by the things falling in the class. In the actual act or process of classification, we (correctly) classify things based on their properties and the *similarity* or *resemblance* of those properties with others of a kind or class.

There are at least two things to observe right away with the meanings of these terms. First, in order to correctly classify anything at all, one must already somehow be cognitively in possession of the concept of the class used to do the classifying in the first place. Second, one must already be in possession of, or understand the rule, by which one can discern and compare properties of the things being classified in order to determine their similarity or resemblance with the things already classified, and to note any dissimilarity.

A *concept* is a general or universal notion. It is a notion that is usually taken to be abstract that somehow applies to many things or events. Concepts clearly define classes. As youngsters, all other things being equal, we grow up learning the meaning of a concept by learning that certain properties of things, such as size, texture, color, and so on, are similar or resemble still other things that can be grouped or classified together. Exactly how we learn this is not clear. What is clear is that possessing concepts assumes that we already possess the means by which to group similar things together. We must already possess a rule that permits us to compare properties of things noting their similarity and resemblance as well as the dissimilarity and lack of resemblance of things that should not be classified together.

Obviously, the notion of a concept is that it is much broader than all the things that any person might experience which may be classified by means of it. The concept *triangularity* is an abstract mathematical notion or object, a universal. We will never actually experience triangularity, though we all experience many particular representations of triangles. There is a real sense in which concepts are ideas in the mind while representations may be "concretely" experienced.

A *representation* is usually defined as a presentation of something that stands for something else. That is, a representation can be a visible or tangible thing that stands in place of something else as an equivalent or that results in an equivalent. Representations can take many forms, such as symbols, pictures, patterns of action. For example, 1's and 0's can represent meaning in many forms such as literature, messages, paintings, actions. Drawings of particular triangles can be taken to represent the idea of triangularity however imperfectly.

3.1.1 Reality and the Influence of Representationalism

As we saw in the last chapter, the history of thought about human intelligence is filled with various theories directed primarily to answering the questions, "What is intelligence, mind, soul?" and "Who has superior intelligence, mind, and soul?"

In general, the concepts of intelligence, mind, and soul were merged, along with the concept reason. Classical philosophers sorted intelligence generally according to speculative (or theoretical) reason and practical reason, with the speculative held to be superior to the practical. With few exceptions, intelligence was held to be a property of humans. With few exceptions, it did not generally extend to the rest of the animal world.

Though there is no widely agreed upon definition of intelligence, nonetheless its boundary it is taken by most professionals these days to be the cognitive domain of human behavior. The cognitive, in turn, is taken to be identical with verbal knowledge, knowing that shows up in intentional or rule-governed linguistic behavior. Knowing anything at all is in turn defined in terms of reason that operates according to rules of logic and grammar. And though knowing has been largely limited to logico-linguistic performances, excluding practical doings and kinds of behavior not related to logic and language, strong arguments and evidence show that such a restriction is not warranted.

But intelligence researchers have often not addressed themselves to the question "What is *knowing*?" They assumed that either the question had already been settled, or that it is one they need not ask. After all, they had already behaviorally or operationally defined the cognitive domain, even if it did have all those problems we earlier pointed out. Nonetheless, the full scope of any study of intelligence must sooner or later address all of the *who*, *what, when where, why*, and *how* of knowing. In many ways, answering the *who* question is putting the cart before the horse. We must first know more about knowing.

During and immediately following the Classical period, some philosophers addressed themselves to that question "What is knowing?" and focused upon even more basic questions such as "What objects or things can we know?" along with the implicit question, "What objects or things are *real*?"

Further implied within these latter questions is that one cannot reasonably be said to know that which is not real. Amid much controversy over questions about reality itself, these questions were entwined with "*How* can we know, or *come to know*, anything at all?" With the birth of the science of psychology in the 19th century, this last question was reduced to theories of learning, defined behaviorally or operationally, which are now central in questions about the nature of intelligence generally.

But all the bundle of interconnected questions related to "What is *knowing*?" turn out to be like a woven fabric: one can not address a single thread without pulling on the others as well. To Plato, the standards for knowing something, as opposed to merely having an opinion or *belief*, seemed to require that *what* one knows, the object of one's knowing has to be objectively real, independent of the one who knows. Otherwise, one could not be said to know anything at all.

Thus the underlying prior question "What is real?" took center stage when he introduced abstract, immaterial, eternal Forms, later known as universals, as the only things that are perfectly knowable because they are the only things that are perfectly real. Knowing, universals, and reality were intimately connected.

3.2 The Continuing Problem with Universals (Concepts): Some History

The question "What is real?" soon turned to the problem with universals. Today, we refer to these as concepts. It is a problem that has persisted to this day. This problem might be stated as follows: How do we experience *universal* or common properties in *particular* things? When we see a triangle drawn on a piece of paper, for example, how do we know that it is a triangle? Why isn't it just a bunch of pencil marks? After all, we do not *see* triangularity the universal. We see particular pencil marks on a piece of paper. But the pencil marks are in the shape of a triangle that are drawn or otherwise represented on some tangible medium, the piece of paper.

On the other hand, triangularity the universal is a Form, a perfect abstract idea. As such, there is no way it can actually be found in those pencil marks on paper. An individual drawn triangle is a particular. And particulars, according to Plato, are always imperfect. They are the things we come to apprehend or perceive with our senses. Yet, we can look at a drawn particular triangle, as imperfect as it is, and see that it represents its ideal perfect Form or idea of Triangularity. We understand that the pencil marks represent a triangle.

Plato

Forms or universals which we experience with reason were contrasted by Plato with those particulars, bound by space and time, which we experience with our senses. It is particular things in the material world, like those pencil marks on paper that we experience with sight, touch, smell, hearing, or taste. But it is only with our faculty of reason that we experience universals or Forms such as triangularity. For another example, just as the name "Socrates" was the name of a particular individual, an existing flesh and blood human, it is Humanity that is the Form, the universal. Yet when Socrates' fellow Athenians saw Socrates the individual, particular human being, they knew he was a human being, a particular instance of the universal Form, Humanity. Yet the universal Form, humanity, was not to be seen in the flesh and blood making up the particular individual who was Socrates.

Even though particulars were conceived by Plato to be imperfect copies or instances of their universal, perfect Form, the problem Plato posed was that we could never actually *know* any instance or particular of any universal. That is because those particulars, those individual instances that are imperfect images of universals, like Socrates in flesh and blood, are fleeting and illusory.

For Plato, one cannot *know* anything with one's senses because they are always imperfect. At best, one can only have beliefs or opinions about sense objects. A drawn image of a right triangle is imperfect. The drawn lines are never perfectly straight. The angles are never perfect. Moreover, it can be erased or otherwise simply fade with time. Socrates the person was flesh and blood that must, sooner or later, die. On the other hand, one cannot see, touch, smell, taste, or hear perfect and eternal universals. One can only apprehend them with reason.

Plato's reasoning about this was generally as follows: Our senses can deceive us, so we can never be certain of anything we experience or become acquainted with by way of our senses. It is only the mind or reason, devoid of all trickery of the senses, that can provide us with any certainty and hence, real knowledge. And real knowledge is only knowledge of universals, never of particulars in the world.

Plato answered the bundle of interwoven questions, "What is real?" "What can we know?" and "How can we know anything?" in the following ways:

1. The perfectly real or *being* [not *becoming*] is the perfectly knowable.
2. To be perfectly knowable is to be an object of intelligence (reason), not an object of the senses.
3. To be an object of intelligence (reason) is to be a universal or Form. That is, it is to be an essence or common quality grasped by reason in a concept.
4. To be universal is to be unchanging. That is, it is to be permanent or stable and enduring. Essences do not change.

5. Pure thought is consistent. That is, the essences or Forms can be unified under the Form of Forms which is the One or the True or the Good or the Beautiful.
6. But to be an object of the senses [sight, touch, smell, taste, or hearing] is to be an appearance. It is the way an essence appears to one. It is a *becoming*, not a being.
7. A becoming is changing, an appearance, thus it is imperfect, it cannot be a universal. To change is to be non-permanent or non-stable and non-enduring. Appearances all change.

One of the first things to notice is that Plato's view of knowledge renders it impossible to *know* anything about the material world. We can only have opinions or beliefs about it. Our only relation with the material world is through our senses. As such, we can never have any certainty about the material world precisely because we cannot apprehend it with reason. We cannot apprehend it with reason because the material world is not perfect. It is fleeting, always changing, and never stable. There is nothing permanent or enduring about it. Only properties of perfection, permanence, and endurance apprehended with the mind, with reason, can give us certainty, and hence, knowledge.

Again, for Plato, knowledge was directly tied to these concepts of reason, perfection and certainty. Knowing was conceived by him to be a pure, *non-sensory* relation between reason, found in the subject (the person) and the eternal Form or universal that is the object. This knowing relation between a subject and object also assumes many other things. It assumes that *the capability of reason in a subject, a person, is itself non-sensory.*

Somehow, reason itself in a person *is separate from the body* of that person. The mind and the body are separate. The mind or reason could not be part of the body because the body is an imperfect, sensory-laden material thing that is non-enduring. Yet the object of the *knowing* relation, in Plato's theory, is an eternal Form, a universal, perfect thing.

Another assumption by Plato is that the Form or universal, the object of the knowing relation, is also *not* found in the subject. Universals or Forms are not concepts found "in the mind" of persons though it is the mind that puts the subject in contact with the universals or Forms. They do not inhabit the mind. Universals or Forms were thought by Plato to inhabit a different, non-sensory, immaterial, eternal realm. To mix imperfection of the body with the perfection of eternal Forms was conceived by Plato to be impossible. It was a contradiction, an abomination of reason itself as well as the Forms.

Yet another obvious assumption by Plato is that the answer to the question, "What is real?" is found in the Forms. The immaterial, non-sensory, eternal Forms are real, while the material, sensory world is not.

Plato's answer to the remaining question, "How can we know or come to know anything at all" was brilliantly set forth in his Allegory of the Cave. It is in that Allegory that he sorted out other concepts related to his view of knowing. In the process of coming to know, each person is initially shackled by their senses that constantly deceive them into thinking they know when they do not. They are surrounded by shadows of images of physical things that they take for reality. They are in a cave of darkness and cannot get out of it on their own.

In the world of the senses surrounded by illusions, the Cave, we cannot be said to *know* anything. We are in darkness. We can actually only make conjectures and have beliefs and opinions about things because those things are shadows of images of physical things. The things we perceive with our senses are actually many steps removed from the physical things that cast the shadows that we see. The only way we can have any relation with these shadows and images and physical things is by way of our senses. Using only our senses, since we do not yet know how to reason, we are tricked into believing those shadows of images of physical things are real. Yet none of it is actually real.

Once we get help from one who *knows how* to reason, however, we then slowly and painfully enter the Intelligible World of the mind, of reason. In the intelligible world, using reason, we can have knowledge which brings understanding. The intelligible world is a world governed by reason, not our senses. It gives us knowledge and understanding, not belief, opinion, or conjecture.

The sensible world is a world of becoming; the intelligible world is a world of being. In the intelligible world we are more truly human beings. Moreover, the Intelligible World is the world of mathematicals and the Forms. It is by means of reason using the mathematicals that we come to know the Forms, the universals, which are perfect, real and eternal.

The process of coming to know is slow and painful. Breaking free of the shackles of the senses and the world of shadows and illusions is not easy. The process is neither continuous nor smooth. It is a process of painful conversions from less to more adequate states, until reason is free of the senses, including free of the use of images in the mind.

Aristotle

In contrast to Plato, for Aristotle there was no separation between the mind and the body. Indeed, the psyche, mind or soul is conceived by Aristotle to be a substance able to receive knowledge. However, though knowledge is obtained through the mind's capability of intelligence, the five senses are also necessary to obtain it.

Foreshadowing a much more modern and scientific view, Aristotle held that the senses are stimulated by phenomenon in the environment. According to him, memory is the persistence of those sense impressions. As he describes the process, the senses receive "the form of sensible objects without the matter, just as the wax receives the impression of the signet-ring without the iron or the gold" (Editors of Britannica Online 1997).

Contrary to Plato, Aristotle maintained that mental activities were primarily biological, and that the psyche or mind was the "form" part of intellect. He insisted that the body and the psyche form a unity. That is, the body and the mind exist as facets of the *same* being. The mind is one of the body's functions.

Moreover, again in contrast to Plato, Aristotle believed that thinking, reasoning *requires* the use of images. While some animals can imagine, however, only man thinks.

"Knowing (*nous*) differs from thinking in that it is an active, creative process leading to the recognition of universals; it is akin to intuition, it does not cause movement, and it is independent of the other functions of the psyche" (Zuzne 1957 pp. 8–9).

Intellect consists of two parts: something similar to matter (passive intellect) and something similar to form (active intellect). Aristotle says that intellect

". . .is separable, impassible, unmixed, since it is in its essential nature activity. When intellect is set free from its present conditions, it appears as just what it is and nothing more: it alone is immortal and eternal . . . and without it nothing thinks" (Britannia Online 1997).

Later, medieval philosophers continued to address the same bundle of questions: "What is real?" "What can we know?" "How can we know or come to know anything?" They gave credit to Plato for at least offering an explanation of how universal properties of particular things are imperfectly modeled after their universal Forms. They gave credit to Aristotle for offering an explanation of how the human mind acquires its universal concepts of particular things from actual experience.

Though neither Plato nor Aristotle's views on these issues provide adequate explanations, their respective positions were something of a baseline from which later philosophers and scientists worked.

Working upon the stage set by Plato and Aristotle, it was later during the medieval period when three very different, separate and competing views or theories on these issues began to clearly form. These theories are still with us today and have great impact on various theories of both human and animal cognition, as well as theories of intelligence. These competing positions can generally be classified into one of three categories: as Realism, Conceptualism, and Nominalism.

3.2.1 Realists, Conceptualists, and Nominalists on Universals

Each of these views is highly diverse, so sketches below will provide only the highlights of each. For example, few realists of either the Medieval or later periods agreed with Plato's realism on universals. And some later versions of each of the theories can be found to be somewhat overlapping. But it is with the rise of conceptualism and nominalism that the history of thought about intelligence, about *knowing,* took highly complex paths away from realism.

In very general terms, realists are those who assert the objective existence of universals before the existence of particular things. Universals have a posterior, objective reality or existence prior to and distinct from the particular things that are instances of universals. Moreover, universals have an objective reality prior to and distinct from the universal concepts or words in our languages. Those universal or general words *represent* universal concepts.

For realists, the universals themselves, and the concepts and words in language or mind representing them, are two entirely different categories. *Realists never doubted the objective existence of universals, independent of minds that may or may not conceive of them, and independent of languages which may or may not have terms referring to them.* As we saw above, Plato's realism is possibly the best example of this.

On the other hand, conceptualists are those who allow universals only, or primarily, as concepts "in the mind." Conceptualists were an early form of a movement later known as Idealism that survives today in some theories of knowledge, such as coherence theories, mind and intelligence. The conceptualists held and still hold that universals have no reality apart from merely being concepts in the minds of persons. Universals are ideas in our minds that we have conceived or invented to explain certain kinds of common experience.

Nominalists are those who acknowledge only, or primarily, universal *words* (Klima 2004). In general, nominalists deny that universals exist objectively and independently of persons and their languages. Whether universal concepts are in the minds of persons or not, it is terms or words in languages that count. And it is only in languages, according to nominalists, that we can find words having universal meaning.

For nominalists, universals are not found outside languages. Universals are only so many pencil or pen marks on paper or vocalizations when we speak. It is languages, those alphanumeric symbols and signs we write or words we speak according to agreed upon rules that ultimately matter in all things related to intelligence, to the mind. Languages are the necessary interface, the representations, between a subject and a reality that may or may not "be out there" beyond us and our language. In fact, according to some nominalists, it is best that we speak only of language, of representations, not of any reality that may or may not be beyond it.

In our day, it is nominalism that has come to hold a predominant view of intelligence, indeed of all knowledge. Nominalists and conceptualists gave up the concept of objectively existing facts, existing independently of either persons' minds or languages, as a basis for the truth of knowledge claims. Indeed, the notion of objective truth was actually given up entirely. Perhaps the most extreme version of this is found in recent theories of the origins of mathematics, which has been described by an early proponent of nominalism as "a game played according to certain simple rules with meaningless marks on paper."[1]

3.2.2 Theories of Knowledge and the Scope of Intelligence

Since Plato and Aristotle, philosophers and some scientists have argued for many theories of knowledge, mind and intelligence.[2] Strictly speaking, those theories are attempts to get clear on the content as well as the boundaries or limits of intelligence. These are the Classical questions, "What can we know?" and "How do we know or come to know anything at all?" "Are there some things that are beyond human knowledge?" "Are there things we cannot ever know?"

The limits of intelligence have both a breadth and depth, but also a kind of ceiling as well as floor. Various philosophers and scientists over many centuries have addressed these or similar questions, giving different answers.

It is unfortunate that psychologists who tend to dominate the sciences of intelligence have given short shrift to the work of some philosophers. In

---

[1] There is some dispute about the authenticity of this quote, attributed to David Hilbert. It is quoted in E.T. Bell, *Mathematics: Queen and Servant of Science*, G. Bell & Sons Ltd, London, 1952, p. 21. It has also been quoted in N. Rose, *Mathematical Maxims and Minims*, Raleigh N C, 1988. Whether correctly attributed to Hilbert or not, it is an accurate reflection of formalists' view of mathematics.

[2] This section is intended solely to provide a broad summary of these theories; it is not an exhaustive examination of them and of the issues each theory entails.

some ways, this neglect has resulted in claims by psychologists about intelligence that have no firm basis, or in some cases no basis at all.

On the other hand, it can also be shown that philosophers themselves have tended to neglect major categories of *knowing* (not knowledge) due to nominalist/representationalist/conceptualist biases that developed over many centuries. For example, setting aside Aristotle's early examination of practical intelligence, the prevailing theories barely mention *knowing how* at all. Nowhere does it get a serious, detailed examination until the 1940's with the publication of Ryle's *The Concept of Mind* (1949).

Realism, Coherence, and Pragmatism

Generally, the major theories are divided into Realism, Coherence, and Pragmatism. The latter has recently developed into what is now referred to as Naturalism stressing a purely physical-causal account of mind, intelligence, and knowledge. In its early form naturalist philosophers tended to adopt simple behaviorist, stimulus-response (S-R) causal models to explain knowledge, mind, and intelligence. The most recent manifestation, however, has extended beyond simple S-R causal schemes to adopt multiple-causal theories based upon genetic determinism. This development has led to a host of other problems we will address.

Each of these theories essentially defines the concept of knowledge, but also mind and intelligence. The scope or reach of intelligence and mind are set forth in terms of theories or standards of truth for knowledge claims. Each of the different theories of truth developed by Realists, Coherence Theorists, and Pragmatists, were initially developed as excluding the others.[3]

Realists developed what is called the Correspondence Theory of Truth. The basic idea is that a sentence or proposition is true if and only if it corresponds to how things are in the world. Although the theory goes back at least to Aristotle, it has been given various new formulations, influenced by Wittgenstein, where propositions are true or false (and hence meaningful) if they picture states of affairs or facts.

Conceptualists, or Coherence Theorists, developed what is called the Coherence Theory of Truth. This is the view that truth consists in a relationship between sentences, sometimes called "truth bearers." Truth is *not* a relationship between sentences and facts objectively and independently "out there" in the world. For Coherence Theorists, a sentence or statement is true if it *coheres*, it is consistent with, logically follows or gains support *from other statements or sentences within a*

[3] Only later was consideration given to adopting all the standards, appropriately modified, to evaluate any given knowledge claim.

framework of beliefs or linguistic system. Conceptualists and Idealist philosophers have traditionally held the coherence theory of truth.

In the earlier American Pragmatist tradition, pragmatists emphasized the relationship between truth and successful scientific inquiry. In a sense, pragmatist theories claim that truth is whatever works or is useful in its entirety, based on science. Truth is a property of the end result of successful scientific research. Their understanding of successful scientific research, in turn, is whatever enables us to have control over nature.

American Pragmatist philosophers such as James, Dewy, and Peirce emphasized these connections between truth and science and that which has useful, long-term practical consequences. Some recent American philosophers, influenced by pragmatist conceptions of truth and knowledge, include W. V. O. Quine, Hilary Putnam, and Richard Rorty.

Each of these major theories in turn takes different positions primarily on two major issues: (1) What is real? (2) How do we know or come to know anything? The above standards for the truth of claims reveals deep divisions between them, influenced in part by traditional nominalism/conceptualism, but also by representationalism and debates between rationalism and empiricism.

Additionally, for some realist philosophers, not all objects of human knowledge are found in a *causal* relationship with a subject, the one who knows. For Plato, the Forms or universals existed outside any causal relationship with human beings. Some contemporary realists, allowing for human freedom of choice or decision, also hold that not all objects of knowledge are in causal relationships with a subject. Many philosophers who consider themselves materialists hold that everything is caused.

In essence, the basic issue between rationalism and empiricism concerns the extent to which our knowledge depends upon sense experience. Rationalists argue that there are ways in which we can have concepts and knowledge independently of sense experience. We have *a priori* knowledge such as mathematics, logic, and certain ethical or moral knowledge, according to rationalists. On the other hand, empiricists generally claim that sense experience is the only source of all our concepts and knowledge. All knowledge ultimately arises from our sense experience.

Though there are different kinds of realists, they all largely make two kinds of claims about the "realness" of objects. They make *existence* claims and *independence* claims. In essence, realists claim: (1) That real objects exist; and (2) The existence of those real objects is independent ofpersons and anything anyone happens to say or think about the matter. This last point should be stressed: for realists the existence of real objects is *independent* of anyone's experience, linguistic practices and conceptual schemes.

Moreover, for realists, the standard that a knowledge claim must meet in order to qualify as knowledge is the correspondence standard. For a claim or belief to be true, it must correspond to an objectively existing fact. For most realists, there are *a priori* facts such as mathematical facts and ethical/moral facts. These facts do not at all depend upon sense experience, thus realists are sometimes rationalists.

But they are also sometimes Empiricists in the sense that some kinds of facts do depend upon our sense experience in the world. There is a complicated sense in which the structure of an asserted proposition or claim must mirror (or correspond to) the structure of the objective fact. Those facts may be either *a priori*, independent of sense experience, or *a posteriori*, dependent upon sense experience.

The Language Interface Issue

In addition to the issue of whether or not there exists an objective reality apart from and independent of persons, another central issue that became paramount involves whether or not there is an "interface" of some kind that exists between the Subject (one who knows) and an Object (that which is known).

Because of the almost obsessive focus upon language representational systems, the rise of Nominalism virtually guaranteed that, of course, there must be some language or symbolic/representationalist interface between Subject and Object that filters (or in some cases blocks) the relation between the Subject and any reality that may or may not be out there beyond us all. Whether or not anything "out there" is real, and whether we know anything at all, are taken to entirely depend upon the nature of that interface between Subjects and Objects, if they exist at all.

Language representations were necessary, it was argued, because *meaning* required them. Everything must be represented or labeled with language to be meaningful and to be known. Furthermore, everything can *only* be meaningful or known through its label.

Not only that, the argument went much further to make what amounts to an unsupportable metaphysical claim: The *only existing objects* are those for which we have a (language) label. James was one of the few Pragmatists who lamented what he called the "usurpation of metaphysics by language" (James 1884) entailed by such a view. But it is also the usurpation of *knowing*, of intelligence, by language as well.

The argument stated that for all those streams of sensory information bombarding our nerve receptors at any time, only language as an interface can provide any meaning to it all. Moreover, so the argument went, though we do not have direct access to any reality that may or may not be out there beyond us, at least we do have access to our language reports about what we experience. Language was a necessary condition to make sense of that experience.

In a sense, Coherence theories and Pragmatist theories, based in part upon the language interface thesis, developed somewhat in reaction to and *against* kinds of Realism. Coherence theories, for example, are kinds of *anti*-Realism. Realists never doubted the objective existence of objects independent of persons and independent of languages and concepts. It is that objective and independent existence that gives us facts, and makes facts so stubborn and ineluctable.

Moreover, Realists argue that it is our *direct* acquaintance with facts that ends a potentially vicious regress of justification for our knowledge claims. Facts are out there independent of our language and independent of us. Everything we can justifiably claim to know must rest on the truth of our claims and that truth is ultimately resolved by reference to objective and independent facts.

But for Coherence Theorists, there are no facts or "truth-makers" existing independently of representations or concepts in the minds of persons. Coherence theories, historically an outgrowth of kinds of conceptualism, require a representational interface with the world that may or may not be beyond the knower and the representation or concept. In the end, all we really have are concepts in our minds.

Again, the coherence standard for a knowledge claim is that the claim must *cohere*, that is, be consistent, with other sentences about the world. This is in stark contrast to the correspondence standard which holds that the structure and content of the sentence or proposition must correspond with an objective fact in the world that exists independently of the Subject.

Others argued that clear conceptions of anything, to constitute a knowledge claim, must be related to their practical consequences. Peirce, the father of pragmaticism,[4] claimed that the whole meaning of a clear conception of anything consists in the *entire set* of its practical consequences. A *meaningful* conception must have some experiential, practical value. Meaning was tied directly to a collection of possible empirical observations under specifiable conditions (Burch 2001).

Peirce rejected realist foundationalist theories, holding that science only deals with phenomena, not "reality" which may or may not be beyond our phenomenal experience.

For all practical purposes, however, under some Pragmatist theories, *e.g.* Peirce's theory, it becomes virtually impossible to determine the clearness of a conception. That is because he defined it in terms of the *entire* set of practical consequences for all time. In its traditional sense, the

[4] This is Peirce's own term which he used in order to distinguish his own scientific philosophy from other conceptions and theories that were called "pragmatism".

pragmatic criterion for knowledge is set very high, virtually beyond the reach of any individual or group precisely because it includes the entire set of practical consequences of any clear conception. The idealism that may be evident in such a view, however, should not be misread. Peirce especially tied concepts to practical, *lived* consequences. He did not leave them "in the mind."

Contemporary Pragmatists, especially starting with Quine (1969), have turned to a more stark version of the theory, called Neo-Pragmatism or Naturalism (also *physicalism*). Certainly in contrast to Realists and Coherence theorists, the new Naturalism embeds everything within a physical, causal context. There is no *a priori* knowledge. There are no abstract objects or universals, only physical stimulations at our nerve endings and our observation sentences about those stimulations.

In a sense, the new Pragmatists or Naturalists are harking back to earlier Empiricism, given the perceived failures of Rationalism to satisfactorily explain human mind, intelligence, and knowledge. From the Naturalist perspective, every event has a cause; all our mental events have causes; our knowledge, hence, has a physical cause. Thus it is characteristic of naturalists to look for justifications in psychological processes, where those are reduced to causal physical processes, responsible for the presence of our beliefs and claims to know. Those causal processes, however, differ according to different naturalist philosophers.

As with the Coherence Theorists, for naturalists such as Quine there are no objective facts existing independently of persons. There are only communities of language speakers who share stimulatory events and observation sentences reporting on stimulations.

More recent naturalists argue for an even more determinist theory. For genetic causal determinists such as Pinker (1994), there are specific "grammar genes" that control the brain and behavior through linguistic rules and representations. Specific genes of DNA become the engine that drives everything in human mind, intelligence, and knowledge.

Naturalists argue that it is rational to be guided by the methods of natural science. Quine and others have argued that theory of knowledge and metaphysics should be approached in ways that are continuous with the sciences, or at least not inconsistent with them. In many respects, Neo-Pragmatists share much in common with early Behaviorists in that the fundamental underlying model of their theories is the basic S-R approach to knowledge, mind/brain, and intelligence.

A Postmodern Heritage and Realist Counterargument

An assessment of each of these theories with some representative arguments, demonstrate that Coherence and Neo-Pragmatist Theories, as reactions against kinds of Realism, opened the door very wide to Post-modernism. Developing on the historical waves of Nominalism and Conceptualism, the anti-Realist theories proceeded to break down traditional notions of objective truth, facts, and the possibility of human reason to discover or come to know either.

Each of the major theories of knowledge worked out by philosophers over the last approximately 2,500 years has different standards that a knowledge claim must meet in order to qualify *as* knowledge.

Realists of certain kinds hold that one can know some objects *independently* of sense experience, that we can know certain things *a priori*. Mathematics and kinds of moral knowledge are sometimes taken to fall into that category. This is represented in the above graph where Rationalism and Realism intersect. Plato's theory is probably the best example of this.

But, contrary to Plato, contemporary realists also hold that we can *know* facts and laws about the physical world. These realists hold the correspondence theory of truth. Claims to know must correspond to *objective* facts in the world, independent of persons, languages, and concepts. Moreover, that correspondence must be falsifiable, testable, and replicable. Bertrand Russell's theory is a good example of this.

For example, reflecting a realist point of view, Russell explains what a fact is: "Everything that there is in the world I call a 'fact'" (1948, p. 143). The computer screen in front of me is a fact, the ending of the Civil War is fact; my toothache is a fact. Facts are also characterized by Russell as independent of anyone.

"I mean by a 'fact' something which is there, *whether anybody thinks so or not.* Most facts are independent of our volitions; that is why they are called 'hard', 'stubborn', or 'ineluctable'. *Physical facts for the most part, are independent, not only of our volitions but even of our existence*" (1948, p. 143).

Facts are what make statements true or false. But "Facts are wider than experience" (1940, p. 383). They are wider than experience precisely because they exist independently of persons, of their language, concepts, and their experience.

Thus, facts exist objectively and independently of anyone. He defines "belief" in the following way: "A belief. . .is a certain kind of state of body or mind or both" (1948, p. 145-6). He continues, "A belief. . .is a collection of states of an organism bound together by all having in whole or part the same external reference." There are various kinds of belief, but there are standards that any of them must meet to be true.

"Truth" is defined as a certain relation between a belief and one or more facts other than the belief. When that relation is absent, the belief is false. However, problems result when we try to precisely delineate the relation which must hold. Nonetheless, it is clear that truth is the *correspondence* of belief with the facts that are independent of all of us. Actually believing a statement to be true is neither necessary nor sufficient for it to be true.

In later writings, Russell defined "knowledge" as a subclass of true beliefs. Every case of knowledge is a case of true belief, *but not vice versa*. However, it should be pointed out that though we are largely discussing *knowledge by description* here, that is knowledge set forth in language, he also set forth a theory of knowledge by *acquaintance*.

Contrary to virtually all contemporary philosophy, Russell held that one of the most important kinds of knowledge, (though he perhaps should have called it knowing), *knowledge by acquaintance*, does not require belief. In his 1913 theory of knowledge manuscript (1984) knowledge by acquaintance is *not* defined as a subclass of belief.

Acquaintance was his theory of immediate awareness, the relation of what is *present*— but not represented—to a Subject. With this, he dispensed with the nominalist requirement that there be a representational interface between a subject and an object in all kinds of knowing. This distinguishes knowledge by acquaintance from knowledge by description.

To summarize, conceptualists and nominalists hinged everything upon concepts in our minds and whatever games we play with our words. Realists want whatever we think or claim to be true to somehow correspond with the facts "out there" independent of us all, beyond our languages and beyond our concepts. Ideally, for realists, that correspondence must be replicable and testable. Early pragmatists agreed with realists on the requirement to test claims out in experience; later pragmatists also wanted to reduce all knowledge and belief to stimulations of our neural systems. Conceptualists and nominalists, on the other hand, will settle for whatever story our language rules permit.

In some ways, nominalists and conceptualists gave up the notion of a reality separate from and independent of the subject and posited instead a language-centered conceptual interface between the subject and object and within the subject. In extreme forms these theories hold that there is nothing beyond that interface.

The problem of universals was traded by nominalists and conceptualists for the problem of how the mind represents anything to itself. It survives today in linguistic and cognitive science theories that strive to prove the genetic origins of universals, including the rules of universal grammar

(UG) by which the mind does its representing with the use of universals. This is a continuation of a legacy of both nominalism and conceptualism, of idealism.

3.2.3 Today's Representationalist Myths: Cognitive Maps in the Brain

Both nominalism and conceptualism are evident in varieties of what is today known as *representationalism* that arose during what is called the Linguistic Turn in philosophy. The Linguistic Turn is described as having been born of Analytic Philosophy during the first few decades of the 20th century. It is characterized as having turned every philosophical problem into a problem about language, or at least into a problem *dependent* upon language.

Later, the Cognitive Turn, starting in the 1960's, was supposed to have replaced the emphasis upon language by turning back to a study of the mind. However, the Cognitive Turn, as we will see, has been little more than an extension of the Linguistic Turn. It still sees the mind and intelligence generally as a linguistic/symbolic representational system.

Representationalism is a theory that proposes kinds of linguistic/symbolic "pictures" of the mind, of intelligence. Mental representations are mental "words." Basically, it is a theory that depicts the brain or mind as filled with representations manipulated according to universal grammar rules. Recent variations on this to include vector notation do not diverge from the basic model (Churchland 1995). In essence, it follows the classical concept of mechanical device, a machine where the concept "machine" is identical with "algorithm" and "computer".[5] The brain, mind and intelligence generally are conceptualized as machines that process linguistic/symbolic representations according to universal rules.

The rules of the representationalist model are interpreted by some as "innate" mechanisms. In essence, the innate machine is a central processor that permits cognition to occur in both humans and animals. The notion of "cognitive map", used today by many cognitive scientists and those in artificial intelligence, captures this theory and model quite well. The following is one example of the use of that notion in explaining how animals, in particular honeybees, navigate:

[5] The terms "algorithm", "computer", "machine", "formal system", and "recursive system" are equivalent and identical in meaning.

"Cognitive maps are representations in the brain of the geometric relationships between salient sites in an animal's environment. . .They have a system in their brain for manipulating symbols associated with landmark coordinates. . ." (Hauser 2000, pp. 76–77).

In spite of the obvious intent to meet rigorous scientific standards, such talk of "representations in the brain" appears to be metaphorical. It is metaphorical because there *literally* are no such representations "in the brain." Moreover, technically, animals do not engage in "manipulating symbols associated with landmark coordinates." Honeybees are not symbol-using creatures; they do not have even the concept of a coordinate. Close inspections of the insides of honeybee brains have not turned up any neural counterpart to the mathematical concept of "coordinate" (Esch et al. 2001).[6]

This nominalist approach to the mind and intelligence has led to much confusion between representations *of* a thing and *the thing* represented. In this case, the author's theory appeals to representations of landmark coordinates as a way of explaining the remarkable navigational abilities of bees to locate food, return to the hive, tell his fellow bees about it, then return to the food site.

The author takes *his* theoretical representations to be cognitive maps located inside the brains of bees. Yet there are no such representations there. This confusion between a representation of something and the thing represented has led to collapsing levels of inquiry and fallacious inferences based upon that collapse. Though bolstered by much laboratory and clinical work, no matter how much empirical data is used to buttress such arguments, conclusions based on fallacies cannot be either valid or sound. As noted by Geach (1957), no experiment can either justify or straighten out a confusion of thought.

Setting aside the issue that good science does not permit appeals to metaphors, the representationalist strategy and its progeny, the computational theory of mind and intelligence, are actually attempts to *avoid* problems, not solve them. But in the representational attempt to avoid the problem of universals, it has unfortunately succeeded in introducing far more serious ones in addition to the original problem of universals which it has not disposed of.

Though the nominalist/conceptualist/idealist strategy was initially an attempt to side-step the realist's problem of explaining how universals can

[6] Latest research shows that the honeybee's navigational capability is due to retinal image flow on the way to their destination. They have what amounts to a visually driven odometer. Moreover, bee dances convey information about the direction of the food source and the total amount of image motion *en route* to the food source, but they do not convey information about absolute distances.

exist objectively and independently of the person, the representationalist strategy has actually introduced far more problems. In many ways, the nominalist, conceptualist, and idealist have merely placed themselves many steps removed from the same problem, introducing even more serious problems along the way.

For example, we might ask the nominalist: How can written or spoken *words* or "representations in the brain" have any *universal* reference? If we are speaking of written words, we are faced with the same problem above with drawn right triangles. A written word may merely be pencil marks on a piece of paper. We are still left with the question, "How can that written word have universal meaning?" By parity of reasoning, the same issue applies to spoken words, which may otherwise be just so much noise.

Moreover, some might argue that words, in both natural and artificial languages, can mean whatever we want them to mean. Once we are committed to a set of rules, however, of course those rules in part also determine meaning.

But we are no closer to answering the original question: "Why do universal words *have* universal or common meaning applied to many things, some applying in principle to an *uncountable* infinite number of things—as in the right triangle example?" Further: "How can the Pythagorean Theorem possibly hold for all *possible* right triangles?" There are, in principle, an *uncountable* infinite number of right triangles. How can the Pythagorean Theorem possibly hold for them all? How do you *know* that the Theorem applies to them all?

The nominalist can reply that it is all a matter of stipulating a definition. "It's all a matter of semantics." Yet that will still not answer the original question. We are still left with the problem of how a *word* in this language in this time and place can correctly apply to an *infinite* number of things for all time and place.

The nominalist has not solved a problem by stepping many places back from the realist's position and placing language in between him/herself and a reality that may or may not be beyond it. The nominalist has only introduced even more problems.

Likewise, for the conceptualist/idealist who holds that universals are concepts in the minds of persons, we can ask the same question, "How can a concept in the mind of one or many persons have *universal* reference?" "How can a concept such as triangularity hold for all triangles for all space and all time?"

Moreover, "How did the concepts get there in the minds of persons (or animals) in the first place?"

3.3 The Innate Versus Emergence Arguments

In many respects, the study of intelligence today has been claimed as the province of linguists and psychometricians. The former have also devoted themselves in the last several decades to "innate" determinist explanations for grammar or syntax,[7] to some degree following the same line of arguments found in *The Bell Curve*. It is syntax that makes language possible, they argue. Thus, it is syntax, in their view, that makes intelligence possible.

More recently, proponents of this argument have turned to the science of genetics, including physics, chemistry and evolutionary theory to make their case. Though the root argument was earlier advanced by Chomsky (Chomsky 1972), its latest advocates include Pinker and Bloom (Pinker and Bloom 1990) among others.

Based on claims that there is a universal grammar (UG) shared by all natural languages, it is their view that language is an evolutionarily *unique* development in human beings, not found in other animals. It should be kept in mind that since a prevalent view of intelligence is that it is *g*, an essentially logico-linguistic measure, it follows on their view that if language is genetically determined, then intelligence is genetically determined as well.

3.3.1 The Genetically Encoded Syntax Argument

As the term "innate" is often used ambiguously, we should get clear at the outset what these proponents of genetic determinism of intelligence mean by it. At the core of the innatist arguments is the claim that syntax is *genetically encoded* in human beings; it is not acquired.

There is purportedly an innate cognitive structure in a certain location in certain molecules in the brain that determines what basic structures and processes will be found in the syntax of any natural language. The notion of "innate" in their theory refers to that claim that there is a specific physical location in the brain, the "syntax module" or "grammar genes" that does syntax processing.

[7] Syntax refers to the set of rules allowing language speakers to encode semantic information and generate an infinite number of sentences. Semantics refers to the meanings of words.

Attributing language structures to specific genetic structures appears to be a category mistake of the most obvious sort. However, the innatist proponents seriously contend that the structures of Universal Grammar (UG) are found in specific nerve and DNA cells.

On one level, the underlying model of this particular argument is the same as the *g*-theory central processing unit, a machine view of intelligence. It is the model of a centralized control unit, a determinist causal agent, that drives everything related to language, hence also everything related to intelligence. Consistent with the single-capacity *g*-theory of intelligence, it is also a top-down, linear, functional-block oriented, serial processing model.

An opposing explanation of origins of language is the *emergent* theory. This view holds that syntax is not the result of specific syntax genes or "module" in the brain, even though some elements of syntax may very well be based genetically. Rather, the rules of syntax largely emerge in a highly distributed, parallel fashion as a consequence of how language develops evolutionarily (Schoenemann 1999). On this view, syntax largely develops as a set of cultural conventions that allow communication of our semantics, or meaning.

Setting Pinker and Bloom's argument almost on its head, one emergent theorist put the relationship as follows: "The grammatical and syntactic regularities that are found across languages occur because of the universals of semantic cognition, not universals of syntax and grammar" (Schoenemann 1999, p. 310).

More recently, advocates of the innate argument have proposed that language and its structures are found within molecular structures of the brain. They have proposed that the actual chemical workings of certain molecules produce kinds of sentence structure. They have utilized the language of physics and chemistry to attribute physical and chemical properties to language, including describing "weak" and "strong" forces between words to explain the language behavior of verbs and adjectives just as physicists explain the behavior of subatomic particles.

Setting aside the glaring reductionist fallacies involved in these proposals, we should look at just how much of actual language behavior could be explained this way. Among other things, in addition to faulty reductionism, the use of obvious metaphors such as "sentence molecules" begs many serious questions.

3.3.2 Nonverbal Communication: Beyond Alphanumeric Symbols and Vocalizations

A fundamental issue between the innatist and emergence theories revolves around the meaning of "language" in the first place. Among other things, depending upon how the two theories differ on the meaning of "language", it would also follow that their respective views of the meaning of "intelligence" will differ as well.

The innatist genetic determinist arguments hold that "language" is defined in terms of the rules of grammar and words as linguistic units. Their definition *excludes* nonverbal communication means, such as gestures. The rules of grammar are a recursive set of rules that allows humans flexibility to express meaning in alphanumeric written or spoken sentences.

Human language as defined by the innatists is held to be far more flexible than found in any other communication system. It allows humans to generate in principle an infinite number of sentences of any desired complexity. Moreover, language in this sense permits us to generate sentences that go beyond the immediate present and to refer to events in other places at other times. It can be used to imagine, fantasize, and to describe events that have never existed and never will. In this sense, Chomsky, Pinker, and Bloom claim that language as they define it is entirely unique to humans; it is not found in lower animals.[8]

It should be pointed out that the innatists often appeal to distinctions between *communication* systems and *languages*, arguing that though lower animals communicate, they do not possess language. In a sense, this is an attempt to do a definitional end run around any opposing argument. It is a way of defining a problem out of existence. Again, they stress that language is a unique development among human beings having no real counterpart in the animal world.

However, most natural language dictionaries, and even more technical ones, do not agree with the innatists' definitional distinctions between communication and languages. The term "communication", strictly defined in an ordinary college dictionary (American Heritage 1993) is:

"The exchange of thoughts, messages, or information. . .Interpersonal rapport."

To communicate is:

[8] It is this infinite capacity that leads some philosophers such as Daniel Dennett (1994) to claim the intelligence superiority of humans over all other forms of life.

"To convey information about; make known; impart. . .To reveal clearly; manifest. . .To have an interchange, as of ideas. . .To express oneself in such a way that one is readily and clearly understood. . .To be connected, one with another. ."

Consistent with the meaning of "communication", "language" is defined as:

"The use by human beings of voice sounds, and often written symbols representing those sounds, in combinations and patterns to express and communicate thoughts and feelings. . .A nonverbal method of communicating ideas, as by a system of signs, symbols, gestures, or rules. . .Body language; kinesics. . .The manner or means of communication between living creatures other than human beings"

Given only one of its usages, the word "language" refers to only one kind or means of communication, though it is broadly understood as largely identical and equivalent with "communication". Nonetheless, the innatists demarcate the meaning of "language" such that it is reserved for human beings, largely due to the recursive rules which permit infinite variety in sentence formation and meaning.[9]

But analysis of the variety of means by which both humans and lower animals communicate shows that this narrow demarcation is not justified. Moreover, the delimitation of "intelligence" to language behavior on the part of humans is even less justified.

Gestures

In opposing the innatist arguments, some emergent theorists point to 3-dimensional gestural communication systems among lower animals as well as humans as evidence of a broader scope of the meaning of "language". In their view, language not only includes 2-dimensional alphanumeric symbolic written and spoken systems with recursive rules, but extends far beyond those to include intentional signs, cues, gestures and other intentional patterns of whole-body performances, movements, and nonverbal communication behavior.

For example, not only are manual gestures part of human present-day speech patterns, cross-cultural studies have shown the *spontaneous* emergence of sign languages among deaf communities everywhere (Goldin-Meadow and Mylander 1998). *Sign languages that do not rely on either spoken or written words are genuine languages with grammars.*

[9] This also permits them to questionably include numerosity in the use of language.

In some ways even more interesting, studies of congenitally blind people have shown that they "would gesture while they spoke regardless of whether the listener was sighted or not" (Corballis 1999).

These studies strongly suggest that gestures are coupled in the brain to the act of speaking, even though the innatists do not include gestures in their definition of "language" and do not account for them in their theories of grammar. Nonetheless, there are individuals who use only sign language, who cannot speak, who manage to learn and teach subjects as abstract as mathematics.

For example, at Gallaudet University, where students use only American Sign Language (ASL), all subjects are taught in ASL, including mathematics, chemistry, philosophy, poetry. They do not use spoken or written words and sentences at all. On the innatists' theory, there is apparently no way to account for this. Gestures are not language, according to the innatists, yet persons using only gestures are somehow learning and teaching languages such as mathematics, chemistry, philosophy, and poetry.

In the opinion of many experts, gestural communication is as natural to the human condition as is spoken language. If there is such a thing as a universal grammar found in a specific "syntax module" in the brain, as Chomsky, Pinker, and Bloom claim, they have not tied it to the use of signs in sign language, but have limited it solely to spoken and written alphanumeric languages defined by universal grammar.

In some ways, following the digital machine model, the innatist arguments have essentially demarcated language in terms of two dimensions. They have done so while leaving out three-dimensional gestures along with other behavioral manifestations that are often found in actual language behavior. They have excluded these three-dimensional means of communication from their meaning of "language" altogether. Elements of real human (and animal) communication, analogical (as opposed to two-dimensional) signals, including voice pitch, loudness, posture, facial expression, gait and gesture, all found in actual language behavior, are excluded (McNeill 1992, p. 4).

Though there are different theories about the relationship between the origins of language in the spoken and written sense, and gestural communications, one theory is that language actually emerged from manual gestures rather than from vocalization in the evolution of *H. sapiens*. Since higher primates, including hominids, are largely visual animals, one theory is that voluntary communication using the hands was an early preadaptation of apes that were to become hominids.

Gestural communication is silent, thus the survival of hominids in hostile environments was better assured by using it rather than vocalizations. Moreover, gestural communication is fundamentally *spatial* communication. It would therefore have been a more efficient way of communicating the presence of dangerous predators.

Others argue that manual and vocal communication developed in parallel. Armstrong and his colleagues (Armstrong et al. 1995) have argued that the basic elements of syntax are intrinsic to gesture. Drawing on evidence from primatology and anthropology, they have set forth a theory that language emerged *through visible bodily action*. This would make sense only if the origins of language are largely found in semantic (meaning) structures, rather than being generated from a central syntax processing unit in the brain.

From Manual Gestures to Whole Body Performances

At least some theorists of the evolution of semantic (meaning) complexity adopt realist assumptions for their underlying model. Sounding like Russell, they assume that there are features of the real world which objectively exist regardless of whether anyone perceives them or even has a word for them. Among the most basic realist assumptions necessary is that in an unavoidably limited view of all of reality, there is a degree of honesty required of any organism to survive.

But even more fundamentally, in accordance with evolved neural systems, all organisms divide up the world differently. Those divisions are what Schoenemann (1999) means by "cognitive categories" or "semantic units" that essentially drive the evolution and development of language. Thus the notion of "increased semantic complexity" refers to the increase in the number of cognitive categories, their perceived interrelationships, and the number of signs that can be used.

Both humans and animals use *signs* to stand for cognitive categories they use to carve up and make sense of the world and their experience in it. Citing Peirce's (Hartshorne et al. 1931-1958) definition of "sign" as anything which stands for something else for somebody, Schoenemann makes the obvious point that human beings are not the only sign users in the natural world. Generally, the term "sign" refers to any nonverbal means of communication. This includes physical gestures, facial expressions, cries, laughter, and the like.

The use of signs by animals to communicate according to some cognitive category has been demonstrated in numerous animal studies such as ape language studies (Premack and Premack 1972, 2003); monkey calls (Page 1999; Griffen 1984, 1992); bees (Esch et al. 2001); and possibly in prairie dogs (Soussan 2004).

Schoenemann and others (Edelman 2004) argue that most cognitive categories are *independent of, and exist prior to, language itself.* Brain mechanisms involved in processing meaning are not limited to language but evolved and emerged relative to highly complex interactions, reactions and transactions with the environment and experience. In turn, the ways we experience the world depend upon evolved sensory systems, upon our evolutionary history.

Earlier arguments that animals are merely reactive organisms are also not supported. An example of this, as it pertains to infranimals, is Griffin's notion of animal communication as the "Groans of Pain" (GOP) reactive interpretation. However, as pointed out by Page, this view is no longer tenable:

"If indeed a vervet monkey, for example, is just expressing an emotional state when it yelps at the approach of a predator, why would it matter whether there is an interested audience for this yelping? . . But the monkey emits alarm cries only if there is another monkey in the vicinity who might benefit from what the signaler 'has to say'. ." (Page 1999, pp. 125-6).

There is ample evidence and strong arguments to support the contention that cognitive categories exist *independently* of and *prior* to language, as the audience effect of the vervet monkey cries suggest. Indeed, evolutionary theory itself strongly supports the argument that communication would not even have evolved if such categories did not already exist, prior to the development of language. All the evidence points to words and sentences, as well as their grouping and manipulation according to rules of grammar, as a process *secondary* to thought.

The entire debate on this particular issue between innatists and emergent theorists is reminiscent of remarks by James about the "usurpation of metaphysics by language," (James 1884) mentioned earlier. Under the influence of nominalism, the view that all we have of what we may call "reality" is the language we use to describe it, came the belief that everything must be labeled with language to be known or experienced. With this also came the belief that the *only existing objects* are those for which we (already) have a language label. In essence, it is this nominalist view that has led to the claim that there is only language-mediated *knowledge that.*

It is also the view reflected in the innatist arguments that language centers in the brain provide us with ready-made language categories by which to process everything we experience in the world. It is the view that humankind already has all the concepts, ideas, and rules needed to know or understand anything because those concepts, ideas, and rules are already found in language.

Yet humans invent new words and new concepts to refer to things that do not exist in the physical world, such as mathematical concepts. In our ever-expanding understanding of kinds of space, for example, we invent new concepts and words to characterize and describe those spaces. These concepts and words are not already given in our language or our thought.

3.3.3 Evolutionary Argument against Innatists

The innatist argument for a "syntax module" somewhere in the brain, along with the corollary argument that the origin of language is unique in human beings, with no homologies to other animals, does not hold up. While innatist proponents claim that language is an evolutionary development, many tenets of the theory are not consistent with evolutionary biological facts. Their theory has been roundly criticized by foremost scientists in human evolution and anthropology.

Among other problems is the very definition of universal grammar (UG), or syntax, itself. Though syntax is supposed to be a set of rules genetically encoded somewhere in a syntax module in the brain, the description given of these shows that they are not actually rules at all but features so general in scope "that they are really nothing more than descriptions of our cognitive semantic universe" (Schoenemann 1999, p. 311). This is problematic for innatists, because it is also consistent with the emergent view that grammatical rules in any natural language are cultural inventions.

Moreover, innatist theory has been highly criticized for ignoring a number of important evolutionary principles. These include such principles as (a) that evolutionary change likely occurs through incremental steps; (b) that these steps most likely build on prior adaptations; (c) that behavioral change drives genetic change; and others (Schoenemann 1999).

It is important to recognize these principles and how the innatist theory and explanation conflicts with or ignores them. Again, given that the innatist argument is built in part on the presumption that human language is a *unique* phenomenon, "without significant analog in the animal world," (Schoenemann 1999), it is difficult to see how it could be explained by evolutionary theory at all. If the development of human language shares no homologies with features found in other organisms, then it likely defies *scientific explanation*. As one noted theorist points out:

"If this is the case, its existence cannot be explained in the scientific sense of the word because scientific explanations (as opposed to religious or mythical ones) are descriptions of the interplay between the laws of the natural world" (Cartmill 1990).

In spite of this rather astonishing implication of their theory, Pinker (1994) has nonetheless argued that syntax has *no* homologies in related species. However, substantial research has shown that both chimpanzees and baboons do in fact have cognitive structures permitting abstract thought (Fagot et al. 2001). This is significant in the case of baboons because they are part of a different primate family that split from the family that gave rise to apes and then humans some 30 million years ago. It has been known for some time that chimpanzees have also demonstrated abstract thought.

In the case of chimpanzees, our nearest related species, experimentation has shown that chimpanzees use *logical* structures, including equivalence and order relations, when grouping objects both within sets of objects as well as between sets. They showed a higher level of logical organization than monkeys do (Fagot et al. 2001; Pot 1997).

Moreover, there are marked similarities between the brains of humans and brains of chimpanzees with respect to language. Researchers at Mount Sinai School of Medicine, Columbia University and the National Institutes of Health found that a region of the brain thought to control language is proportionately the same size in humans and chimpanzees.

This finding apparently disproves a long-standing theory that the brain section, the *planum temporale* of the left hemisphere, was enlarged only in humans. The study showed that 94 percent of the chimpanzee brains studied demonstrated the same asymmetry (Gannon and Holloway 1998). That region of the temporal cortex is also known as Wernicke's area, thought to control language comprehension, but only in humans.

Though it is not clear what the interpretation of the finding should be, one author of the study believes chimps may converse using a sophisticated array of facial, body and hand gestures, perhaps augmented with grunting or other vocalizations Holloway 1998). Others from the same study stated:

"If both chimps and humans have an enlarged *planum temporale*, their common ancestor probably had the feature as well, though the brain region may not have acquired its language functions until humans split off from other primates 6 to 8 million years ago. Finally, it may well be that the *planum temporale* is not involved in language in either chimps or humans."

Again, it has been evident for some time that chimps have the cognitive structures underlying "argument relationships." As Schoenemann notes, "to argue that this is not evidence of homology is to argue that humans and chimps *independently* evolved underlying cognitive structures that allow them to mark the same semantic features" (Schoenemann 1999, p. 314).

That argument, however, contradicts established principles of evolutionary biology. Among them, that evolutionary change most likely occurs through incremental steps and that these steps are most likely to build on prior adaptation. Quoting Jacob, Schoenemann notes "Evolution does not produce novelties from scratch. It works on what already exists. . ." (Jacob 1977).

He also notes that other uniquely human adaptations, such as *bipedalism*, show clear homologies with related species. Moreover, the principle of building upon previous adaptations is evident in the evolutionary development of cortical circuitry co-opted from circuitry for non-verbal movements, permitting the development of speech.

To argue that syntax alone has no homologies is to defy evolutionary biological explanation. Moreover, a corollary to the evolutionary biological principle that behavioral change drives genetic change is that "if some behavioral change is beneficial, we are therefore provided with an explanation for why it evolved."

Nonetheless, Pinker and Bloom "argue that "need" is an insufficient explanation for the existence of syntax in natural language. . . [reflecting] a Lamarckian perspective on evolution" (Schoenemann 1999, p. 316; Pinker and Bloom 1990).

According to Schoenemann, Pinker and Bloom beg the question at issue of how much of syntax is genetic:

". . .much of human behavior is clearly "Lamarckian" in this sense. If some mechanism is needed by a language in order for that language to effectively communicate certain aspects of the reality of its speakers, then a mechanism will be invented or a convention will be settled on to accomplish this" (Schoenemann 1999, p. 316).

To go to such lengths as Pinker and Bloom have gone to deny homology in the face of such convincing evidence "belies a profound and unreasonable bias," states Schoenemann.

One might also conclude that it belies a remarkable insistence on an anthropocentric view of the natural universe, with human intelligence uniquely placed at the pinnacle of an intelligence hierarchy, divorced entirely from all other living things beneath it.

3.3.4 Cognitivism, Mechanism, and "Innateness": How the Mind Does Not Work

In addition to the above linguists and psychologists, some theorists in other fields hold that universal ideas or concepts are innate. These theorists appeal not only to "innate knowledge," but also "innate mechanisms" as explanations to solve the problem of universals. "Universality is often a telltale sign of an innate mechanism at work" (Hauser 2000, p. 24) as one theorist says. Yet invoking the concept of "innate knowledge" or an "innate mechanism" is just another way of attempting to sidestep the original problem.

As the above assessment of such arguments shows, appeals to "innateness" often turn out to be *pseudo* explanations masquerading as solutions. One can still ask, "How does the supposed innate knowledge or mechanism *make* universals have universal reference—especially those universals that have reference to an infinite number of things for all time and all place?"

One attempt to answer this is yet a further appeal to natural selection. One theorist writes: "When organisms, including humans, encounter recurrent themes or statistical regularities, natural selection builds such information into their brains" (Hauser 2000, p. 23). Yet, though an appeal to natural selection may be a partial answer to the problem of some universals, it will not work at all with others, such as triangularity. The universal triangularity is not reducible to a "recurrent theme" or a "statistical regularity" though particular representations of it may be. We are still left with the original problem of universals.

In a further attempt to explain the concept of "innateness," one author states:

> "What is innate . . .is the mechanism for learning about a specific domain of knowledge, not the knowledge itself. Thus, the learning mechanism filters the experiences, guiding the organism to attend to some events in the environment but not others. . we are born with a learning mechanism that allows us to recognize objects such as hockey pucks and coffee mugs and make predictions about their behavior." (Hauser 2000, p. 23).

Thus contrary to earlier philosophers who wrote of innate knowledge, this author claims that what is innate is not the knowledge itself but the mechanism for learning about a specific domain of knowledge.

Innate Learning Mechanisms

Though this author does not explain what he means by "learning mechanism" he describes to some degree what it does. The purported learning mechanism filters experiences and guides the organism. The author does not provide any explanation of how, exactly, it does this, though his description is reminiscent of earlier philosophers' attempts to explain the same things with appeals to a process of "abstraction."

But just as with such appeals that beg the question, so does the appeal to inscrutable "learning mechanisms." Any so-called process of abstraction, for example, *assumes what it seeks to explain, i.e.* filtering experience and guiding the organism. We are still left with the original question. On the surface, this appears to be a *pseudo* solution as well.

Representationalism, as found in current conceptualist/idealist and nominalist strategies to explain the mind and intelligence, in particular the problem with universals, is a continuation of a failed nominalist and idealist tradition. A view of the mind or brain as having representations of the world and manipulating those representations according to innate mechanisms or rules begs far more questions than it answers.

It is yet another attempt to avoid the problem while introducing even more problems in the process. In particular, the appeal to "innate knowledge" has historically been proven to be a *pseudo*-solution to the problem of universals. That is because it is no solution at all, but a barely disguised attempt to avoid the problem.

The Classical Computational View of Mind and Intelligence

Representationalism, born of nominalism and conceptualism, in turn gave birth to the computational view of mind and intelligence. Together, they became the program of the new cognitive sciences born along with early artificial intelligence. More specifically, representationalism merged with the program of cognitivism that is pervasive today in theories of mind and intelligence.

Though cognitive science may be any theory about how the mind or intelligence works, cognitivism is the view that all mental phenomena—all phenomena included in intelligence —are fundamentally cognitive. In the accepted sense of "cognitive" that meant that all mental and intelligence phenomena involve thinking.

As Dreyfus explains (1995), the notion of thinking was then expanded under cognitivism far beyond its traditional boundaries involving only thoughts and reason. It was expanded to include perception, as well as skills and emotion. Thinking was traditionally conceived to be in a separate category from perception, skills and emotion. However, under cognitivism it was expanded such that skills, emotion, and perception were described and explained as *unconscious* "thinking," which seems to be a metaphorical stretching of the meaning of the word.

It is important to stress the underlying classical mechanical view of thinking extended to include perception, skills and emotion that these developments entailed. Fundamentally, thinking (or reasoning) was interpreted according to propositional or sentential logic categories. These in turn were treated as Boolean categories and operations, encoded as strings of "0"s, "1"s. This included the familiar "all-or-none" concepts, including the operators "and", "or", "either-or", "not", used in machine logic. Thus, all thinking, including the extended categories of perception, emotion, and skills, was represented or encoded as "0"s and "1"s and manipulated according to traditional Boolean operators.

This underlying theory of computation, now known as the classical theory, developed over many decades, following the works of logicians such as Gödel, Turing, Church, Kleene, Post, among others. It has been enormously successful in laying the foundation of computer science generally and developments in artificial intelligence.

However, for reasons I will later, it is a theory that may be necessary but is no longer sufficient for the foundation of modern scientific computation, where the appropriate algorithms are based on the *real* numbers. The classical theory of computation is also inadequate for a science of natural intelligence. From a mathematical point of view, the contemporary representationalist *cum* classical computational theory of mind and intelligence, born of earlier nominalism and conceptualism (idealism), is based in formal logic, having little relation with the continuous concepts of real and complex numbers. It is combinatorial rather than analytical. Natural intelligence, however, requires a mathematical analytical (real number) approach.

In part, the classical computational view underlies those representationalist theories that appeal to "innate mechanisms" and "innate knowledge" to explain knowing universals (or cognition and knowing in general), especially that found in babies and animals. It is the point of view taken by earlier theorists such as R. Gregory and W.V.O. Quine, and later theorists such as P. Churchland, and others.

Representationalists, even those that diverge from the traditional sentential categories, nonetheless follow an underlying mechanistic, classical computational model. What is wrong with this computational theory of mind and intelligence is, among other things, the underlying theory of computation. It is a discrete and linear, top-down, logic- and knowledge-based serial approach to computation that cannot handle actual continuous nonlinear natural intelligence found in even the simplest things both humans and animals know how to do.

Missing Practical Intelligence

But there are other problems as well. The traditional representationalist *cum* conceptualist (idealist) approach to mind and intelligence leaves out at least two major categories of knowing. In the history of serious thought about human intelligence, stemming back to the Greek philosophers who addressed practical intelligence sorting it from the theoretical and speculative knowledge held in highest esteem, later theorists tended to leave it out altogether.

As earlier points of view on the questions "What can be known?" "How can we know or come to know anything at all?" diverged into the realist, conceptualist, and nominalist camps, those in turn later congealed into at least 3 major theories of know*ledge*, not know*ing*. With few exceptions, these major theories did not address knowing how or immediate awareness at all.

These theories became the Realist, Coherence, and Pragmatist theories of knowledge. They all assumed a representationalist, propositional foundation of all human knowledge. It was knowledge that had to be mediated by language, either written or spoken.

The influence of nominalism, combined with the later representationalist and computational theory of mind and intelligence, became so strong that most philosophers *never questioned the underlying assumption that all knowing can be represented in public natural or artificial languages.* The propositional view of knowing and intelligence became the predominant view.

Moreover, with few exceptions, each philosopher who identified him or herself as in the Realist, or Coherence, or Pragmatist camp fundamentally rejected any notion of *immediate awareness*. It is rejected precisely because of the notion that it is "immediate" and not "mediated" by representations. There can be nothing cognitive that is *immediate* precisely because, so the doctrine held all cognition is mediated by language, by representations.

Furthermore, *coming to know* was excluded as not a proper question for philosophers. "Coming to know" was assumed to be identical to learning, held to be mostly a psychological matter.

Each of these theories assumes that knowledge, embedded solely within propositional or language structures constitute the full scope of intelligence. Where they had major disagreements usually concerned competing explanations of the sources of our concepts and knowledge.

Rationalist Sources of Innate Arguments

The representationalist strategy that eventually developed into the computational theory of mind followed in part from what is called the Rationalist Tradition. This is a position that holds that at least some propositions can be known *a priori*, not based on experience, and other propositions can be known when they are deduced from *a priori* propositions.

Mathematics is traditionally thought by Rationalist to be known *a priori* and by deduction. Some rationalists hold that ethics is also known the same way, while still others hold that metaphysical claims such that God exists, we have free will, and our mind and body are distinct substances, are all known *a priori* (or by intuition) and by deduction (Markie 2004).

But some Rationalists hold that we have *innate* knowledge or we have innate concepts. It is in these arguments that one finds some parallels with those advocating the computational theory of mind and current innate arguments for universal grammar.

Traditionally, innate knowledge has been explained as knowledge that is independent of any experience and is simply somehow *in us* as part of our rational natures. This is either because of our capacity for pure reason, or because we are born with certain structures in the mind. However, experiences may trigger a process by which we bring this knowledge to consciousness, though the experiences do not provide us with the knowledge itself. It has in some way been with us all along.

Earlier Rationalists argued that we gained the knowledge in an earlier existence. This is the Platonic argument. Others argued, as contemporary computational mind theorists, that it is part of our nature through natural selection. The appeal to "innate mechanisms" is an appeal to innate knowledge (or an innate "learning mechanism") following this same rationalist tradition to some degree.

The innate concept (universal) argument is much the same as the innate knowledge argument. Some argue that some of our concepts are not gained from experience. Though sense experiences may trigger a process by which these concepts are brought to consciousness, experience does not provide the concepts themselves. They are simply part of us. Since we do not experience perfect mathematical objects, such as triangularity, nor do we experience immense numbers, these have been argued to be "innate" concepts. Our knowledge of them may be "innate" as well, or so the argument goes.

Yet such appeals to "innateness" are bogus explanations. In the first place, even with the example of such mathematical concepts as triangularity, not all otherwise perfectly healthy individuals have the concept. Yet if such concepts were innate, in the sense that they are part of us as rational beings or because natural selection somehow selected for them in us, then it would be reasonable to assume that all human beings would have them. Yet not all do.

Moreover, if "innate" mechanisms are somehow in us for some of the same reasons, just waiting for the right experience to somehow "spark" or give rise to the knowledge that goes with them, we are left with explaining exactly how that occurs as well as explaining how it is that not all otherwise healthy human beings have such a mechanism or the supposed knowledge that goes with it.

While appearing to be an explanation, the appeal to "innate" mechanisms or "innate" knowledge is in fact something on the order of an *ad hoc* attachment to a theory to provide the appearance of an explanation where there isn't one. It is a fraud of an explanation.

Lastly, if the appeal to "innate" mechanisms or "innate" knowledge is a disguised appeal to a genetic explanation (as advocated by some psychologists, linguists and intelligence researchers), that has even more serious problems attached to it than the historical "innate" arguments.

For reasons already touched upon earlier, we now know enough about the human genome to know that genes just don't work that way. Not only are genes *not* primary in determining what a person is, becomes, does, and knows, there is no genetic counterpart to mathematical or any other concepts in the mind.

In spite of prevalent and, unfortunately popular notions to the contrary, the state of our knowledge and understanding of our own intelligence, including how we know or come to know, is not advanced enough to provide scientifically respectable explanations for these serious problems of natural intelligence.

3.4 Summary

We explored the meaning of intelligence and scope of cognition by looking at the development of realism, nominalism (conceptualism), and the rise of representationalism. We also looked at major theories of knowledge since cognition is largely defined in terms of *knowing*. We saw that from the time of Plato, philosophers addressed themselves to the question "What is knowing?" and focused upon even more basic questions such as "What can we know?" along with implicit questions such as "What is real?" These questions are in turn related to others such as "*How* can we know, or *come to know*, anything at all?"

Over time, the latter questions were largely reduced to psychological theories of learning. Moreover, earlier concepts such as practical intelligence were dropped with the rise of language representationalist and the classical computational theories of the mind. None of the major theories of knowledge: realism, coherence, and neo-pragmatism, addressed *knowing how* to do anything.

As only one of many such issues, the problem of universals has come down through the centuries as a particular sticking point in most theories of knowledge. This in turn has affected theories of intelligence and the mind. The problem of universals can be summed up in the following question: How do we experience *universal* or general, common properties in *particular* things? When we see a triangle drawn on a piece of paper, how do we know that it is a triangle? Why isn't it just a bunch of pencil marks? Responses to this particular problem vary dramatically among various theories of knowledge and intelligence.

In general, realists are those who assert the objective existence of universals. By "objectivity," they mean that these exist independently of human beings, of human language, and of human concepts. Nominalists are those who primarily acknowledge only universal *words*. Conceptualists (later known as idealists) are those who primarily allow universals only as concepts of the mind. Both nominalists and conceptualists hold that universals have no reality apart from merely being concepts in the minds of persons or common words in languages.

Given the influence of nominalism, representationalism quickly became dominant during the Linguistic Turn in philosophy during which every philosophical problem was turned into a problem about language, or at least into a problem *dependent* upon language. Though the later Cognitive Turn was supposed to have replaced the emphasis upon language by turning back to a study of the mind, it still sees the mind and intelligence generally as a linguistic/symbolic representational system manipulated by universal rules of grammar.

With the rise of neo-pragmatist theories, there was a reduction of knowledge and all things related to the mind and intelligence, especially language, to causal theories. This was accompanied by representationalist theories that propose kinds of linguistic/symbolic "pictures" of the mind, of intelligence. Mental representations became kinds of mental words.

Some recent psychologists, linguists, and other theorists hold that universal ideas or concepts are innate. These theorists have appealed not only to "innate knowledge," but also "innate mechanisms" as explanations to solve the problem of the origins of language, universals and to explain intelligence in terms of language acquisition.

As noted, some theorists hold that there are innate structures that act as a central processor driving the engine of language and intelligence. That central processor, they say, is found in "grammar genes" in a specific location in the brain. Intelligence, these theorists claim, is confined to logico-linguistic/symbol structures of spoken and written language. It excludes anything *unrelated* to language.

Closely examined, however, their arguments have been found to be fallacious, begging the questions at issue and in many respects were found to contradict established evolutionary principles. Moreover, where "innate" is taken to refer to genetically encoded mechanism brought about by natural selection, their arguments betray a fundamental misunderstanding of genetics as well, while still not explaining the problem of universals.

Some of the arguments for innate molecular structures taken to be the origins of language betray fallacies of reductionism that collapse biological as well as levels of explanation without justification; a confusion of correlation with cause; and given to unwarranted and unscientific use of metaphor in place of evidence and argument.

We concluded that invoking the concept of innateness turns out to be another way of attempting to sidestep the issues. The appeal to inadequately explained or supported innateness is a *pseudo* explanation masquerading as a solution. One can still ask, "How do the purportedly innate structures *make* universals have universal reference?

The problem with universals remains unresolved. Representational and computational theories of mind and intelligence leave out much of our intelligence. They leave out a concept of natural intelligence that includes other animals besides us, and by their very structure, such theories do not recognize the intelligence of *knowing how* and immediate awareness.

4 The Intelligence of Doing: Sensorimotor Domains and *Knowing How*

A recurring question in intelligence studies which we addressed to some degree in the last chapter is where the human mind and intelligence come from. In much of the research literature, the term "mind" often gets interpreted or defined as "consciousness" or "conscious awareness", without clarifying what those concepts mean, though they often get explained in nominalist terms.

For example, "consciousness" often gets defined as "consciousness *that*" such and such is the case, meaning that the one who is conscious can say or otherwise indicate that they are aware. Likewise, "awareness" gets defined the same way, as "awareness *that*" such and such is the case.

Neither of these are equivalent to "consciousness *of*" or "awareness *of*" which may be outside explicit, straight-forward verbal or linguistic indicators or parameters of a subject altogether. Determining whether or not a subject is conscious or aware may have to be sought another way.

Words such as "categorization", "mappings", "abstraction" and "discrimination" also get used in proposed explanations, yet those often leave us with far more questions than answers.

We looked at arguments that the genesis of intelligence, identified with verbal intelligence, is found in language centers in the brain where the biological bases of universal grammar are supposed to be found. For all the reasons and evidence considered, and more we did not, those arguments do not hold up. They do not hold up in part because they commit too many fallacies, beg too many questions, base too many claims upon assumptions known to be false, and violate fundamental scientific principles of evolution.

4.1 The Intelligence of Doing

Humans tend to be good at thinking and doing; and thinking is a kind of doing. It is an old adage that if you really want to know what someone thinks, look more at what the person does and listen less to what they say.

Extended to intelligence, the obvious principle is that intelligence is disclosed more in what one does than in what one says.

What humans know how to do is far more pervasive and fundamental than their verbal intelligence; indeed, knowing how is fundamental to all intelligence, including the verbal. Moreover, it is not reducible to the verbal. You cannot bring someone to know how by telling them how. Knowing how showed up long before hominids became language speakers; it still shows up after we are born before each of us becomes language speakers.

Though we do not know what animals are thinking or even if some of them think at all, we know that they are also good at doing a lot of things, including tricking humans. As noted in the Introduction, from my earliest years trying to keep wildlife from stealing crops and chickens, I learned to carefully watch them and never underestimate their intelligence.

On a broader evolutionary scale, some have posed a much larger question: How did inanimate matter evolve into creatures like ourselves who can think? Among many other things, some biologists have proposed that given the overwhelming odds against survival, mammals eventually learned the trick of providing nests for their offspring. Nests made it safer for offspring to learn by trial and error without being punished for being wrong. The nest made it possible to get things wrong without dying (Cohen 1998).

This trick enabled mammals not only to have fewer offspring because more of the ones they had survived, but it also meant that the mechanisms for survival no longer had to be entirely built into their genome. Parent mammals taught the trick of making nests to their offspring who in turn provided protective nests to their own. The trick of providing nests for offspring eventually led to a new kind of intelligence that included learning and teaching (Cohen 1998).

In some ways, this larger evolutionary perspective provides a more informed way of looking at intelligence than examining brain structures. As Cohen notes:

"Mind isn't just a matter of sophisticated brain structure. The cultural context that passes on tricks through learning and teaching is crucial, particularly for the most important and apparently unique features of human beings: imagination, creativity, and morality."

Mammalian *know how,* knowing how to build nests, knowing how to teach their young, and the young knowing how to learn enabled greater survival. Understanding the emergence of the intelligence of doing requires that we look not only at sophisticated brain structures, but also the context in which those structures find themselves, along with acquired processes of teaching and learning. We have to look at the processes of reaction, interaction, transaction among all these components of the patterns of emergence.

4.1.1 A Two-Pronged Approach to Intelligence Inquiry

Though the brain evolved and is not artificially designed, intelligence is a product both of evolution and intentional design. Teaching is a deliberate process to bring about changes in the intelligence and behavior of the one being taught. Moreover, persons can seek ways to teach themselves. They can deliberately set out to shape their own intelligence in ways that the natural world would not.[1]

Thus a theory of natural intelligence cannot be reduced to the biology of the brain. It must look more broadly at the context and experience within which intelligence emerges. To be explicit: there are at least two major categories of phenomena to which a science of intelligence (where we are specifically addressing human intelligence) must attend. One category includes the brain and its context; the other category includes the person and the person's experience in the objective world. The former major category is solely from an objective science point of view; the latter category is addressed with both an objective science point of view as well as the phenomenal view of the person in relation to an object.[2]

Moreover, the latter category requires viewing the phenomenal experience of the person within a set or matrix of relations obtaining between Subject (S) and Object(s) (O). Not all objects within such a set will be physical; many will be artifacts, including ideas. For example, the accumulated knowledge of a discipline may constitute an artifact which becomes an object in the relation with a Subject. As such, these require a different level and kind of inquiry than object-level science as performed in a clinical laboratory.

This chapter will focus on the structures and nonlinear cognitive elements employed in active natural intelligence. In particular, it will focus upon the neurobiological architecture of the sensorimotor system and awareness and how kinds of intelligent doing emerge from those.

[1] Humans can also intervene in natural brain and intelligence processes by artificial means such as drug inducement. I will not explore those possibilities here.

[2] The word "object" is used in the broadest sense to refer to any term of thought, where "term" is not limited to its language sense as a word. Objects can include physical or abstract terms, including patterns.

Though much of the research into intelligence assigns motor-related behavior to the non-cognitive, extensive neurobiological and related research reveals highly complex *cognitive* behavior at very primitive levels. Moreover, it will address to some extent how those sensory and motor areas are related to highly abstract thought structures as found, for example, in mathematical cognition, though most of these issues will be addressed in a later chapter.

Indeed, within a larger biological and evolutionary perspective, knowing how and cognitive awareness at primitive sensorimotor levels is necessary to account for our very survival. These findings call into question the definition of "cognition" and "intelligence" as usually circumscribed by language centers of the brain and evidenced solely or even primarily in verbal behavior. More broadly, I want to focus on certain intentional aspects of everyday experience and knowing exhibited or disclosed in normal human and animal doings.

Fallacies to Avoid

Clearly, psychological matters are involved in our inquiry. However, the natural intelligence domain of inquiry is neither equivalent nor identical with the psychological. Nor is it equivalent or identical with the neurological.

Instances of the difference can be shown in part with the distinction between learning and coming to know. Though there is an obvious relation between the two, the concept "coming to know", if only because it includes the concept "know" cannot be reduced to the concept "learning" and behavioral change. The concept "know" implicitly includes normative standards such as the concepts "fact", "truth" and "justification". Moreover, it includes standards for performance that are included in our understanding of *knowing how* to do something.

The concept of learning does not necessarily include any of these. One can learn much that is false; one can learn bad, ineffective, and even fatally incompetent ways of doing things. But it cannot be said (at least not without contradiction or nonsense) that one comes to *know* anything which is false; one cannot be said to *know how* to do a task yet consistently fail or perform badly and ineffectively when doing it.

Moreover, though it may be tempting on reductionism grounds, physics and neurophysiology must not be assumed as sole *premises* in theory of natural intelligence. We certainly need to understand neural and physical theories and methods in order to fully understand natural intelligence. These are necessary and must be part of any complete causal theory of natural intelligence. However, necessity is not sufficiency. To assume physics and physiology as sole *premises* results in kinds of material reductionist, or naturalistic fallacies (Russell 1984, p. 51).

One example of such fallacies is to argue that *because* we know certain empirical facts about the retina, it is therefore the case that we can solve the problem of kinds of intelligence we obtain when we observe things with our eyes. Knowledge of the retina will not answer questions about the knowledge or intelligence we obtain by using our eyes.

Among other things, this is faulty reductionism; it is also equivocation and yet another example of a category mistake. Intelligence gotten by using our eyes cannot be found by surgically opening up the retina. This is similar to the claim that we can find the explanation of a man's moral values by performing an autopsy on him.

Moreover, it should be added that the other side of this fallacy includes the assumption that *because* we use our brains to think, that there are neural networks [connections] in our brain when we think, we can therefore solve the natural intelligence problem by operating on the brain or submitting it to fMRI (functional magnetic resonance imaging) to observe its neural processing when we think.

But we will not find thought, reasons, logic, or universals in brain tissue when we open the skull or observe its functional MR images. Though such fallacies are readily passed off as explanations in some circles these days, they are rejected here.

Scientific inquiry into intelligence is broadly multi- and interdisciplinary, drawing upon concepts, methods, and research strategies spanning many disciplines from physics, chemistry, biology, neuroscience, psychology, and anthropology. It also requires the normative discipline of epistemology, the study of knowledge, knowing, belief, and theory of evidence and justification.

Epistemological analysis and inquiry proceed differently from psychological and neurophysical analysis and inquiry. Its objects, methods, and categories differ in that there is a greater reliance upon logical and mathematical analysis, seeking critical examination of prior hidden assumptions and underlying models, then evaluating those against a full spectrum of evidence and standards. It is usually (though not always) less involved in direct data collection and analysis though those are certainly taken into account. In many senses, epistemology is more a meta-level inquiry than are other areas of inquiry, particularly psychology.

It is important to stress these differences between kinds of inquiry because of the deeply ingrained material and genetic determinism that has pervaded American society. As earlier noted, this is the view that the answer to all serious questions about intelligence or life itself can be found at the physical, material or genetic level. Yet, to paraphrase a noted reproductive biologist, one can sit brain tissue and DNA in a test tube forever, and it will not come to life; it will not do anything but die and decay (Cohen 1998; Stewart 1998). DNA code alone must be plugged into a highly complex, multi-leveled network of support for life to eventually emerge.

Certainly, as neuroscientists assert, our sensations are functions minimally of our sense organs and our nervous system. However, this is not *primitive* knowledge, but *a scientific inference which we make*. Hence it cannot form sole *premises* of scientific inquiry into intelligence, though we must include that knowledge in our quest to understand the emergence of natural intelligence. For example, we must seek to understand how visual processes emerge to become seeing and observing, and how these and other senses emerge as awareness (both "awareness *of*" and "awareness *that*") and perception found in the intelligence of everyday *knowing how*.

A scientific explanation of natural intelligence must seek to provide a causal account of relationships among many domains. These include the neurological and physical domains, but also contextual and highly subjective domains as well. The aim of a science of intelligence is to explain how intelligence emerges from interrelationships among components in all those domains. In general, I will refer to the contextual and subjective domains as the domain of *experience*, when referring to human and animal intelligence.

4.1.2 Cognition, Consciousness, Awareness

We must use knowledge of the most primitive sensory and sensorimotor components in order to determine where the organism enters the circle of cognition. Neither the concept of cognition nor the concept of intelligence is taken here as identical with the concept of consciousness, as the latter term is defined in many neurological and other research communities today. That term is often defined in terms of ill-defined and question-begging processes such as *categorization* and *abstraction*. Or it is defined in nominalist terms that presumptively require a language interface between a subject and object.

For example, technically categorization is a process of classification. Both terms already imply the existence of one or more concepts since a concept is a necessary condition to the existence of a category. Likewise, concepts and categories are necessary conditions to the process of classification. Classification is a process of sorting things based upon similarities among properties of the things sorted. Thus the process of classification *already assumes* the possession of a rule of similarity (a concept and a category) to permit the process to begin in the first place. To define the term "consciousness" in terms of "categorization" begs the question. It assumes what it seeks to explain.

Any definition that assumes what it seeks to explain ends up not explaining anything. The process of categorization assumes more basic levels by its very definition. In spite of this, some theorists apparently believe they have provided the most primitive levels of explanation of consciousness, when they have not (Edelman 2004). Additionally, the following terms are used in similar inadequate ways: "concept", "discrimination", "identification", "detection", and "abstraction" resulting in questionable analysis and explanation.

Moreover, relations between "consciousness", "awareness" and "cognition" are often left very unclear. Edelman (2004) proposes that consciousness (hence the mind) arises as a result of the interaction of two parts of the neural system differing in their anatomical structure and evolutionary history. Primary consciousness, being aware of things in the world, arises from reentrant processes that connect perceptual categorization and value-laden memory.

On his theory, "higher consciousness" requires that one make the distinction between self and non-self; it also requires semantic ability, the assignment of meaning to a symbol. In essence, this is actually linguistic ability, though he apparently thinks one only has that when one can master a whole system of symbols and a grammar.

Though not directly significant for purposes here, one should note that Edelman's self and non-self requirement contradicts empirical research on the formation of self-identity and some basic logical principles in children. These occur before they have developed semantic ability as he defines it (Piaget 1950, 1971, 1972, 1990; Piaget and Inhelder, 1956; Gruber and Voněche 1995).

Thus according to Edelman being aware of one's self is possible if the brain is capable of perceptual categorization, memory, learning and self/non-self discrimination. Regions of the brain that function to define self within a species include the amygdala, the hippocampus, the limbic system, and the hypothalamus. Regions that define non-self include the cortex, the thalamus and the cerebellum. We will return to all this below.

In effect, an adequate theory of cognition in relation to natural intelligence, specifically the intelligence of doing, must provide adequate definitions of the above terms and answer at least the following questions:

- What are the most fundamental primitives of the sensorimotor system that comprise the first step into cognition?
- What are the cognitive parameters of awareness?

- What are the sensorimotor primitives of awareness?
- Specifically, how are these related to concept formation?
- Are these primitives still there in the most outer reaches of abstract reason and thought?
- What is the relation between cognitive awareness in humans and animals?

Indeed, what is the scope of cognition, the length, breadth, floor and ceiling of natural intelligence?

We can start by focusing upon a broader concept of subject experience, where "experience" is broadly defined in terms of awareness and doing.

4.2 The Science of Awareness

A bit of history: before the advent of objective scientific measures and the growth of the neurosciences, the test of any direct or immediate[3] awareness of even conveniently sized objects was traditionally some version of Descartes' (1596–1650) introspective method of doubt or some other appeal to the possibility of illusion or error in perception. It is worthwhile to retrace a few historical steps here since our present understanding of the problems and some of these concepts directly follow from them.

A close look at Descartes' method of doubt alone shows that these tests and methods were directed largely to what one is justified in believing. They were largely directed to *knowledge by description* (*knowledge that*) by closely reflecting upon what evidence or reasons one has for believing something. They were not actually directed to *knowledge by acquaintance,* the kind of knowing that comes from first-hand physical, immediate awareness and experience.

Russell (1912) had initially introduced the distinction between *knowledge by acquaintance* and *knowledge by description* primarily to provide a foundation in actual immediate physical experience and awareness for the latter.

Yet over many decades various philosophers argued against the notion of direct or immediate awareness as a kind of knowing or cognition, as well as the distinction between *knowing by acquaintance* and *knowing by description,* precisely by appealing to the same tests of the possibility of illusion, hallucination, and other sources of error. They allowed *mediate*

[3] The term "immediate" means there is no language interface between the subject who is aware and the object of that awareness. Sometimes the word "direct" is used. "Mediate" awareness means there is such an interface.

awareness ("awareness *that*"), the kind a person can verbally report upon, but even then one could not refer to internal states such as pain to prove it.

The major conclusion drawn from their arguments seemed to be that the only kind of knowledge we can actually have is knowledge by description. It is the only kind that can be publicly checked out, because it can be represented in propositions, without resorting to private introspection or personal subjective experience of individuals.

However, the very framing of the distinction and the argument in the first place, by comparing knowledge by *acquaintance* with knowledge by *description,* led to these and other fallacious arguments against *knowledge by acquaintance* and certainly against immediate awareness. These philosophers mistakenly thought that knowledge by acquaintance should meet the same kind of standards as knowledge by description. They would not allow a non-propositional knowing.

But beliefs are not necessary conditions to a subject's *knowledge by acquaintance* or immediate awareness. Neither is justification for beliefs, or reflecting about beliefs, as those are the province of knowledge by description. Any justification for knowledge by acquaintance or immediate awareness has to be found in what one discloses in one's actual experience, especially in what one knows how to do. One cannot look to verbal structures for such justification because it is not found there. It must be looked for in the structures of actual experience, especially the structures of *knowing how*.

Moreover, since the logical possibility of kinds of error or hallucination is *always* present, the framing of the distinction and the argument led as well to some philosophers and others pitching the objective foundations of *all* cognition, reason, and knowledge altogether. These are largely known as postmodernists.

Again, though Russell had initially introduced knowledge by acquaintance in relation to knowledge by description primarily to provide a *foundation in actual physical experience* for the latter, it is clear that in the arguments against it propositional *knowledge by description* was given priority over any direct or first-hand experience or awareness.

Given the influence of nominalism and representationalism, that priority is not entirely unexpected. *Knowledge by description* was and still is the standard by which all epistemological categories and intelligence are evaluated and tested.

Immediate awareness will always fail the propositional *knowledge by description (knowledge that)* test. But that is not the test that should be used.

4.2.1 Cortical Structures and Information: Neural Bases of Awareness and Intelligent Doing

We should clarify the use of certain terms here as well as get clear on the overall anatomy and physiology of the senses and movement (Seeley et al. 2002).[4] Trivially, the sensorimotor system of the human is that system that is of or relating to both the sensory and motor system.

The sensory system includes the somatosensory system which is the sensory system of somatic sensation. Somatic sensation consists of various sensory receptors that permit the experiences of touch, pressure, temperature, pain (which includes itching and tickles), muscle movement, joint position (including posture and movement), and facial expression.

The somatosenses include the cutaneous skin, kinesthesia (movement) and organic senses which have to do with information from the organs, such as stomach pains.

The sensorimotor area is that area of the cortex including the precentral gyrus and the postcentral gyrus and combining sensory and motor functions.

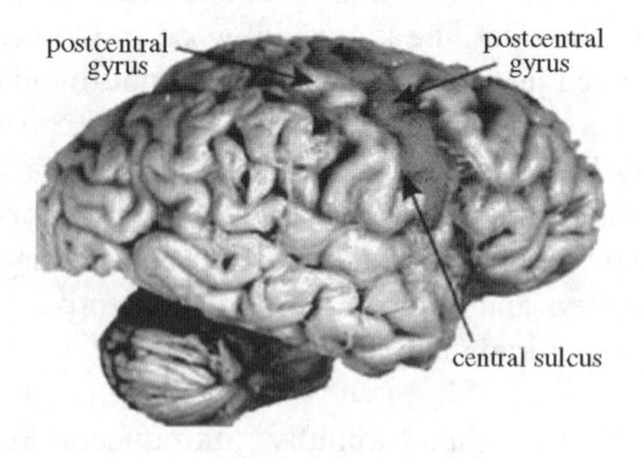

Fig. 4.1. Brain Showing Postcentral and Precentral Gyrus

The primary motor area is a group of networked cells in mammalian brains that controls movements of specific body parts associated with cell groups in that area of the brain. This area is linked by neural networks to corresponding areas in the primary somatosensory cortex.

The motor pathway consists of the corticospinal tract which originates in the pyramidal neurons of the motor cortex. The cell bodies in the motor cortex send long axons to the motor cranial nerve nuclei of the midbrain, pons, medulla oblongata. The bulk of these fibers extend to the spinal cord.

[4] Much of the information in this section relies upon Seeley et al. 2002; Hernegger 2005; and Edelman 2004.

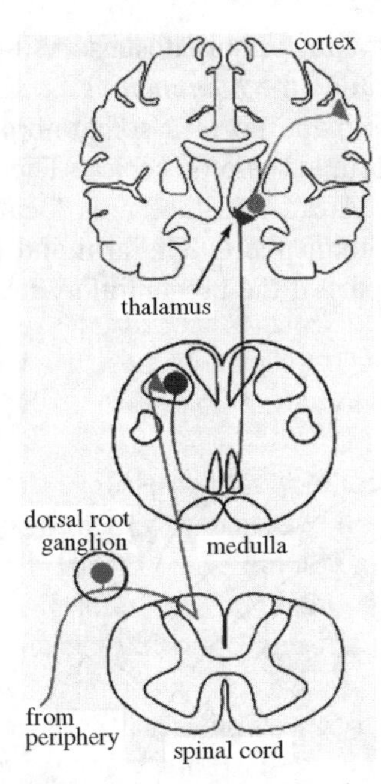

Fig. 4.2. Basic Somatosensory Pathway

Sensation enters the periphery *via* sensory axons. All sensory neurons have their cell bodies sitting outside the spinal cord in a clump called a dorsal root ganglion. There is one such ganglion for every spinal nerve. The proximal end of the axon enters the dorsal half of the spinal cord, and immediately turns up the cord towards the brain.

These axons are called the primary afferents, because they are the same axons that brought the signal into the cord. (In general, *afferent* means towards the brain, and *efferent* means away from it.) The axons ascend in the dorsal white matter of the spinal cord.

At the medulla, the primary afferents finally synapse. The neurons receiving the synapse are now called the secondary afferents. The secondary afferents cross immediately, and form a new tract on the other side of the brainstem. This tract of secondary afferents will ascend all the way to the thalamus, which is the clearinghouse for everything that wants to get into cortex. Once in thalamus, they will synapse, and a third and final neuron will go to cerebral cortex, the final target (WUSM 2005).

The primary somatosensory area in the human cortex is found in the postcentral gyrus. Areas of this part of the brain map to certain areas of the body depending upon the amount of somatosensory input from that area. This somatosensory map is called the *homunculus*.[5]

The motor homunculus is made up of a somatotopic representation of different body parts in the primary motor cortex. The leg area is located close to the midline with the head and face area located laterally on the convex side of the cerebral hemisphere. The arm and hand motor area is the largest and occupies the part of the precentral gyrus located in between the leg and face area.

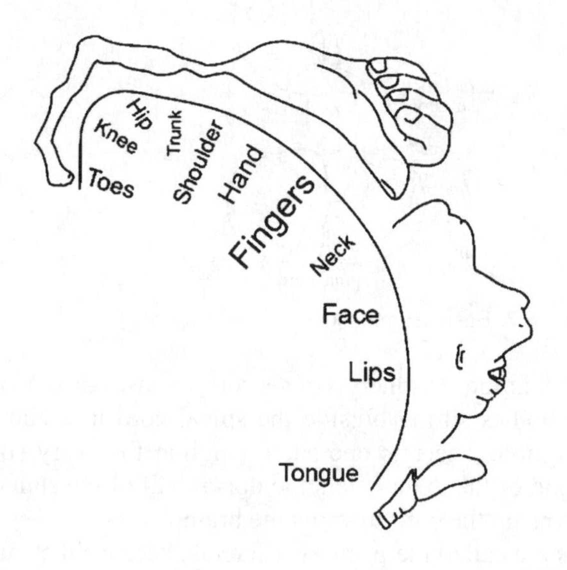

Fig. 4.3. Homunculus

A sensory system is part of the nervous system consisting of sensory receptors, neural pathways, and those parts of the brain which function to process information. The sensory systems include those for vision, hearing, somatic sensation, taste, and olfaction.

Receptive fields have been identified for the visual, auditory, and somatosensory systems so far. These systems code for four aspects of a stimulus, including type, intensity, location, and duration.

[5] Latin: *little man*.

Reticulo-Thalamo-Cortical (RTC) System

Organisms store and analyze information in the cortical network, centralizing controls in the reticulo-thalamo-cortical (RTC) system. Complex interactions, reactions, and transactions with the environment are enabled by the neocortical network. The primary and secondary sensory areas of that network represent the peripheral sensory receptor system in the cortex. These areas continue the functions of analysis and filtering of information from the environment.

As part of the sensory system's filtering function, the visual system analyzes differences in light, colors, movement, shapes and contours. It is important to note that the filtering function is the means by which the sense qualities are *selected* before the act of seeing can take place. The cortical sensory detectors are the carrier of code for the sense qualities which have to be decoded into information in order to be meaningful.

A viable theory of intelligence must be clear on the mechanisms of percept (sense) formation. This includes not only the biology of the brain, but also the context or environment in which the brain finds itself. Moreover, such a theory must become clear on *concept* formation, without a conceptual "collapse" or dissolving of, the crucial differences between percepts and concepts. "Percept" and "concept" are two different categories of mind and intelligence; they are not one.

4.2.2 How Concepts (Universals) Get Formed: A Global Map Theory

Again, consciousness usually gets defined in terms of kinds of awareness. One example of this is Edelman's (2004) definition of "primary consciousness" as the "state of being mentally aware of things in the world, of having mental images in the present" (Edelman 2004, p. 9). This is then used by him in explanations of how cognition arises. He explains that this kind of consciousness can be possessed by both humans and animals, but one of its defining features is that it is not accompanied by any sense of a socially defined self. Setting aside references to the self and non-self condition, we should look at his explanation of how perceptual categories are formed.

According to his theory, perceptual categorization in mammals is carried out by interactions between the sensory and motor systems in what Edelman calls "global mappings." This is a dynamic structure containing

> ". . .various sensory maps, each with different functionally segregated properties, linked by reentry. These are linked in turn by non-reentrant connections to motor

maps and subcortical systems such as the cerebellum and basal ganglia" (Edelman 2004, p. 49).

Global maps "sample the world of signals by movement and attention" and then categorize the signals as coherent through reentry and synchronization of neuronal groups (Edelman 2004, p. 49). Consisting of both sensory and motor components, this is what he says is the main basis for perceptual categorization in higher brains.

But he goes on to explain that this categorization by itself cannot generate generalizations across signal complexes. Generalizations require that the brain map its own activities that are represented by global mappings, "to create a concept—that is, to make maps of its perceptual maps" (Edelman 2004).

"Higher order cortical maps in the prefrontal, parietal, and temporal areas are likely to carry out this construction, which might correspond to a 'universal,' a concept. . .generalization arises by abstracting certain features of such mappings by means of higher order maps" (Edelman 2004, p. 50).

Edelman continues by elaborating upon the function that memory performs in these processes, and outlines the interactions of the three major parts of global neural systems. These include the thalamocortical maps; the subcortical organs concerned with temporal succession, including the hippocampus; basal ganglia; and cerebellum; and the ascending value systems. The latter are neuronal structures which have different neurotransmitters; from nuclei of origin they each send axons up and down the nervous system in a spreading fashion. The value systems directly affect learning and memory as well as control bodily responses necessary for survival.

Though not all of it is described here, Edelman's Theory of Neuronal Group Selection is impressive. Concept formation is intended to precede language. The perceptual system and memory in part drive the formation of concepts. Those structures that drive concept formation are those that categorize, abstract features, and recombine patterns in different kinds of global mappings.

However, even at this point, one must raise some serious questions regarding his explanations. I have already alluded to some of these above. But in addition, we must question his use of *categorization* and *concept*; his appeal to a purported process of *abstraction*; and his resting the process of "sampling" on attention (with movement).

Notice that he uses the notion of *categorization* as somehow temporally and morphologically prior to the notion of *concept* in the formation of consciousness. Somehow, categorization by global maps sampling the world of signals by movement and attention occurs prior to the formation of any concept or universal.

However, this contradicts the usual understanding of category. The process of categorization, indeed the process of sampling as well, requires a *prior concept of category and rule for sampling.* His explanation of this process appears to be begging some questions, that is, assuming what he seeks to explain.

Additionally, he says that generalizations, maps of perceptual maps, arise by "abstracting" certain features of such mappings by means of higher order maps. The use of the notion of abstraction is yet another can of worms. Purported processes of abstraction have a very long and dubious history in inquiry into the operations of the mind going back many centuries. And appeals to "abstraction" are commonly found in material reductionist arguments in explanations for how generalizations supposedly arise. Additionally, these are tied to an equally troubling concept of induction that often gets defined as "learning".

The Bogus Process of Abstraction

Possibly one of the best recent descriptions is given by Quine (1966; 1969) a naturalist philosopher who sought neural explanations for the formation of beliefs. His description of abstraction fits well with Edelman's use of the term.

As found in both Edelman and Quine, the process of abstracting requires that one *single out for attention from other features of the environment* some *single* feature given in sense experience. This process may be repeated many times to then form generalizations based upon those experienced features. That is, it is the method of induction, he claims, that is the psychological process of *abstraction* from experiential events to general concepts or universals.

It is specifically these claims regarding the doctrine of abstractionism, the "lynch pin" which supports what Edelman claims is the process of forming those generalizations or "maps of perceptual maps" that give rise to concepts or universals. Generalizations arise by "abstracting" certain features of such mappings by means of higher order maps.

But the process of abstractionism breaks down when we carefully analyze the formation of any kind of concept, including logical, arithmetical, and color concepts by this process. In order for one to attend to "certain features" of such mappings, they must already have some prior rule by which to do so.

Appeals to abstractionism always assume a (usually hidden) prior rule. The process of abstracting "certain features" depends upon a *prior rule of similarity* of some property that the one doing the abstracting discerns that the features have in common. Otherwise, there is nothing to direct the process; there is nothing to select some features over others. There has to be some rule holding the purported process of abstracting together such that it results in selecting certain features and not others.

Apparently, Edelman wants the higher order maps to fulfill that function. This is supposed to be made possible by recategorical properties of memory and the redundancy built into nonlinear interactions in multidimensional networks of neuronal groups. "Such interactions allow a non-identical "reliving" of a set of prior acts and events. . ." (Edelman 2004, p. 52).

Yet this will not work. The formation or learning of concepts or universals as Edelman describes still *presupposes cognizance of*, not merely *discrimination of*, features. That is, such an explanation *already assumes* the concept, universal, or intelligence it sought to explain. That concept, universal, or intelligence is in the purported act of abstracting itself. The acts of memory re-categorizing the same features will not dissolve this problem as the memory itself must act according to the same or a like rule. No matter how many times a feature gets recycled or reentered and recombined by memory and global maps, it does so by virtue of a prior rule.

Moreover, it should be pointed out that Edelman deals solely with perceptual concepts and their formation. He has left aside the formation of abstract concepts entirely.

A Spurious Sense of Induction: The Appeal to "Sampling"

Edleman's explanation is based upon a faulty appeal to "sampling" which is actually a concept of learning that is in turn based upon an entirely spurious sense of induction. There are *two* senses of the concept "induction" which must be distinguished from one another. There is induction *as statistical form of argument* and there is induction *as process*. Edelman is using the latter. That is, he takes induction to be a process of concept formation in which one somehow derives concepts or universals from sampled sense data.

This sense of induction can be traced back to Francis Bacon (Anderson 1960), who presented induction as a way of discovering truth. For Bacon, through the supposed "process of abstraction" from particulars, generalizations about the world can arise. That is, induction as process or learning takes place.

But this is an unsupportable notion of induction because the so-called process of abstraction already assumes what it purportedly seeks to establish. It is yet another instance of begging the question. Induction as process is erroneously taken to be a kind of logic of discovery when in fact it is a logic of verification (of what one already assumes to be true).

True induction is in fact a kind of statistical inference or reasoning which is involved in determining whether theory is supported by data. That is, it is a kind of *statistical* argument.[6] When induction is taken in the sense as a statistical form of argument, it rules out other spurious senses of the kind involved in Edelman's appeal to sampling as a way of forming concepts or universals.

A Problem with Attention

Another problem with Edelman's explanation is that he starts the process of global mapping with *attention*. He says, global maps "sample the world of signals by movement and attention." Thus, consciousness, specifically "primary consciousness," which he defines in terms of awareness ("being mentally aware of things in the world, of having mental images in the present"), can only arise from the level of attention. If I have accurately interpreted his claims, attention (along with movement) is the starting or entry point for consciousness.

But there are serious problems with this position, in addition to his concept of consciousness and its relation to awareness and the numerous faulty assumptions underlying his use of the terms "category", "concept", and the process of "abstraction". There is substantial empirical evidence showing that there is a great deal of *cognitive* awareness prior to the level of attention.

Yet on Edelman's theory, this is not tied in to his theory of how consciousness or concepts (universals) arise. Levels of preattentive selection and how they are related to attention should be closely analyzed.

4.2.3 Primitive Awareness

Admittedly, there is debate over how early the attentional filter operates across the sensorimotor process. However, experimental evidence shows

[6] Through induction, one makes an inference from *some* instances of a collection to *all* instances of that same collection. The conclusion makes a claim which goes beyond the premises, making the conclusion only probable and not logically necessary. It is induction *as statistical inference* which is the true sense of induction, not induction *as process*.

that preattentive analysis *precedes* the first storage of information and conscious perception. It has a latency period of about 60 ms.

During the preattentive phase, the RTC and the stimulus excite primary arousal of the activation system itself and the sensory fields. The body and its senses become aligned with the stimulus *via* the sensorimotor paths of the reticular brain stem. In general, the function of the sensory system during the preattentive phase, including the sensory fields of the cortex, is to analyze stimuli so that the sensory system can filter the stimuli and align the filtered sense qualities with the stimulus.

According to experts (Wolfe 1996), preattentive orientation proceeds *subconsciously* (which appears to be interpreted as the absence of "consciousness *that*" such and such is the case) at the level of the nervous system. It is only when sensory perception is attained that *attention* can then focus upon information as an object with which it can operate.

Only when this level is reached does preattention make the transition to the *conscious* attention of a cognitive system. And this appears to be the line between *non-cognitive* neurophysical activity and *cognitive* neurophysical activity, according to some experts (Kunimoto 2001).

Some appear to argue that only at the attention phase, interpreted as "consciousness that" or "awareness that" does a subject's neurophysical activity become (or is held to be) cognitive. This is the level at which subjects can give responses in language or by some other acceptable public indicator that he or she is aware of a stimulus. Any activity below this is held to be non-cognitive.

However, there is substantial evidence we will review of cognitive immediate awareness *below* the threshold of attention. If that is the case, then the intelligence research community must radically revise its understanding of the cognitive domain and the place where we enter it; the entire scope of natural intelligence must be enlarged to accommodate the facts. Moreover, Edelman's theory of consciousness, among others, would seem to require revision.

Scientific Definitions of "Awareness"

As noted, earlier attempts by theorists to address the problem of awareness in general tended to rely upon introspective reports of inner, subjective experiences. Perhaps among the earliest examples is found in Descartes' method of doubt, referenced above, but these are also found much later in James (1890) and certainly in Russell (1984). However, as we also noted, these reports are more appropriate to test and evaluate *knowledge by description* than acquaintance or immediate awareness.

Though theorists and researchers today use more objective methods, they still tend to identify *attention* as the starting point for intentional *cognitive* activity. This is due in large part to the fact that attention is usually interpreted to mean one's conscious "awareness *that*" such and such is the case and subjects can explicitly verbally or otherwise indicate in a straight-forward, conscious "awake" fashion that they are aware. The concept of attention is tied to subject linguistic reports of experienced stimuli.

Obvious problems and disagreements with this have been widely reported in cases of subject extreme disability, such as coma or persistent vegetative state.[7] With the exception of looking at some measures of awareness under surgical anesthesia, I will not directly address those issues here since there are ample problems even under the most ideal conditions.

Possible Subject Bias

Among other problems, one drawback with any experimental inquiry relying upon introspective or otherwise subjective reports of one's inner experiences is, of course, that subjects' reports of their inner or private experience may very well be influenced by bias.

Among other things, any given subject may claim to be *unaware* of a stimulus unless they are completely confident in their response, or they may claim to be aware on the basis of just about any sensation. Individual subjects may tend to determine whether or not they aware on the basis of their own private criteria for awareness. Thus such reports cannot be used to precisely define awareness in general. They certainly cannot precisely define "immediate awareness".

Subject bias can also be found even in experimental studies on awareness without introspective reports. In sensory discrimination tasks, for example, there is evidence that subjects can sometimes be systematically underconfident. That means that they may systematically claim *not* to see stimuli that they have partially or even entirely seen (Bjorkman et al. 1993; Kunimoto et al. 2001; De Becker 1997).

Moreover, with objective definitions based on correct *versus* incorrect identifications by the subject, subjects making an incorrect identification may nonetheless still have some awareness of the stimuli.

Even with objective definitions based on chance and greater than chance performance, issues of whether perception of a stimulus can occur without awareness will not be resolved because they are insensitive to

[7] News reports of the Terri Shaivo case are abundant as this is being written, March 2005.

subjects' phenomenal experience. There are other approaches as well with similar or even more complicated problems.

Among the more promising objective measure of awareness appears to be offered by Kunimoto (Kunimoto et al. 2001). They have suggested that the general concept of awareness should not be viewed in terms of two mutually exclusive states, *awareness* or *unawareness*. It should be viewed as a continuum of states ranging from unaware through an infinite number of partially aware states, to complete awareness.

Awareness of and Awareness that

However, Kunimoto et al. have not clearly distinguished between "awareness *that*" such and such is the case [tying awareness to "that" clauses or linguistic reports] and "immediate" awareness which is not tied to such reports. This is so even though their stated concern is with subliminal awareness. This distinction should be factored into any continuum, with a clear map showing where the two categories lay on it.

Their proposal is to measure awareness in terms of subjects' ability to discriminate between correct and incorrect responses using a metric provided by Signal Detection Theory (SDT). "Awareness" is operationally defined such that a subject is aware if and only if confidence is related to accuracy (with the metric greater than zero). The approach uses both subjective reports for assessing awareness by analyzing confidence reports with techniques developed in SDT to eliminate response bias.

Because Kunimoto's operationally defined concept of awareness ties awareness to subject reports of their own inner states, and though this method may overlap in some ways with our concerns, it does not directly address *immediate* awareness in its fullest sense.

It is apparently more addressed to "awareness *that*" than immediate awareness which would occur in the absence of language and reflection *about* the awareness. Nonetheless, we will shortly return to the Kunimoto study.

4.2.4 Experimental Evidence of Immediate Awareness

The scope of immediate awareness should be viewed as a spectrum of primitive sensory relations extending from the preattentive phases to attentive, layered in complex succeeding integrated network "sheets" throughout the sensory system, including somatosensory, and sensorimotor system, as well as the imagination.

Sorting such a hierarchy of primitive relations of the entire spectrum of awareness, including those of immediate awareness, and showing where they lay on the continuum, poses a challenge that has yet to be undertaken. Evidence for *cognitive immediate* awareness ("awareness *of*" in contrast to "awareness *that*") activity during the *preattentive phase* has been empirically shown or strongly suggested in a variety of research studies (Li et al. 2002; Colombo et al. 1995; Näätänen et al. 2001).

The following list generally describes a number of those research studies, but the list should not be considered exhaustive.

- Kunimoto (Kunimoto, et al. 2001) using their method described earlier, conducted four subliminal perception experiments using the relationship between confidence and accuracy to assess awareness. Their operational definition of "awareness" was applied to simple visual tasks. Subjects discriminated among stimuli and indicated their confidence in each discrimination response. Subjects were classified as aware of the stimuli if their confidence judgments predicted accuracy and were classified as unaware if they did not. In the first experiment, findings indicated that subjects' claims that they are "just guessing" should not be accepted as sufficient evidence that they are completely unaware of the stimuli.

Their experiments tested directly for subliminal perception by comparing the minimum exposure duration needed for better than chance discrimination performance against the minimum needed for confidence to predict accuracy.

The latter durations were slightly but significantly longer, suggesting that under certain circumstances people can make perceptual discriminations *even though the information that was used to make those discriminations is not consciously available*. "Consciously" again means "consciousness *that*."

Kunimoto has stated that the major contribution of their research findings may be methodological in that they have shown how to dissociate perception from awareness in that people "can discriminate among stimuli at better than chance levels even with displays so brief that their confidence is unrelated to their accuracy."

They also propose that their operational definition of "awareness" could be applied to auditory perception, memory, or other cognitive tasks to determine the extent to which various types of performance are carried out with or without awareness.

- Li (Li et al. 2002) conducted experiments testing rapid *visual* categorization in the absence of awareness. Subjects were asked to respond to masked and unmasked natural scenes when they contained an animal. Subjects also rated their confidence in perceiving the contents of each masked image. For a majority of the scenes, masking effectively prevented awareness of the stimuli, as indicated by the fact that confidence ratings did not predict categorization accuracy. For the same scenes, however, subjects responded significantly above chance level to the presence of animals.

In the same experiments, *motor* responses started to reflect *correct* categorizations at the same time for masked and unmasked stimuli. This indicated that early responses in "normal" (unmasked) visual categorization probably also rely on the first milliseconds of stimulation. Similar results were obtained with simpler displays for which stimulus and mask contrast could be controlled. In those cases the earliest motor responses to "perceived" and "unperceived" targets showed virtually identical distributions.

According to the researchers, these experiments showed that information about the first milliseconds of visual stimulation can propagate throughout the visual system, unaffected by later changes, and determine behavior *even when it is not (or not yet) available to consciousness.* Again, "consciousness" here refers to "consciousness *that*."

- Repp (2001) conducted research on finger-tapping which revealed an internal mechanism which guides motor actions in response to subliminal changes in stimuli. Through a total of five experiments, subjects were assessed in terms of sensorimotor coordination, phase correction, timing adjustment of a repetitive motor activity to maintain synchrony or some other intended temporal relation with an external sequence of events. They were also tested in terms of phase resetting which is a more dramatic timing adjustment that immediately restores synchrony after a large synchronization error.

In each test, subjects correctly altered their motor actions in response to subliminal changes in stimuli even *without* a conscious perception of change. *Repp concluded that the brain agent guiding the motor behavior is below the perceptual threshold.* At some level, the brain is much more sensitive to timing information than the results of previous psychophysical experiments suggest. This precise timing information seems to be used in the control of actions, *without awareness* (interpreted as "awareness *that*").

- In Colombo (Colombo et al. 1995), researchers tested visual search asymmetries in 3- and 4-month old infants *indicative of a preattentive phase.* Thirty-two infants from each age group were presented with 2 visual arrays to the left and right of midline. The stimuli were constructed of feature-positive and feature-absent arrays, each paired with a corresponding homogeneous array in which no discrepant element was embedded. The visual fixations of infants were measured, showing a "pop-out" effect for feature-present stimuli in both age groups. The results were similar to but not as strong as results found for adults.

As the researchers note, the findings may reflect limitations of infant visual search, the methodology used to assess it, or the difference in the size of the effect between adults and infants. The findings also show evidence of visual quality selection in the preattentive phase for infants.

- In Näätänen (Näätänen et al. 2001), tests were conducted with multiple simultaneously active sources of seemingly chaotic composite signals, with overlapping temporal and spectral acoustic properties, impinging on subjects' ears. In spite of the chaotic composite signals with overlapping temporal and spectral acoustic properties, the subjects' perception is an orderly "auditory scene" that is organized according to sources and auditory events. This allows them to select messages easily and recognize familiar sound patterns, and to distinguish deviant or novel sound patterns.

The data suggest that subjects' ability to organize such impinging signals is based on a kind of "sensory intelligence" [sic] in the auditory cortex.

"Even higher cognitive processes than previously thought, such as those that organize the auditory input, extract the common invariant patterns shared by a number of acoustically varying sounds, or *anticipate* the auditory events of the immediate future, occur at the level of sensory cortex (even when attention is not directed towards the sensory input)" (Näätänen et al. 2001).

- Some studies on "blindsight" or "numbsense" convey how some persons who are conventionally blind or insensible by objective measures can nonetheless discriminate visual or tactual test stimuli correctly with near-perfect accuracy (Weiskrantz 1997).

These patients will insist that they can't "see" or "feel" anything despite objective evidence to the contrary, demonstrating a level of awareness I refer to as "awareness of" not reducible to the subject's "awareness *that*." Subject actual responses *correlated negatively* with their verbal reports.

- Similar studies conducted decades earlier showed that subjects presented with a series of nonsense syllables who were then subjected to mild electric shocks at the sight of certain syllables, soon showed symptoms of anticipating the shock at the sight of "shock syllables." Yet, on questioning, they could not identify the syllables. The subjects had come to know when to expect a shock but could not tell what made them expect it. These findings are similar to other experiments which showed they knew or could identify persons by signs they could not tell (Polanyi 1966; De Becker 1997). The findings also seem to suggest that subjects also knew patterns of *timing* associated with "shock syllables" by signs they could not tell.

- Finally, experiments in human perception show that in spite of "noise" in images and gaps in contours caused by light intensity variations and occlusions, human perception is able to account for these by using an intrinsic process of *line completion* and *grouping of parts into whole entities* (Livingston and Hubel 1988). There is evidence that this entire process is purely preattentive without any top-down (know*ledge* or "awareness *that*") influences. Edelman (2004, p. 37) depicts the Kanizsa triangle demonstrating these intrinsic processes of perception in order to also demonstrate the context-dependency of perception.

The above list of various experiments involves most of the sensory (including somatosensory) and the sensorimotor system. They also involve large numbers of primitives in the preattentive phase and attention system. They show *cognitive* immediate awareness of objects that is not mediated by linguistic units or grammatical rules. In fact, certain of the experiments, as in the blindsight and numbsense experiments, showed that subjects' correct responses *correlated negatively* with their own verbal language reports.

More to the point, the experiments show that humans will often verbally deny *what tests show they have accurately perceived*: their language reports of their own knowing correlate negatively with what they actually know.

Language is not always a valid guide to all kinds of intelligence, to *knowing*. Sometimes it misses facts of intelligence altogether.

Evidence of Awareness Under Anesthesia

Studies showing the incidence of awareness under surgical anesthesia are not directly related to our inquiry due to the nonstandard conditions of such research. However, some of these studies should be mentioned due to the close attention they have paid to defining and examining the parameters of awareness under surgical anesthesia. The obvious alarming possibilities of awareness under those conditions have led researchers to closely examine many variables that others may be inclined to neglect.

Their efforts to closely refine an understanding of awareness and its conditions should be part of our inquiry.

One study specifically reviewed definitions of awareness, its incidence, clinical relevance, causes, and ways to avoid it (Schwender et al. 1995). This study sorted the following stages of intra-operative awareness in patients:

- Conscious awareness with explicit recall and with severe pain;
- Conscious awareness with explicit recall but no complaints of pain;
- Conscious awareness without explicit recall and possible implicit recall;
- *Subconscious* awareness without explicit recall and possible implicit recall;
- No awareness.

Explicit recall is tied directly to patients' verbal reports. There are methodological problems associated with demonstrating implicit recall. However, such recall may be tied to postoperative development of post-traumatic stress, including re-experiencing the event awake or in dreams, sleep disturbance, and avoidance of stimuli.

In spite of methodological problems, the spectrum of stages of awareness is tied to sense perception (pain) and memory as well as possible behavioral profiles post-operatively. Though nonstandard, such an instrument provides a far better way of measuring degrees and kinds of awareness and absence of awareness than seen in some standard research.

A second study examined the extent to which meaningful auditory input can be processed by the brain during surgical anesthesia (Aceto et al. 2003). They examined whether patients may be able to remember some information implicitly after anesthesia as well through a "dream-like process" that is subconscious awareness. The details of the experimentation are too lengthy to repeat here. However, we should look at definitions of "awareness" they used.

They sorted one kind of awareness where that is defined as ability to recall (obviously "awareness that") events that occurred while patients were considered to be unconscious. This was sorted from *subconscious* awareness, a state in which information registered by the brain does not enter consciousness.

In postoperative interviews, none of the patients were able to recollect explicit memories of intra-operative events. However, their study showed that auditory information may be processed and retained in the form of implicit memory during deep general anesthesia (Aceto et al. 2003, p. 633). One patient reported a dream associated with a tape played while under anesthesia. Their findings included that auditory information perceived during anesthesia "was incorporated into this form of dream" and related to the implicit memory system.

Even under induced subconscious states such as those brought about by various kinds of anesthesia, patients reveal degrees of awareness. Though considerably more research must be done to document those levels of awareness, the evidence in some experiments shows degrees of cognition as well.

What the Experiments Show

Cognition is viewed by a number of cognitive scientists, neuroscientists, and intelligence researchers as starting with the attention system and continuing on to "higher" levels, that are one way or another aligned with language. But all of the experiments cited above showed some deeper level of *cognitive* awareness, below the attention system threshold, that *correctly* affected subjects' overall behavioral responses.

For example, in Kunimoto et al. 2001, their experiments showed that people can make perceptual discriminations even though the information that was used to make those discriminations is not consciously available. In this as well as in the remaining experiments, "consciously" means "consciousness *that*."

Moreover, some of the experiments, including Kunimoto et al. 2001, also show that there is in fact a *negative correlation* between subjects' own verbal judgment (*knowledge that*) about their own awareness and their awareness as actually measured in the experiments. Kunimoto et al. 2001 found that subjects "can discriminate among stimuli at better than chance levels even with displays so brief that their confidence is unrelated to their accuracy." In other words, their responses correlated negatively with their verbal confidence levels. Their cognitive awareness transcended their verbal cognition.

The same finding is evidenced in Li et al. 2002. Even though masking effectively prevented awareness of the stimuli and confidence ratings did not predict categorization accuracy, subjects nonetheless responded significantly *above chance level* to the presence of animals. Subjects' *motor* responses started to reflect *correct* categorizations at the same time for masked and unmasked stimuli.

Li et al. concluded that information about the first milliseconds of visual stimulation (the preattentive phase) can propagate throughout the visual system, unaffected by later changes, and determine behavior *even when it is not (or not yet) available to consciousness.* Again, "consciousness" here refers to "consciousness *that*."

Repp 2001 concluded that the brain agent guiding the motor behavior is below the perceptual threshold and that the brain is much more sensitive to timing information than the results of previous psychophysical experiments suggest. This precise timing information seems to be used in the control of actions, *without awareness* (again, "*awareness that*").

Näätänen concluded (Näätänen et al. 2001) that the data suggest that subjects' ability to organize impinging signals is based on a kind of "sensory intelligence" [sic] in the auditory cortex.

Weiskrantz (1997), Polanyi (1966) and De Becker (1997) all cite experiments or findings that show subjects' actual intelligent responses correlated negatively with their verbal reports.

This evidence among much more supports the argument that the circle of cognition is larger and deeper than previously thought. This is so as it pertains to not only vision, but also the psychomotor and entire sensory motor parts of the brain.

Among many things, these and related experiments tend to show that not just the brain but the entire central nervous system (CNS) of the human is an intelligence system designed to "home in" on indicators present to them long before they are aware *that* they are doing so. It is this capacity of the primitive sensory and sensorimotor systems shown in some of the experiments which permitted subjects' cognitive awareness and in some cases appropriate motor response before they are even aware *that* they are aware.

4.2.5 Primitives of the Preattentive Phase of Awareness

The preattentive phase in human perception appears to be the most fundamental primitive level of cognition. The experiments above including other strong evidence exists showing that it is at that point that natural intelligence actually begins, at least in the human.

The research literature defines preattentive processing (of visual information) as that in which visual tasks can be performed on large multi-element displays in less than 200 to 250 msec of eye movements (Treisman and Hayes 1992). In certain visual experiments, the subjects accomplished search task in time less than 200 msec, suggesting that certain information in the display is processed in parallel by the low-level visual system and that visual information is processed unattentively.

Visual Fields

As Hernegger explains (2005), there are several visual fields (prefrontal, supplementary, and parietal fields); the same is true of the other sensory systems. There are also several hand-arm fields in the immediate vicinity of the visual fields. This proximity suggests a coupling of eye-hand-arm control by the RTC activation system.

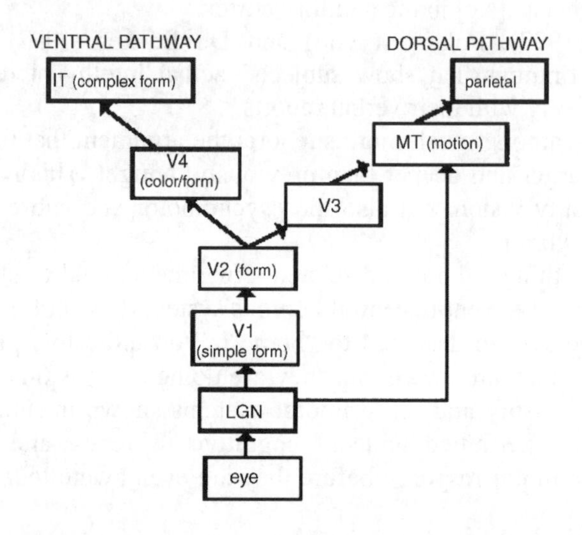

Fig. 4.4. Organization of the Visual System

The entire visual system is usually divided into distinct processes. These include distinctions between *object* and *spatial* vision that are related to different neural pathways. One pathway is the dorsal stream that processes visual information to the parietal cortex, and a ventral stream that processes to the temporal cortex.

Preattentive vision operates in parallel segregating objects from a background. This is familiarly known as the "pop-out" effect. If background features permitting segregation of the object are not available, then a serial inspection is conducted by attentive vision.

The organization of the visual system is both parallel as well as hierarchical. Parallel pathways transfer different kinds of information from the retina to the LGN (lateral geniculate nucleus) and cortex.[8] The

[8] The striate cortex: The lateral geniculate projects to the striate cortex, or Brodmann's area 17, which is located in the occipital lobe at the back of the brain. The striate cortex, also called area V1, is the most highly developed cortical structure in humans. Most of the

pathways recombine in the cortex. After recombining, two pathways emerge as shown in the graph, the dorsal, *magno*-dominated pathway flows to the parietal cortex. This pathway involves space, movement and action. A ventral, *parvo*-dominated pathway flows into temporal areas and is primarily concerned with object identification and perception (Lamme and Roelfsema, 2000).

Properties of visual information such as shape, color, and size that can actually be called a part of *unattended* input are processed preattentively without needing focused attention. Again, preattentive processing is information without focused or selective attention.[9] Because the processing occurs so quickly it obviously occurs in the absence of language and in the absence of reflection about the process.

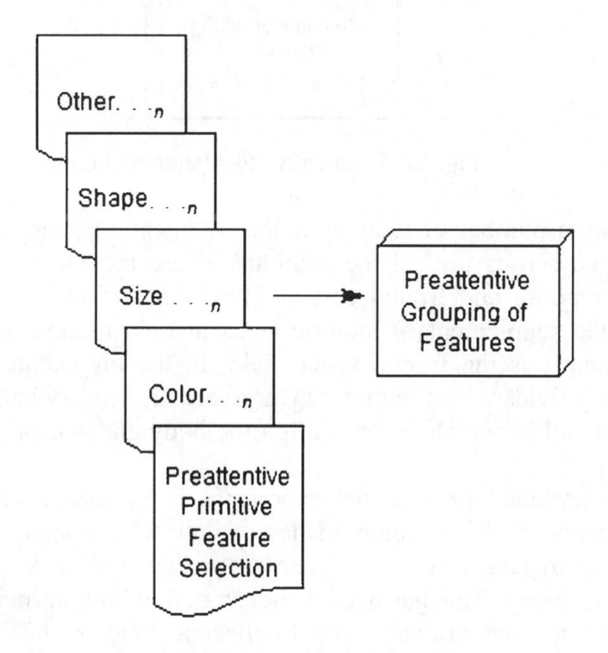

Fig. 4.5. Preattentive Feature Selection and Grouping

Experimentation shows that preattentive processing of features acts to deploy attention. Not until sensory perception is attained can attention

cortex is organized into layers, and the striate cortex is the most laminated region in the brain with at 6 to 9 discrete layers.

[9] There is either some confusion or sheer carelessness in the use of these terms in the research literature. The use of qualifiers such as "focused" or "selective" to modify the word "attention" appears to be redundant since "attention" is defined in those terms.

focus upon information *as an object* with which it can operate; only when this level is reached does preattention make the transition to the *conscious* attention of a cognitive system.

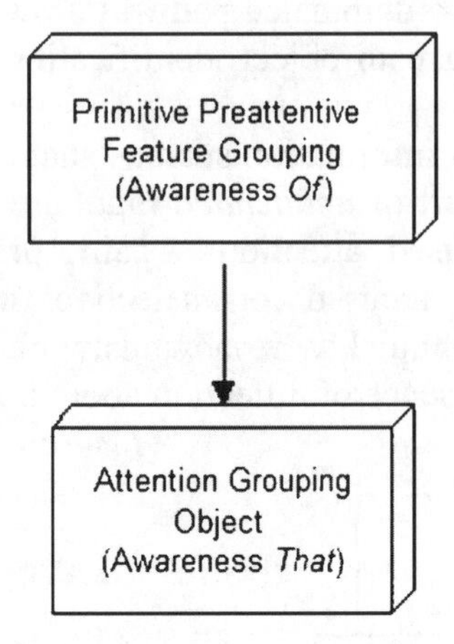

Fig. 4.6. Preattentive to Attention Grouping

There are a number of fields of selective attention that serve to align the body and senses toward the stimulus. These include neurons in the parietal, temporal, and frontal cortex. They also include neurons in the region of the supplementary motoric areas in field number 6. The best-known example is the frontal visual field. In the immediate vicinity of these sensory fields with attention functions are the sensory hand-arm field and the like, all of which serve to align the body and sensory systems to the stimulus.

Though normal human visual experience is by means of conscious, aware percepts (such as seeing shades of yellow in a sunrise), a visual stimulus that bypasses visual awareness *can nonetheless be transformed into a motor output*. This has been demonstrated in both normal as well as pathological subjects (Lamme and Roelfsema 2000, p. 571). The very segregation of visual pathways into ventral and dorsal streams may underlie differences between conscious perception and visually guided action bypassing consciousness.

Preattentive and Automatic Processes

Issues have arisen because of the similarity between *automatic* processes and preattentive ones, and the need to distinguish preattention from attention. By definition, both kinds of processes occur in the absence of attention. Additionally, even if they are different and distinct processes, a recurring issue is whether automatic and preattentive processes reflect the same underlying processes. Usually, neuroscientists sort both the preattentive and attention processes from automatic processes in a variety of ways.

Treisman (Treisman et al. 1992) argued that there are significant differences. First, automatic and preattentive processes become independent from attention in different ways. Though automatic processes are developed through extended practice, preattentive processes are governed by mechanisms that are either innate or acquired early in life.

Moreover, in the same source, Treisman argues that automatic and preattentive processes "are functionally different because automatic processes support skilled behavior while preattentive processes support low-level perceptual functions such as feature detection."

Primitive Preattentive Features, Processes and Cognition

Though there is no doubt that automatic processes support skilled (intelligent doing) behavior, so do preattentive processes as well. This is evident in the experiments cited above as well as other studies on intelligent performances (Frank et al. 2004).

It should be stressed at the outset that the preattentive processes of awareness are obviously in the absence of language. This was demonstrated explicitly in a number of the experiments above, especially Kunimoto, et al. Moreover, preattentive processes are understood as well to be in the absence of attention, usually taken to be the starting point of cognition. *Yet much cognitive activity is evident preattentively before attention as all of the above experiments show.*

Moreover, the activity of preattentive feature selection cannot be the activity of classification, though researchers may describe it, in some metaphorical sense, as classification. It does not proceed by comparing properties of objects based on a principle of similarity. There are, in most preattentive processing, no conjunctions of features, including no conjunctions of like features with which to compare properties. Moreover, preattentive processing takes place in new-born babies, seeing features for the first time.

Additionally, it should also be emphasized that though much of the research on the preattentive phase of awareness is related to preattentive processing of *visual* information, preattentive awareness in general is not limited to the visual system. Though we know more about vision than other sensory systems, the above experiments involved the auditory, tactile, as well as the visual system. The preattentive processes are found in auditory, tactile, olfactory, and other senses of the somatosensory and sensorimotor systems as well.

The peculiar characteristic of preattentive processing in the visual system is that one's visual system processes just *single* features of an item such as hue or form of an object. These are a limited set of visual features or properties that are detected very rapidly and accurately by the low-level visual system. These features "pop-out" and are detected in parallel, immediately, within approximately 200-250 msec. This is usually contrasted with attentional processes that can handle conjunctions of features, are usually thought to be conducted in a serial manner, taking more time.

For example, one can preattentively process the color of an apple, but cannot preattentively process a *conjunction* of features such as color, shape, size, and so on. If an object is made up of a conjunction of unique features, then the current view is that these objects cannot be detected unattentively.[10] A conjunction target item is composed of two or more features. A search for such an object is called a "conjunction search".

An example of an experimental preattentive task is the detection of a red circle in a group of blue circles. The target object, the red circle, has a visual property "red" that the blue distractor objects do not. It "pops-out" in the array. All non-target objects are considered distractors. An experimental subject can usually tell immediately at a glance whether the target is present or absent. The visual system identifies the target through a difference in hue because the red target is in a space filled with blue distractors.

The primitive features identified that are used during the preattentive phase (but of course not necessarily in each case) minimally consist of the following: color, orientation, motion, size, curvature, depth, vernier offset (small departures from the colinearity of two line segments), gloss and, perhaps, intersection and spatial position/phase.

As Wolfe notes (Wolfe et al. 1996, 1998)[11] there may be a few other local shape primitives to be discovered because the primitives of preattentive shape processing are not entirely known. The problem is a lack of a widely agreed upon understanding of the layout of "shape space." Shape or form appears to be the most problematical primitive feature in the preattentive phase. For example, simple color space is a two-dimensional plane or it could be three-dimensional if the surface has luminance. As

---

[10] Though there is some evidence in the research literature which contradicts this.

[11] Moreover, there are differences in how each of the primitive features is actually processed in the preattentive phase. However they are in fact processed, they are used to intentionally guide attention to some object.

Wolfe notes, it is not clear what the "axes" of shape space might be. But preattentive processing of “shape space,” whatever we take that to be, enables us to then make sense of objects we attend to, and to make sense of a whole lot of other properties of things, including motion.

The following table is a partial list of preattentive features taken from a recent research survey (Healey 2005):

Table 4.1. Lists of Some Preattentive Features

List One	*List Two*
length	binocular lustre
width	stereoscopic depth
size	3-D depth cues
curvature	3-D orientation
number	lighting direction
intersection	texture properties
closure	artistic properties
color (hue)	direction of motion
intensity	
flicker	
line (blob) orientation	

Treisman (1991) was among the first researchers to document preattentive processing, determining to a large extent which visual properties are detected preattentively. She called those properties “preattentive features.” Additionally, she hypothesized how the human visual system actually performs preattentive processing.

Triesman’s and other researchers essentially demonstrated two different ways that preattentive and nonpreattentive tasks are performed. Though their findings have recently met with some challenge (VanRullen et al. 2004), they demonstrated that preattentive tasks are performed bottom-up in parallel. Such processing takes place very quickly. On the other hand, nonpreattentive tasks are performed serially, taking far more time.

Conjunction targets are processed serially, thus nonpreattentively, and take more time. An example would be a red circle target, made up of two features, red and circular. Healey (2005) presents an experiment in which one of these features is present in each of the distractor objects, red squares and blue circles. By its very design, the experiment denies the visual system of a subject a unique visual property to search for when trying to locate the target. When a viewer searches for red items, the visual system always returns true because of the red squares in each display; moreover, a search for circular items always returns blue circles.

According to Healey, numerous studies have shown that this target cannot be detected preattentively. Viewers must perform a time-consuming serial search through the displays to confirm its presence or absence. The experiment shows that such a target search is conducted by attention processes.

In contrast to Kunimoto's method, Triesman and other researchers measured preattentive task performance in two different ways: by response time and by accuracy. In the response time model viewers were asked to complete target detection quickly while still maintaining a high level of accuracy, with the number of distractors in a scene repeatedly increased. If task completion time is relatively constant and below some chosen threshold, independent of the number of distractors, the task is said to be preattentive. If the task were not preattentive, viewers would need to search serially through each display to confirm a target's presence or absence.

In the accuracy model, a display is shown for a small, fixed exposure duration, then removed from the screen. Again, the number of distractors in the scene varies across trials. If viewers can complete the task accurately, regardless of the number of distractors, the feature used to define the target is assumed to be preattentive.

Preattentive Feature Integration

Based upon extensive experimentation, Triesman and other researchers compiled the above list of features that are detected preattentively. Moreover, Triesman proposed a *Feature Integration Model* for preattentive processing to explain how that process works.

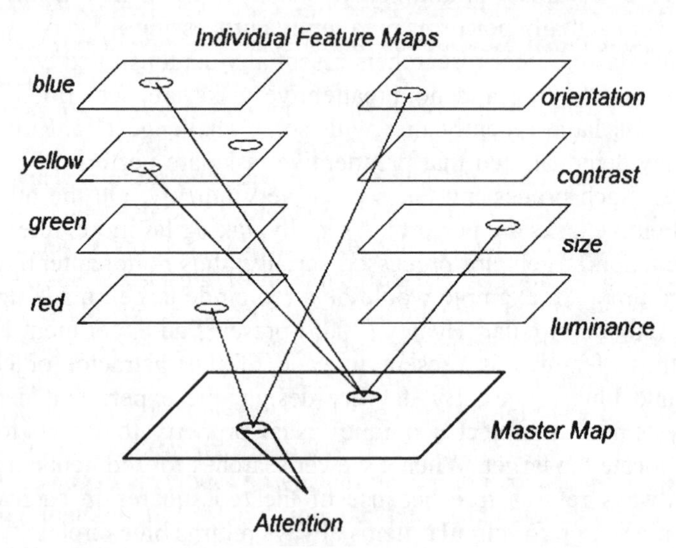

Fig. 4.7. Triesman's feature integration model

In the above figure, individual feature maps are accessed to detect feature activity. This is conducted in parallel by the preattentive process. However, focused attention acts through a serial scan of the master map of locations.

Triesman's hypothesis was proposed to explain how preattentive processing occurs. If the target has a unique feature, one can simply access the given feature map to see if any activity is occurring. Feature maps are encoded in parallel, so feature detection is almost instantaneous. However, a conjunction target cannot be detected by accessing an individual feature map.

As Healey's examples show, activity there may be caused by the target, or by distractors that share the given preattentive feature. In order to locate the target, one must search serially through the master map of locations, looking for an object with the correct combination of features. Focused attention requires a larger amount of both time and effort because of the requirement for a serial search.

Feature Integration Theory led to the view that our perception of the visual world relies on the parallel extraction of a set of preattentive features which is then followed by the serial integration of these features into a coherent object. The integration of features is performed by visual attention. In visual search experiments, the detection of preattentive features can be performed in *parallel*, whereas searching for more complex stimuli such as feature conjunctions requires a *serial* examination by some form of attentional spotlight (Wolfe et al. 1989; Treisman and Gelade 1980).

Possible Dichotomy of Visual Discrimination

Thus we are also left with the notion that there exists a dichotomy of visual discrimination tasks, with parallel/preattentive discriminations at one level, and serial/attentive discriminations at another.

However, more recent theories of attention have posed challenges to this dichotomy. VanRullen et al. (2004), provide evidence that challenges the assumption that preattentive and parallel processing are two equivalent ways to refer to the same subset of visual discrimination tasks. According to them, experimentation with dual tasks reveals that some tasks can be preattentive but not parallel, and parallel but not preattentive.

Though they do not directly challenge the notion of preattentive features, their results show that such features are not limited to early cortical representations and that there is in fact a hierarchy of preattentive features. They agree that preattentive features can be processed in parallel, but with constraints. Their experiments showed that the preattentive (dual) tasks that result in parallel search seemed to rely on neuronal selectivities present in early visual areas, such as orientation and color. Those that result in serial search probably rely on higher level neuronal selectivities such as color and orientation conjunctions, animals, and faces.

They propose that a feature or stimulus is preattentive if there exists a neuronal population selective to that feature or stimulus, independent of the cortical area involved. ". . .neurons in V1 are well known for their selectivity to orientation (in columns) or color (in blobs). . .Similarly, neurons selective for simple conjunctions of color and orientation. . can most likely be found in areas beyond V4" (VanRullen et al. 2004).

Detection and Attention to Faces

Interestingly, a number of visual search studies demonstrated that facial information such as identity and expression is not registered in parallel even though one of their own researcher's dual task experiments indicates that gender facial information *is* available to our visual systems even when attention (in dual task) is not (Brown et al. 1997; Reddy et al. 2002). Thus, they also propose the following general principle:

> "[T]hat the extent to which our preattentive (dual task) features can be discriminated in parallel is an inverse function of the receptive field size of the neurons that represent this feature. At higher levels of the ventral hierarchy, only very few "features" can be processed in parallel, and the corresponding stimuli must be well enough separated to avoid having a target and a distractor falling within a single receptive field" (VanRullen et al. 2004).

They also speculate that with receptive fields covering the entire visual field, there could be no parallel processing "even when the target is defined as a preattentive [dual task] feature." In the case of pop-out that nonetheless occurs for features that are not discriminated without attention, they are left with determining the mechanism that underlies the parallel search, assuming that attention is limited and cannot be deployed simultaneously across the entire search array.

Healey (2005) and others have pointed out that if the low-level visual system can be harnessed during visualization, it can be used to draw attention to areas of potential interest. Indeed, it is this low-level visual system, including preattentive processing of features, that is fundamental for *knowing how* to do anything at all and for survival itself, as some of the above experiments involving predator images indicates.

But this same low level visual system, including preattentive and attention processes, has become the focal point of interest to software designers and those engineers who seek more biologically inspired models for artificial intelligence. Expanded to the complete sensory and sensorimotor awareness system, these processes should also be a focal point of interest to those who seek to understand the full scope of intelligence, especially the intelligence of doing, *knowing how*.

4.3 Primitive Intelligence of Moving and Touching

Some of the above listed experiments involve the visual sense of motion as well as visually guided action. What is called the Middle Temporal complex (MT+) region of the inferior temporal sulcus, consisting of Middle Temporal (MT), Medial Superior Temporal lateral (MSTl) and Medial Superior Temporal dorsal (MSTd), has multiple regions specialized in different aspects of motion perception. It is motion perception that extracts the three-dimensional structure of the world, defining the edges and forms of objects.

Motion perception involves what are called the "what," "where," and "when" pathways and sharing of information. While some neurons are good at determining the direction in which an object is moving, they cannot identify the object.

Some cells in layers of the visual system are sensitive to orientation and also to motion in particular directions. Parts of (Medial Temporal complex) MT+, MSTl and MSTd sense when *objects* move; others sense when *you* move. Moreover, different patterns of optic flow are produced in your retina when you move in different directions and the neurons in MSTd recognize these different patterns. Analyzing the above experiments, it is fairly easy to see that this enormously complex system is involved in them all.

All the regions of the brain that guide a variety of movements are involved as well. There are multiple representations of space in the posterior cortex that makes all this possible. The Lateral Intra Parietal (LIP) region represents locations of objects that you intend to look at and may reach for.

The Medial Intra Parietal (MIP) region represents immediate *extra-personal space*, which is the space you can reach to, guiding arm movements. The Anterior Intra Parietal (AIP) region represents the shape information we need in order to grasp objects. And the Ventral Intra Parietal (VIP) region represents the *near space* used to guide the head, mouth and lips during feeding. This region receives visual and tactile information from the face (Vilis 2002).

Touch discriminative ability depends on a variety of touch receptors coding millions of stimuli. This is part of the somatosensory system, including multiple types of sensation from the body - light touch, pain, pressure, temperature, and joint and muscle position sense, all of which may be involved in highly complex interrelations with one another. Each of these kinds of sensation is carried by different pathways; each has different targets in the brain, and each cross one another at different levels.

The primary motor cortex (also known as M1) works in association with premotor areas to plan and execute movements. Premotor areas plan actions together with the basal ganglia and refine movements based upon sensory input which requires the cerebellum.

4.3.1 Multiple Spaces of the Senses, Images and Probing

Natural intelligence systems use the body to *attend* to physical things outside it. We attend to things outside our body, *from* our body, and we can feel our own body *in terms of* the things outside it *to* which we are attending.

There is a spatial direction to the primitive relations of the preattention phase and of attention. There is also a sense in which we can make an external physical thing function as a *proximal term*[12] of immediate awareness. That is, there is a sense in which *we extend our kinesthetic bodily intelligence to include that object we attend to*, by extending our body with instruments, such as mechanical probes, that we use to attend to an object or set of objects.

Moreover, we do this as well with purely abstract objects, including images. We can use images in our minds as proximal terms to probe or to stand in for physical objects (Kosslyn 2002),[13] as well as previous events in time. Some recent research has shown how subjects use images to reinstate a context from the past as aids in remembering past events (Quiñones 2005).

Additionally, there is increasing evidence of the effects of motor imagery-based mental practice for activating cerebral and cerebellar sensorimotor networks. The evidence suggests that mental practice may be an effective substitute or complement to physical practice to activate compensatory networks for motor rehabilitation (Lacourse, et al. 2004).

---

[12] The terms "proximal" and "distal" are borrowed from anatomy, but can be used to analyze the structure [or anatomy] of our intelligent doing, *knowing how*.

[13] According to Kosslyn, using images this way also causes the same effects on memory and the body as occur during actual perception, but the two functions are not identical.

Though I provide an example of physical probing in a medical surgical task, I hope it is clear that I do not limit probing in general to physical spaces and with sense [physical] objects. Human beings also probe abstract spaces, as mathematicians and logicians clearly evidence. Both Russell (1903) and Gödel (1964a, 1964b), for example, have pointed out the close analogy between how we know reality with our senses and how we know abstract objects of mathematics with our minds. They both recognized the significance of immediate awareness in mathematical knowing.

One who performs a complex task must know how to coordinate movements of their body, using most if not all their senses, by a kind of "cognitive indwelling" in the external physical thing that is functioning as a proximal term of their *knowing how*. There is a sense in which all the things we use in a physical task performance, such as tools or other implements as probes become extensions of our body, specifically, extensions our bodily natural intelligence.

From an intentional phenomenal (first-person) point of view, the senses along with imagining, touching and moving form a multi-relational, multi-layered set, with touching, imagining and moving on a higher level than any of the other senses, except sight. All these must be examined within the context of spatial relations each sense has with the body and spatial relations of moving and touching.

A clarification of all this should help to understand a little better the significance of the nature of *probing*. With respect to the particular senses, seeing, hearing, feeling, tasting, smelling, there are a number of principles pertaining to these which require explanation over and above the spatial relation each sense has with our bodies. Firstly, not all of the senses are on the same primitive *epistemic* phenomenal level. The epistemic level is that pertaining to intentionality. In physiologically unimpaired persons, the sense of sight may take some priority over the other senses, and it is clear that the space of our visual experience is not identical to the space of the other senses.

For example, visual space is *binocular* space, while the space of the other senses, for example smell, is not. But as already noted, we still have limited understanding of the space of all the primitive features processed during the preattentive phase of the visual system. We do not yet have a complete understanding of "shape space."

Additionally, there are different representations of space in visually guided actions. The multiple representations of space in the posterior cortex, used to guide a variety of movements such as grasping and reaching, and feeding, are mapped on several forms of egocentric frames of reference and are derived from several modalities of sensory information such as visual, somatosensory, and auditory (Vilis 2002).

Moreover, the MT+ complex helps to extract the three-dimensional structure of the physical world, to define the form of objects, to define relative motion of parts of objects, and a variety of other facets of moving objects.

The senses of sight, hearing, feeling, smelling and tasting are epistemologically sorted from *touching*, (specifically discriminative touching) which is *not* identical to mere tactile feeling. The former is clearly intentional while the latter is not.

Eccles (2002) cited an excellent experiment effectively showing the difference between the two. The experiment showed the effect of silent thinking on the cerebral cortex, in which a subject was:

> ". . .concentratedly attending to a finger on which just detectable touch stimuli were to be applied. There was an increase in the rate of cerebral blood flow (rCBF) over the finger touch area of the postcentral gyrus of the cerebral cortex. These increases must have resulted from purely *mental* attention because actually no touch was applied. . ." (Eccles 2002).

These senses are also sorted from *moving*, which is treated as a complex sensory *and* somatosensory-motor phenomenon in the neurophysiological literature (Berthoz et al. 1995), involving the above different representations of space in visually guided actions. I include moving with touching as primitives at a level higher and distinct from though including the other senses, including tactile feeling.

In part, this is because the concept *touching* is clearly *bodily* intentional, in the sense that we use our bodies cognitively to index in kinds of space when we touch. On the other hand, mere tactile feeling is not used this way.[14] Moreover, the space of feeling and the space of touching are not identical. For example, intentional touching is not clearly always in Euclidean three-dimensional space because of the relation of imagining, including *anticipatory* imagining to it, as we will see below. However, mere tactile feeling [such as feeling a pin prick] clearly is in Euclidean space. A pin prick is felt *here*, *now*, *in this space that I can physically point to.*[15]

The senses and the concepts *touching* and *moving* are enormously complex concepts. One way of distinguishing the senses from touching

[14] Substantial empirical research has established this claim, including that of Berthoz et al. 1995. Also see Gardner 1993, especially references included under bodily-kinesthetic intelligence.

[15] There is an enormous diversity and complexity of kinds of space characterizing the primitive features in the preattentive phase and the senses, as well as touching and moving. Thus I have chosen to limit the discussion here.

and moving and other primitive epistemic relations, including imagining, is to note that the objects of the senses, that is particular sights, smells, tastes, or configurations of these, are exactly that. They are particulars which occur "now" with the subject. They are not universals or generalizations publicly accessible to anyone, though in principle the same particular may be experienced by more than one person.[16]

The intentional concept of *touching* is more complex in that it entails a deliberateness with the body which is not found with the senses *per se* and where imagining is involved there may be abstract universals which may be experienced by more than one person. A thorough analysis of the epistemic structure of *touching* requires an analysis of probes, their intentional use, and epistemic relation to our body.

It should be noted that our scientific knowledge and understanding of human moving [movement generally, or whole-body displacement] is quite limited. We do not as yet even understand how moving is stored in the memory, how we spatially image or reconstruct a trajectory path [path integration] in our minds, or how we "home" in on a target, objective or goal with our bodily movements.

Moreover, what we know of the human use of the fingers as probes to explore or come to know the texture and shape of objects has much in common with results of scientific neural experimentation with the rat trigeminal system. We know that rats rely on rhythmic movements of their facial whiskers *much as humans rely on coordinated movements of fingertips* to explore or come to know objects in their proximal environment. The trigeminal system is a multilevel, recurrently interconnected neural network which generates complex emergent dynamic patterns of neural activity manifesting synchronous oscillations and even chaotic behavior (Nicolelis et al. 1995).

There is an epistemological sense in which touching requires that one intentionally *heed* and *focus* upon the object of touching with one's body, whereas one can experience with one's senses without that kind of intentionality. Moreover, this intentional heeding and focusing will differ epistemically in its structure depending upon whether or not one has unaided visual access to the object(s) to which one is heeding or focusing. It will also differ depending upon whether or not one is touching the object directly with one's body [for example with a hand or finger], or if the touching is mediated by an instrument used as a mechanical probe of some kind.

[16] That is, these particular objects of the senses necessarily have a temporal relation with the subject who is having particular sensations, but only in principle can two subjects experience the same particular object of the senses, such as a particular color or taste.

For example, if one is probing with one's fingers the interstices of a surgical incision not visually accessible, the intentional structure of that *coming to know,* the probing, differs from the digital inspection of a wound which is visually accessible. This is so in part because the structure of the former requires more complex relations of imagining. As with any kind of exploration, such probing requires *continuously forming images of the object* that is the particulars making up the configuration of the inside of the incision, as touching proceeds.

Moving which has epistemological significance is clearly intentional and requires a focal heeding with one's body, as does touching. However, it is not clear whether the space of intentional moving is equivalent to or identical with the space of intentional touching because of the relation of imagining and especially imagining which is anticipatory to the latter.

Moreover, intentional touching is obviously a more *close*[17] relation than moving which can be more distant. With touching, the body is clearly used indexically in a very close, concrete way with proximal objects. With moving, the indexical function of the body may be more abstract because it can involve objects that are imagined and anticipated which may be at great distances from the body.

The same kind of heeding or focusing found with touching and moving is not found with experiences of the other senses precisely because of the unique digital or indexical use of the body, and the spatial relations of touching and moving with the body. Moreover, though touching may involve any part of our body, the fingers *as digital indexes* are pivotally involved in an epistemic sense, as a means of directing our *coming to know* an object of touch.

As already noted, there is an epistemic sense in which we extend our body to include the object presented, to which we attend, *from* the [sometimes imagined and anticipated] focal and subsidiary configurations of particulars we are aware of with our fingers. And as our analysis of probing shows, the relation of imagining is also pivotally involved in touching, in tactile efforts to come to know objects of touch. Not only are images formed of configurations of physical particulars presented, but images are also formed of abstract configurations of particulars *anticipated.*

[17] I emphasize that the terms "close" and "distant" as epistemic relations are defined in relation to proximity with the human body. The human body is the ultimate instrument of all our external knowing.

4.3.2 Smoothness and Timing in Intelligent Doing

The entire scope of *knowing how* can be classified according to performance standards as well as the number of paths and termini that define a task. Standards for intelligent performance include such notions as *smoothness* in the actual doing or execution of a task as well as timing and other conditions as well.

Timing in the actual performance of a task or method is usually one sure sign that a person knows how to do what they are doing. Timing is directly related, usually, to the actual demonstrated physical smoothness of the performance.

In F.C.S. Bartletts' (1958) early study of thinking and skills acquisition, he noted that the movements of a novice, a person who does not know how, are *oscillatory*. That means that the person's movements exhibit wavering between extremes, a "jaggedness" or discontinuity between movements that do not exhibit the controlled, continuous movements of one who knows how.

As a skill or task is mastered, however, a learner develops timing and controlled movements so that the oscillations are reduced and the actions flow continuously. That is, the actions, the movements, of one who *knows how* are smoothly connected, both sequentially (spatially) and temporally.

The smoothness condition alone is demonstrated in a variety of ways and, depending upon the task being performed, can be captured well in time-lapse photography with a digital camera. When one closely examines and analyzes someone (or an animal) performing a task, the sensorimotor system and much more of the entire central nervous system is involved. One's sense of sight, smell, even taste, hearing, and certainly touching and moving is fully integrated in a dynamic, continuous performance.

According to Maccia (1973), the smoothness condition in *knowing how* not only distinguishes *knowing how* from an accidental "happened to be," it also distinguishes a procedure, a way of doing something, from a performance, a doing of something. It is evident that one can know a procedure for doing something without being able to do it, such as Olympic judges.

Moreover, the more a person or animal practices or does a task, the better their sensorimotor system becomes "tuned" to do it the next time. The task itself becomes so familiar that the very perceptual and conceptual clues and cues which characterize the doing of it, especially if these are redundant, become part of the "conceptual and perceptual repertoire" of one who knows how. This is a clear demonstration of all that variability and plasticity found in the mapped connections between the senses through the thalamus to the region of the somatosensory cortex.

So little attention has been paid to *knowing how* that we do not have adequate research establishing a coherent, consistent and complete theory showing how it is integrated within cognition generally. Human performances alone, let alone animal performances, need extensive study and research to delve into the most fundamental levels of those components which can give rise to the expertise in performance that we witness everyday in certain fields.

Because kinds of intelligent doing show up in the smoothly timed, highly context-dependent patterns of moving and touching by one who knows how, these must be incorporated within a broader theory of natural intelligence. The emergence of smoothly timed and context-sensitive patterns from the complex dynamics involved in interactions of very large numbers of all the sensory and motor components and relations between them must be explained.

Explanations of this kind of intelligence, however, must be on many levels, not reduced to the neurological and biological. Some researchers (Kunimoto et al. 2001) have insisted that at least with respect to the components of awareness, the phenomenal experience of subjects must be incorporated into the research. Their experiments included controls on subject verbal responses. There are other ways, however, how this might be done which can be concretely demonstrated by analyzing the interactions among those components of very complex tasks.

Limitations of Computational Models of Awareness: Selection without Classification

One of Edelman's (2004) arguments against computer models of the brain and mind is that the "mapped connections from the sense of touch in the hand through the thalamus to the region of somatosensory cortex are variable and plastic." As he explains, those regions of the somatosensory cortex that map the fingers shift their boundaries as a result of excessive use of even one finger.

His point is that there is a highly variable context-dependency involved in the dynamic circuit variation of the sense of touch. Computer models, especially top-down, logic- and knowledge-based models cannot handle this kind of context-dependency and variability. And he is correct.

Though I will address these limitations more thoroughly in a later chapter, we should briefly review why this is so. As obvious as the distinction may be, we must sort levels of scientific inquiry to make clear some fundamental concepts about the preattentive and attentive phases of awareness. There is the level of the object of inquiry itself. In this case, the preattentive and attentive phases of cortical and central nervous system phenomena are the object of scientific and technological inquiry.

There is the next level, the level of language describing the phenomena, language *about* the object or phenomena of interest. That language sets forth representations of the objects of interest, setting forth (under the best conditions) necessary and sufficient descriptions to fully explain the phenomena. Those representations are not found in the object itself.

We should take care not to confuse the level of the object or phenomena with that level at which we set forth representations of the object or phenomena. We should not confuse the symbol *of* something for *the thing* we symbolize.

The evidence we have reviewed in experiments and some clinical trials supports the view that the preattentive phase of cognition is the most primitive. It is located at the *perceptual* "edge" of the circle of cognition, the most primitive point at which we actually enter that circle.

Because the preattentive phase selects single, unique features, the activity of that phase cannot *literally* be classification. There is a pervasive confusion in the neuroscience research literature between selection and classification. The assumption is that when an organism selects, it has therefore classified.

Yet selection and classification are neither identical nor equivalent. First, selection of single unique features is not classification because it does not proceed by *comparing features* because only single features are selected. Nor does it proceed by selecting properties of objects based on a principle of similarity [among properties]. That is the only notion of classification recognized by scientists adopting the logico-linguistic notion of intelligence, as well as those adopting the computer model of mind. Though researchers may describe what is happening in the preattentive process as classification, they must be doing so in some metaphorical sense.

In most preattentive processing, no conjunctions of features, including no conjunctions of *like* features with which to compare properties is taking place. Moreover, preattentive processing takes place in new-born babies, seeing features for the first time.

It should be mentioned, however, that VanRullen, Reddy, and Koch's research has shown that in *dual* tasks pop-out can occur for features that are not discriminated without attention.

In effect, the fact that the activity in the preattentive phase cannot be a classification process means that it does not work according to classical Boolean operators such as conjunction. This has been empirically verified in a number of experiments, some of which are described above. As such, it follows from this that the preattentive phase cannot be reduced to *classical* machine models of cognition.

It is follows from the empirical evidence presented thus far that the primitive phases of human cognition cannot be understood or modeled with classical top-down, knowledge-based, discrete, functional-block, linear models. These are the same underlying models of human cognition found in classical IQ research and those tests based upon it. It also includes underlying assumptions as well of current computer systems architectures of intelligence.

In sum, neurobiological architectures are not modular, linear, or feedforward. The neurobiological representations of visual modalities alone, including depth, motion, color, and form are entirely unlike those currently employed by conventional computer vision systems. However, we will look more closely at computational problems with natural intelligence in a later chapter.

In the preattentive processing phase alone, we have yet to figure out all the primitives involved and exactly how the process and the interrelations among all the primitives actually work. In the visual system, we have three types of cones that allow us to distinguish between about 2 million colors, but there are probably billions of actual primitive featural relations involved in the preattentive phase and attention system. Again, we do know that however the process works preattentive processing acts intentionally to deploy attention, the place where actual perception occurs.

4.4 Summary

This chapter sought to understand the fundamental structures that give rise to the intelligence of doing by looking at sophisticated brain structures and the context in which those structures find themselves, along with some focus upon the phenomenal experience of human subjects.

Specifically, we looked at the neurobiological architecture of the sensorimotor system and awareness and how kinds of intelligence emerge from those. We found that extensive neurobiological and related research reveals highly complex *cognitive* behavior at very primitive levels.

Where We Enter the Circle of Cognition: Immediate Awareness

The primary objective was to determine from knowledge of the most primitive sensory and sensorimotor components where the human organism most likely enters the circle of cognition. The concept of cognition is not taken here as identical with the concept of consciousness, as it is defined in some recent theories (Edelman 2004). Because his theory is to some degree representative, we critiqued some of Edelman's definitions, including "categorization", "concept", and "abstraction", finding that they assume what they seek to explain.

Additionally, his explanation of concept formation is inadequate. His appeal to the questionable process of "abstraction," which is actually an underlying appeal to a problematic notion of induction renders his explanation unacceptable. The use of the notion of abstraction is yet another example of a process that assumes what it seeks to explain. The formation or learning of concepts or universals as Edelman describes still presupposes cognizance of, not merely discrimination of, features. That is, such an explanation already assumes the concept, universal, or intelligence it sought to explain.

Another problem with Edelman's explanation, along with other theorists, is that he starts the process of global mapping at the level of *attention*. There is substantial empirical evidence showing that there is a great deal of *cognitive* awareness prior to that level, specifically in the preattentive processes.

The research literature on preattentive and attentive processes shows that there is some dispute on when and where the preattentive processes make the transition to attention. Preattentive orientation proceeds *subconsciously,* interpreted as the absence of "consciousness *that*" such and such is the case, at the level of the nervous system. Many researchers hold that it is only when sensory perception is attained that *attention* can then focus upon information as an object with which it can operate. Some appear to argue that only at the attention phase, interpreted as "consciousness that" or "awareness that" (aligned with language) does a subject's neurophysical activity become cognitive.

However, we reviewed substantial evidence in multiple experiments of cognitive immediate awareness *below* the threshold of attention. All of the experiments cited showed some deeper level of cognitive awareness, below the attention system threshold, not aligned with language that *correctly* affected subjects' overall behavioral responses. Moreover, as noted, some of the experiments also show that there is in fact a *negative correlation* between subjects' own verbal judgment (*knowledge that*) about their own awareness and their awareness as actually measured in the experiments.

This evidence shows that the circle of cognition is larger and deeper than previously thought. Minimally, the evidence supports our position that it begins with immediate awareness; this is prior to the level of attention and is not aligned with language. This is so not only as it pertains to vision, but also the psychomotor and entire sensory motor parts of the brain.

Thus I argued that not only do we need to revise our understanding of the scope and depth of the cognitive domain and the place where we enter it, we must also revise our understanding of a network of related concepts, minimally including cognition itself, natural intelligence and learning. If the empirical findings and our interpretations of them are correct, natural intelligence begins with cognitive immediate awareness in the preattentive phase.

Additionally, researchers in those experiments as well as those who have investigated cognitive awareness under surgical anaesthesia recommend a broader spectrum of the concept of awareness. The general concept of awareness is not viewed by many researchers in terms of two mutually exclusive states, awareness or unawareness, but is viewed as a continuum of states ranging from unaware through an infinite number of partially aware states, to complete awareness. This continuum also distinguishes between "*awareness that*" such and such is the case [tying awareness to "that" clauses or linguistic reports] and "*immediate awareness*" which is not tied to such reports.

Primitive Selection and Problems with Consciousness

Arguments were also presented raising issues with the description of the preattentive phase and with the use of the concepts "conscious" and "attention", and distinctions between cognitive and non-cognitive. Because the organism is already making *preparations* and *aligning its senses* with some stimulus during the preattentive phase, this logically implies that the organism is *already directing itself* in ways to attend to some stimulus that it has already *in some more primitive sense* selected to align itself with.

It has to have made such a selection since any given stimulus would be in an environment filled with possibly an infinite number of stimuli from which to select. The preattentive phase is said to precede conscious sensation which is held to occur with the activation of attention, with (again) the problematic use of the term "conscious" tied to "awareness that" such and such is the case. And it is in the attention system combined with the activation system that, so it is claimed, cognition occurs.

The experiments cited earlier, however, reveal that this nominalist framework within which some research is interpreted must be rejected.

To summarize, intelligent doing, *knowing how* in relation with *immediate awareness,* is more fundamental in the total scope and structure of natural intelligence than is *knowledge that*. It is deeper, broader, and more intricately threaded throughout the tapestry of everything an intelligent being does.

Verbal intelligence is a narrow slice of the total intelligence of the person. Indeed, it misses some of the most important and far-reaching properties of natural intelligence altogether.

In the remaining chapters, we will look closely at intelligent doing involving abstract structures and methods as in the doing of mathematics. And we will examine computational models, and more issues and problems related to natural intelligence.

5 Universals, Mathematical Thought and Awareness

In the last chapter, we left unfinished an examination of the issues surrounding the formation of concepts, otherwise known as universals. We examined Edelman's (2004) global mapping theory that sought to explain the formation of concepts by means of interaction of two parts of the neural system, but we found that theory seriously inadequate for a variety of reasons. Among other things, his explanation of the process of formation of concepts does not work. In the first place, it begs too many questions and effectively collapses concepts to percepts. Even given his theory, however, there is no way to explain the formation of abstract concepts as in mathematics.

We also left unfinished the job of examining intelligent doing where that addresses those abstract objects and methods dealing with them as found in mathematics. We also mentioned that there are serious limits on the use of computer models to characterize kinds of intelligence. Specifically, the use of top-down, logic- and knowledge-based models are unable to characterize self-organizing complexity of emergent properties of intelligent doing and immediate awareness.

With respect to the computability issues, we agree with Edelman's criticisms of the top-down computational approach to mind and consciousness, though our reasons for disagreement will extend beyond his. Serious assessment of the evidence concerning preattentive processes alone shows the limitations on computer models of such processes.

Based on that evidence, it appears that activity in the preattentive phase does not work according to classical Boolean operators such as conjunction. This has been empirically verified in a number of experiments, some of which were described earlier. As such, the preattentive phase cannot be reduced to classical machine models of cognition. It cannot be understood or modeled with classical top-down, knowledge-based, discrete, functional-block, linear models. As I earlier noted, these are the same underlying models found in classical IQ research and tests based upon it.

Later, I will examine in much greater depth these and other issues related to the computability of kinds of natural intelligence. We will distinguish between whether or not certain processes of an intelligent organism are computable on the standard Von Neumann digital computer; and whether or not such processes are computable, even in principle. I will argue that there are limitations even in principle on the computability of facets of living intelligent systems.

This chapter will examine several issues. One of those is related to mathematical doing, the kind of intelligent doing or knowing how *par excellence* involving the conception and use of universals. We will examine several different arguments on the nature and origins of mathematical thought. This may require going back over some of the ground we have already covered when we earlier addressed the formation of universals. Mathematic doing, from counting and elementary arithmetic to the far reaches of algebra, number theory, combinatorics, topology, among other branches, is entirely about universals and their patterns.

Involved in this will also be an examination of related issues surrounding immediate awareness and problems related to analyzing phenomenal relations between subjects and objects. This includes addressing the limits of classification and the nature of objects of the awareness relation between a subject and object. This will in turn lead us in following chapters to examine in much greater detail those problems with computational approaches to kinds of intelligence.

5.1 On the Origins and Nature of Mathematical Thought

The concept of "origins" here is not intended to be historical, anthropological, or cultural. It is intended in the most general sense, broadly including assessments of the neurological and epistemological research on the nature of mathematical thought. Since we have already surveyed much of the historical development of some concepts related to this kind of intelligence, we will not repeat that here. That development already includes an examination of how different theories of knowledge developed and how the nature of reason and logic were viewed. Needless to say, mathematical thought and reason is implicit within that development.

Given the increased interest in the origins of mind, specifically where the concept of "origins" is interpreted to mean *genetic* origins, we should clarify the nature of a peculiar kind of fallacy that continues to rear its head in such discussions. It is called the "Genetic Fallacy."

The Genetic Fallacy

Recall that no matter how much data one might present, or how strongly committed one might be to an argument, if the argument is based upon a fallacy, then it cannot be *valid*. If the argument is not valid, it cannot be *sound*. Therefore reasonable people must reject the argument.

A recurring fallacy in discussions of the origins and nature of mathematical thought is the "Genetic Fallacy." In general, this is a fallacy of irrelevancy. That means that it is an argument employing reasons and evidence that are irrelevant to the objective truth, validity, and soundness of an idea or argument.

In the genetic fallacy this involves appealing to the origins or history of an idea, or in this case, the entire discipline of mathematics to make unwarranted claims. It is fallacious to either support or reject an idea based on its origins or source, including its past, unless its past in some way directly affects the truth, validity and soundness of the idea or argument. The genetic fallacy is committed when one nonetheless evaluates an idea or argument based upon where (or from whom) the idea or argument originates, rather than evaluating the argument or idea on its content and objective merits.

It should also be mentioned that the genetic fallacy is usually committed by those who are otherwise known as reductionists. As noted earlier, though there are valid reductions in the sciences, especially sciences such as physics and chemistry, reductionist arguments in the life sciences are sometimes fraught with fallacies and logical incoherencies. In essence, these are arguments advocating the view that the biological origins of a thing are the most scientifically revealing features of a thing; that some facets of living things, such as intelligence, are really as primitive as their origins.

The discipline of mathematics has been around for thousands of years. During that time, just as in the sciences, mathematicians have established formal, proven methods and techniques to objectively evaluate the truth, consistency, coherency, and pragmaticism[1] of their abstract ideas, concepts, theorems, principles and laws. It is those established proven methods and techniques that determine the nature and worth of the ideas of mathematics. It is not persons who happen to be mathematicians; or persons who happen to be cognitive scientists.

Thus, in contrast to arguments committing the genetic fallacy, the nature and value of mathematical and scientific ideas, concepts, laws,

[1] This is Charles Sanders Peirce's term, which meant the long-term consequences of testing out an idea in the world.

theorems, can be objectively determined and evaluated by established methods and techniques. The human origins or history of the ideas are irrelevant.

It appears to me that when closely examined, among other problems, the overall argument proposed by Lakoff and Núñez (2001) commits the genetic fallacy. Their arguments commit other fallacies as well, but it is the genetic fallacy in particular that we should examine first since it is directly related to their fundamental thesis about all of mathematics.

We should start with some of their recent arguments on the origins and nature of mathematics and assess these in terms of the genetic and other fallacies we have pointed out in earlier assessments.

5.1.1 A Postmodern View: The Body Shapes Development and Content of Mathematics

Lakoff and Núñez (2001) aver that they are not mathematicians. Nonetheless, they state that it is their intention to analyze mathematical ideas and theorems for meaning and truth, among other things:

"This book is about mathematical ideas, about what mathematics means-and why. It is concerned not just with which theorems are true, but with what theorems mean and why they are true by virtue of what they mean."

They state they are cognitive scientists who are studying the origins of mathematical ideas. They also state that they have looked at those ideas, not as mathematicians but as cognitive scientists, and come up with what they say is a plausible explanation for where mathematicians" ideas come from. Their theory is a theory of what they call "embodied mathematics"; mathematics comes from the bodies of mathematicians. Their theory, they claim:

". . .is the first attempt at a rigorous methodology for Mathematical Idea Analysis-a cognitive analysis of the structure of mathematical ideas, of how those ideas are rooted in everyday bodily experience, what cognitive mechanisms they use, and how they are related to one another."

Where do those abstract concepts and ideas of mathematicians come from according to Lakoff and Núñez?

"Most ideas are unconscious, and that is no less true of mathematical ideas. Abstract ideas, for the most part, arise via conceptual metaphor-a mechanism for projecting embodied (that is, sensory-motor) reasoning to abstract reasoning."

Among other things, setting aside the almost paradoxical reference to unconscious ideas; the lack of clarity as to what "unconscious" means; and what sensory-motor reasoning amounts to, these authors conclude that abstract ideas of mathematics are born of the sensorimotor regions of the brain that generate conceptual metaphors.

"The authors believe that understanding the ideas implicit in mathematics-especially the metaphorical ideas-will demystify mathematics and allow it to make more sense. Moreover, understanding mathematical ideas and how they arise from our bodies and brains will make it clear that embodied mathematics is all that mathematics is-the only mathematics we know or can know."

The authors' position in general is that such abstract ideas developed by mathematicians, even the most esoteric ideas, are little more than metaphor:

"conceptual metaphor plays a central, defining role in mathematical ideas within the cognitive unconscious-from arithmetic and algebra to sets and logic to infinity in all of its forms: transfinite numbers, points at infinity, infinitesimals, and so on. Even the real numbers are constituted by metaphorical ideas coming out of the way we function in the everyday physical world."

Those mathematicians in particular who the authors label "romantic" or Platonic realists are especially bothersome because according to Lakoff and Núñez they are suffering from the delusion that those ideas actually have something to say about objectively existing reality, "truth", and the universe. But according to Lakoff and Núñez, mathematics has nothing to do with reality. Again, it is all metaphor.

"Mathematics is not built into the universe. The portrait of mathematics has a human face."

These authors base their theory on the concept of a technical use of the term "metaphor" which is "a grounded inference-preserving cross-domain mapping — a neural mechanism that allows us to use the inferential structure of one conceptual domain (say, geometry) to reason about another (say arithmetic)" (Lakoff and Núñez 2001). Summarizing their point of view, one might possibly state their argument as follows:

- Ideas of mathematics originate in the minds of mathematicians;
- Ideas of mathematics arise *via* conceptual metaphor;
- Conceptual metaphor is a mechanism for projecting "embodied reasoning" of the sensorimotor system to abstract reasoning;

- Therefore abstract ideas of mathematics are all metaphor. And:
- Mathematical ideas are not "built into" the universe, but are"embodied" in brains and bodies of persons.

On its face, this is an argument based upon the genetic fallacy, wholesale fallacious reductionism, along with a great deal of ambiguity and equivocation, among other things. It is a relativistic postmodern reconstructionist view born of nominalism and conceptualism.

Though the authors claim they are dedicated to reason and love mathematics, they reduce mathematical reason to natural selection and biological origins. In effect, they "animalize" reason, rendering it subservient to natural selection. Based upon their reductionist assumptions, they have ignored the power of mathematical reason, indeed reason of any kind, to be independent even of its origins.

Moreover, though they ground their position by positing the existence of a neural metaphor mechanism, a "grounded inference-preserving cross-domain mapping" permitting us to use cross-domain inferential structures, in point of fact, there is no evidence of any such neural mechanism. In fact, the neural mechanisms necessary and sufficient to construct coherent and unique percepts, let alone concepts, from a large set of ambiguous and highly segregated sensory and motor signals are basically unknown (Tononi and Edelman 1998).

Their arbitrary and ungrounded assertions about this purported neural metaphor mechanism has no evidence to back it up either in neuroscience or physiology. This so-called neural mechanism is evidently designed to support an unsupportable argument.

Additionally, contrary to these authors' postmodern reductionist point of view regarding the objectivity of mathematics (that it has none), given all the clearly objective and repeatable tests to which many mathematical ideas have been put over at least the last 2,500 years, it is clear that the natural world, indeed the universe at large, is remarkably in agreement with many of those ideas.

Any argument against these authors' view could cite the fact that mathematical ideas, theorems, concepts have been objectively and repeatedly tested, both in abstract problems in mathematics itself as well as in applied empirical problems in demanding fields such as physics, chemistry, biology, and engineering sciences, among others. One could note that mathematical ideas, just like ideas and concepts of any other hard-core discipline, (physics immediately comes to mind), must "test out" in the real world, "out there" independently of any of us.

There is not only a consistency and a coherence requirement that mathematical arguments and assertions, concepts, ideas, and theorems must meet; there is also a pragmatic requirement for mathematical concepts and ideas.

Mathematical ideas, concepts, theorems, and principles are fruitfully utilized in all serious research and all knowledge-generating disciplines in the world. Indeed, it is mathematics that gives form to all our knowledge of the world because it gives form to laws and law-like statements used to describe, explain, and predict.

Conceptual Metaphors and Begging the Question

Moreover, even setting aside the absence of any evidence in neuroscience or physiology to support the authors' questionable use of "conceptual metaphor" at the level of the sensorimotor system, it also seems to beg many questions. Unlike the literary definition of "metaphor" the authors claim that conceptual metaphor is a basic cognitive function and that it is by means of this function that mathematical ideas arise.

What does "conceptual metaphor" as a function or process amount to? The authors say that it functions to "project embodied reasoning" of the sensorimotor system to abstract reasoning. However, the sensorimotor system does not *reason* in any understandable sense of that word. The sensorimotor system does many things, but reasoning is not one of them.

What does the notion of "projection" amount to? The authors do not adequately explain this, however, the explanation of metaphor entails that that humans "naturally see common traits in subjects which are factually distinct." Though the authors utilize the notion of conceptual metaphor to explain how abstract ideas of mathematics are formed, even assuming it does exist, their concept of metaphor *already entails* the explanation they seek. That explanation is that humans tend to see common traits in subjects which are factually distinct. Hence, according to the authors, spatial and other metaphors abound in mathematics. Their explanation, in addition to committing the genetic fallacy among others, also begs the question.

It is difficult to make sense of the notion of "embodied mathematics." It is not clear that Lakoff and Núñez mean anything other than that the discipline of mathematics, just as all other disciplines of any kind, originates from persons who occupy bodies.

One is almost astonished with such a line of argument. But even though individual persons who are mathematicians may be the originating source for mathematical ideas, those ideas do not remain there. Their objectivity is found in the formal expression and the objective tests to which these ideas have been subjected, sometimes over millennia.

Mathematical concepts, theorems, and principles are not identified, except usually for historical purposes, with those persons who may have conceived of or discovered them. The ideas are certainly not identical with some "embodiment" either in the persons who discovered or conceived of them or those who may come to know them.

Lakoff's and Núñez's arguments represent an outgrowth of the coherence theory of knowledge combined with earlier conceptualism. These have been combined with spurious biological reductionism. Recall that for coherence theorists and conceptualists, there are no facts in the world "out there" independent of subjects (persons). Early conceptualists held that all we have are concepts "in our minds."

Coherence theorists essentially held that we could not escape the self-referring network of our sentences. Neo-pragmatists held that there is no escape from the circle or "web of our beliefs" that are reducible to stimulations of our nerve endings and observation sentences based upon those. And now Lakoff and Núñez argue that we are embodied minds, and there is no escape from our bodies such that we can know anything at all "out there" beyond those.

5.1.2 The Language Causal Argument: Language Shapes the Development and Content of Mathematics

Perhaps more problematic in many ways are the innatist *cum* nominalist arguments regarding numerical concepts, extending to mathematical concepts in general. This is the issue whether or not the ability to develop numerical concepts depends on language ability which in turn is genetically based in language centers of the brain. Linguists and those psychologists convinced that thoughts in general are inseparable from words we use to label or "name" them are convinced that numerical cognition depends upon language.

Generally, this argument is also a genetically determinist argument, reducing language and hence mathematics to biology. The argument is that numerical cognition depends upon knowledge of the vocabulary of counting words[2] and the recursive capacity of syntax and morphology.

The innatist language argument is that children's initial counting is embedded in natural language as a result of their learning more and more count words. This argument is advanced by some linguists and like-minded psychologists, including Pinker and Bloom (1990); Bloom (1994), Pinker (1994).

---

[2] The counting words are the natural numbers.

Recalling Quine's (1969, 1978) learning-as-induction argument, their explanation is that as this learning progresses children *infer* that there are more count words. Then with enough actual experience they also *infer* that the counting numbers are discretely infinite. This learning process is supposed to be inductive. All those inferences are purportedly made based upon so many experienced instances or particulars, the numbers.

However, this is induction in the spurious sense, in which generalizations are supposedly reached based upon experience of so many particular instances. The problem with this notion of induction (and learning), as well as the argument, is that it assumes what it seeks to establish. The recognition that one counting number follows another, after another, and after another, on and on. Then the recognition that the series goes on *infinitely*, *assumes* cognizance of the rule allowing the numbers to continue. And it assumes cognizance of infinity.

Hauser et al. (2002) argue that the mathematical idea of discrete infinity is *derived from* the recursive capacity of human languages that is unique to humans. That is, they argue that there is a simple, direct causal dependence of mathematical knowledge upon the recursive generative capacity of natural language.

Yet there are numerous experiments showing that number concepts and language concepts are not only neurologically distinct, number concepts *cannot* be reduced to natural language concepts. As some researchers assert ". . .it is one thing to hold that language facilitates the use of numerical concepts and another that it provides their causal underpinning" (Gelman and Butterworth, 2005).

Gelman and Butterworth (2005) argue that numerical concepts have an ontogenetic origin and a neural basis that are *independent* of language. Excluding the non-mathematical use of number terms and notation, numerosity is the cardinality denoted by numbers, including both approximate and exact values.

Their research cited a number of neuroimaging experiments showing the distinct cerebral circuits that underlie observed language and number dissociation. They also cited research showing very rapid learning of new vocabulary for abstract concepts of numerosity among relatively primitive groups. That rapid learning supports the conclusion that the learners already possessed the concepts of numerosities and the idea of discrete infinity:

"the English counting system is almost always instantaneously mastered by Warlpiris . . . independently of formal Western-style education.' . . .the Oksapmin of New Guinea, a group who used [sic] use a fixed number of sequential positions on their body as "count words"...Within 6 months, the Oksapmin had introduced a generative counting rule..." (Gelman and Butterworth 2005).

Additionally, the above researchers' rejection of the language thesis, that language is necessary for mental representation and manipulation of numerosities, is also supported by numerous neuroimaging studies.

The Neurological Evidence

Those parts of the brain involved in numerical processing are in the parietal lobe, not in the language areas of the brain (Cipolatti and Harskamp 2001). Indeed, there is evidence that numerical and linguistic processing work at cross purposes:

> "One neuroimaging study has found that activity in Broca's area is depressed relative to rest during numerical tasks, suggesting that numerical and linguistic processing are even in opposition" (Presenti et al. 2000).

Based on these and other arguments and evidence, these authors conclude that understanding recursive infinity and mathematical thought in general are not derived from language at all. Their conclusion is consistent with a number of arguments one can find going back in the mathematical and philosophical literature many centuries. Specifically, one can find testimony to this effect from mathematicians themselves (Hadamard 1945), based on much introspection.

However, setting introspection aside, there is a growing body of brain-imaging evidence that supports dividing numerical abilities into exact calculation and approximate calculation. Under functional magnetic resonance imaging (fMRI) and event-related potentials (ERPs), different portions of the brain have been shown to activate depending upon whether a task involves exact calculation or approximate calculation.

In general, the evidence shows that *approximation* mathematical tasks involve the following areas of the brain (Dehaene et al. 1999):

- Bilateral parietal lobes: The banks of the left and right intraparietal sulci, extending anteriorily to the depth of the postcentral sulcus and laterally into the inferior parietal lobule;

- Right precuneus; left and right precentral sulci; left dorsolateral prefrontal cortex; left superior prefrontal gyrus; left cerebellum, and left and right thalami.

Research shows that most of these fall outside traditional perisylvian language areas. They are instead associated with visuo-spatial and analogical mental transformations (Dehaene et al. 1999).

On the other hand, fMRI showed that exact calculation involves the following:

- A large left-lateralized activation in the left inferior frontal lobe;
- Smaller activation was found as well in the left cingulated gyrus, left precuneus, right parieto-occipatal sulcus, left and right angular gyri, and right middle temporal gyrus (Dehaene et al. 1999).

The angular gyrus is thought to have functional links with posterior language areas such as Wernicke's area, because it is assumed to be involved in mapping *visually* presented inputs onto linguistic representations (Horwitz et al. 1998). For example, research has shown that one side-effect of damage to the left angular gyrus, is a condition called *anomia*, in which subjects have difficulty finding words for objects and confrontive naming difficulty. These individuals have difficulty naming objects and describing pictures, for example.

Moreover, lesions involving the angular gyrus, or when damage occurs between the fiber pathways linking the left inferior parietal lobule with the visual cortex, can also result in Pure Word Blindness. This is due to an inability to receive *visual* input from the left and right visual cortex and to transmit this information to Wernicke's area so that auditory equivalents may be called up. Such patients are thus unable to read and suffer from *alexia*[3] (Joseph 2000).

Thus the most recent conclusions based upon neuroimaging experiments regarding numerical functioning are the following:

- Arithmetical *approximation* relies on nonverbal visuo-spatial cerebral networks (S. Dehaene, et al. 1999).
- Arithmetical *exact* calculation is arguably language-dependent in that the areas, together with the left angular gyrus and left anterior cingulate, may constitute a network involved in the language-dependent coding of exact addition facts as verbal associations (S. Dehaene, et al. 1999).

[3] Alexia is a neurological disorder characterized by loss of the ability to read or understand the written word. It is a complex visual disturbance resulting from disease in the visual-association areas at the back of the brain.

However, the evidence for exact calculation being language-dependent relies upon interpretations of a portion of the brain that is not generally considered a classical language area, the left angular gyrus (S. Dehaene, et al. 1999; Gelman and Butterworth, 2005).

All characterizations of the left angular gyrus make reference to visual functions because it apparently functions to map *visually* presented inputs onto linguistic representations. Thus the conclusions remain unsettled as to whether or not there are any language-dependent numerical functions.

Indeed, there is some evidence showing that there may be experimenter intervention in some fMRI experiments in the form of asking questions of subjects, hence leading subjects to process arithmetical operations via language centers when they otherwise would not. If so, findings from these experiments would be questionable.

5.1.3 Thinking in Patterns and Images

Mathematicians generally regard their entire discipline as a science of *patterns*, a spatial concept, whether one is referring to numbers, functions, geometric forms, sets, or any combination of these. Popular conceptions of mathematics as essentially about number, or numerosities, are mistaken. However, one finds this conception pervasive, even among those cognitive scientists and others researching the nature of mathematical thought. For example, one finds the following:

"the numbers generated by counting seem to be the foundation of mathematical thought" (Gallistel and Gelman 2000).

But mathematics ceased to be the science of number as of about 500 B.C. Up to that time, it was almost entirely arithmetic, a series of techniques for counting, measuring, and accounting. From about 500 B.C. to 300 A.D., however, mathematics eventually came into its own as a discipline with Greek mathematics, especially Thales. It was Thales who intoduced assertions of mathematics that could be logically proved by *formal argument.* It was Thales who introduced the idea of the theorem, the bedrock of mathematics (Devlin 1994).

Thus, investigating thought processes associated with counting the integers and reals will be enlightening on many levels, but it will not, in the end, say much about mathematical thought. Investigations of mathematical thought would require focusing upon spatial formal thought processes or patterns *about* patterns consisting of very high-level abstractions, functions such as permutations, transformations, and deductions, among other things.

Nonetheless, to the degree that research into those mental or brain processes involved in counting may shed light upon any other mathematical thought process, it is illuminating that the research tends to support the view that such thought is non-verbal.

Visual and verbal systems of the brain are separate systems, though related in some complex ways. Above, I cited neurological evidence showing that arithmetical *approximation* relies on nonverbal visuo-spatial cerebral networks while arithmetical *exact* calculation is arguably language-dependent. Though this last point is disputed, it appears that the areas of the brain that are activated during exact calculations, together with the left angular gyrus and left anterior cingulate, may constitute a network involved in the language-dependent coding of exact addition facts as verbal associations (S. Dehaene, et al. 1999).

As I also noted above, however, this evidence relies upon interpretations of a portion of the brain that is not generally considered a classical language area, the left angular gyrus (S. Dehaene, et al. 1999; Gelman and Butterworth, 2005).

Mathematical Thought and Space

However, while extant neurological studies appear to give much attention to the issue whether or not mathematical thought (or just numerosity) is language-dependent, they give short shrift to the issue of the relation between mathematical thought and spatial reasoning.

It appears that the visual system contains separate subsystems for mental imagery and mental rotation. These are central to understanding mathematical thought, especially the formal properties of symmetry and regularity of patterns.

Mathematicians study transformations of objects, including rotations, translations, reflections, stretchings and shrinkings of objects (Devlin, 1994). They investigate whether or not a given spatial figure is symmetrical, which means that after undergoing some transformation, the figure itself is left invariant. Examples of symmetrical figures (non-dynamic in these examples) are the circle and snowflake. One can rotate or reflect a circle and still get the same circle; one can rotate a snowflake one-sixth of a complete rotation, and it will look the same.

In investigating symmetry alone, mathematicians want to determine the symmetry group for any figure. That is the collection of all transformations that leave that figure invariant. Combining two transformations, they seek to establish a third; if successful, they determine that there is a pattern in the operation of combining the two to get a third. For circles, that pattern reveals the arithmetical laws of associativity, identity, and inverses.

Likewise, mathematicians investigate properties of sphere packing, determining the most efficient mathematical structures that would completely fill a space. In the same way, mathematicians investigate tiles to determine what shapes can be stacked together to completely fill a space.

In all this, mathematicians rely upon spatial thought processes as well as imagination. They must be able to image in their minds what happens under certain transformations of objects. Moreover, they image mechanically manipulating the objects themselves in order to "see" in their minds what happens under those transformations.

Moreover, topology, that area of mathematics that studies properties of figures that are unchanged by stretching and twisting the surface, relies heavily upon mathematicians' abilities to spatially image the effects of these functions in their minds. This is supported by substantial evidence from studies of mental rotation and mental scanning of objects (Kosslyn 1980; Shepard and Cooper, 1982; Knauff et al. 2002). In a sense, objects or figures, even though abstract and not physical, are imagined as things to manipulate in certain ways (Trojano et al. 2000).

Recent research including functional brain-imaging techniques has shown that mental imagery makes use of much the same neural substrates as perception in the same sensory modality. Spatial operations on mental images as well as those operations on visually presented material, share the same neural substrate (Trojano et al. 2000; Kosslyn et al. 2001).

That is, the research supports the view that analysis of visual space in perception and imagery has a common neural basis in the parietal lobes. Neural networks involved in spatial transformation might be shared by several cognitive functions, including visuospatial imagery (Trojano et al. 2000).

Not only is spatial imagery and its underlying neural substrate involved, but intentional (imagined) motor tasks as well. Investigations of the neural correlates of motor planning, independently from actual movements, show that the posterior parietal cortex combines somatosensory and visuomotor information; the dorsal premotor cortex then generates the actual motor plan and the primary motor cortex executes the plan (De Lange, et al. 2005).

Space and Theorem-Proving

Given the central place of theorem-proving in mathematical thought, one would naturally expect a number of studies to focus upon deductive processes. And indeed, cognitive psychologists have conducted studies investigating processes underlying different forms of deductive reasoning, including conditional, syllogistic, and relational forms of deductive argument (Braine and O'Brein 1998; Rips 1994; Knauff and Johnson-Laird 2000, 2001).

Predictably, sentential mental proof theories hold that such reasoning takes place as a result of applying language-like rules of inference, while mental model theory holds that such reasoning relies on the construction and manipulation of spatially organized mental models (Knauff et al. 2002). A third view is that visual mental imagery is used in reasoning that is similar to perceptions.

On the neurological level, the sentential mental proof theory would predict that language centers of the brain would be activated when subjects engage in deductive reason; the spatial theory would predict that the cortical areas involved in spatial working memory would be activated; and the visual theory would, of course, predict that the primary visual cortex or nearby visual regions would be activated by reasoning deductively.

Knauff et al. conducted fMRI experiments to investigate the neural substrates of deductive reasoning (limited to conditional and relational forms of deductive arguments), specifically its visual and spatial components. They found that deductive reasoning "is based on spatial representations and processes." Their results:

". . .appear to corroborate the mental model theory of reasoning. Sentential accounts, such as the theory of mental proof, are not supported by the present data" (Knauff et al. 2002).

Their results as to the visual imagery theory of reasoning were left unclear.

"On the one hand, we did not find activation in the primary visual cortex. On the other hand, increased activation occurred in the visual association cortex and parietal regions, although there was no correlated visual input" (Knauff et al. 2002, p. 211).

However, these and many more neurological facts are necessary but not sufficient to both describe and explain the intelligence of mathematical doing. On the neurological level alone, considerably more research must be conducted on all kinds of reason, extending beyond deductive forms.

Moreover, there are other issues to address. Citing only one highly significant issue involving perception and motion of motor tasks manipulating physical or imagined objects, for example, to determine the effects of transformations, how do we perceive structure in motion? We do not yet know the neural correlates of the ability to interpret changing configurations as organized wholes, an obvious necessary condition for mathematicians to deal with transformations of abstract objects and spatial patterns, among other things.

Even if we had all the neurological questions answered, mapping all the neural correlates with mental functions and the highest level abstractions, we will still have the issues of autonomy, independence, and objectivity of reason left unanswered. Those must be addressed at higher levels of reason itself.

5.1.4 The Realism Argument: Reality and Reason Shape the Development and Content of Mathematics

Among the fundamental positions on the nature of mathematics that Lakoff and Núñez particularly objected to is the position of mathematical realism. Historically, of course Plato was the philosopher who set the standard for this position. Though hardly any realists today follow Plato's version, nonetheless some of the foremost mathematicians in the world consider themselves mathematical realists.[4]

In spite of the growth of the coherence and conceptualist view of knowledge and ensuing postmodernism, realism even in mathematics nonetheless experienced a rebound in the 1930's with Gödel's famous consistency and incompleteness proofs. These reaffirmed that the touchstone for knowledge, truth and reality are those objectively existing facts or universals beyond us all and beyond our languages, and also beyond our formalisms.

However, the realist is presented with serious problems explaining how he or she knows or can *verify* the objective and independent existence of universals. How can universals such as facts or truth have an objective reality independent of persons and even independent of language?

Over many centuries, realist philosophers and scientists proposed a variety of theories setting forth standards for verifying the existence of what they hold is real. All those standards, including the famous correspondence standard, seemed to arguably reduce to subjective conditions of an observer. No matter how many statements of purported fact anyone or any group made, all these ultimately seemed to depend upon direct observations by someone.

In 1931, however, Gödel's two famous theorems seemed to prove that there are mathematical facts independent of persons and language

[4] Sir Roger Penrose is probably one of the best known mathematical realists today.

formalisms. In essence, his theorems proved that nobody can set up a formal system and then consistently state *about* that formal system that he or she perceives (with mathematical certainty) that its axioms and rules are correct, and that they contain all of mathematics.

This is so because anyone who says they perceive the correctness of the axioms and rules must also claim to perceive their consistency. But the assertion of the consistency of those axioms is itself not provable in that formal system.

Hence the person who claims to perceive the truth of something that cannot be proved in the system [*i.e.* the assertion of the consistency of its axioms] has to abandon the claim that the system contains all of mathematics. But then we are left with the larger problem of explaining how it is they perceive such truth. How can someone *know* that the axioms are consistent?

Ontological and Epistemological Issues

The history of issues about mathematical thought appears to revolve around two separate but related questions, which we might label the ontological and the epistemological: (1) Are mathematical objects (including numbers and functions) real, in the sense of being independent of humans and their languages? and (2) What are the processes or structures of reason, specifically mathematical reasoning, which permit us to establish whether or not such objects are real? How can we establish any mathematical knowledge?

Gödel was a realist in the Platonic sense. He distinguished between a system of all *true* mathematical propositions, which he called mathematics in the "objective sense," from a system of all *demonstrable* mathematical propositions, which he called mathematics in the "subjective sense."

His Second Theorem demonstrated that no axiom system can possibly fully comprise all *true* mathematical propositions, that is, mathematics in the objective sense. The reason that no axiom system can do this is because of the indemonstrability of the true assertion of consistency of that formal system.

In spite of his rather profound arguments, however, at the time the actual significance of his proofs was largely ignored. The powerful influence of nominalism and conceptualism, combined with other philosophical ideas, gave rise to kinds of *anti*-realism and *anti*-objectivism that we see today.

Structure of Our Inquiry

I argued earlier that intelligence research must adopt a two-prong approach to inquiry into the nature of intelligence. One prong in a fully developed inquiry is the brain level. We must use knowledge of the sensory and sensorimotor system in any fully developed inquiry into intelligence.

However, all our accumulated knowledge of neural functions of the brain is a necessary, but not sufficient level of inquiry. A second prong must also look at intelligence from the level of the person in the context of the person's objective experience in the world.

Essentially, that means seriously analyzing the multiple relations of various kinds obtaining between a Subject (S) and Object(s) (O). Among other things, that is because intelligence is not solely a natural phenomenon occurring at the level of the organism. It also has artificial aspects because teaching and the artifacts of accumulated knowledge are involved. Artifacts are products of human design; these include physical items but also abstract ideas. Note, however, that just because artifacts are involved, that does not mean that they are any less "real" than the brain itself.

Moreover, in addition to artifacts, the term "experience" must be broad enough to encompass both physical as well as abstract objects. The universe within which a fully developed inquiry must proceed includes the self-organizing dynamics of relations between a subject and those large numbers of objects (components) within that universe that include artifactual objects found within the discipline of mathematics.

Above, we found that mathematical knowing, the intelligent doing of mathematics, is not derivable from Lakoff and Núñez "conceptual metaphor" arguments. Those arguments are reductionist and fallacious.

We also found that mathematical ideas and thought are not derivable from genetic language structures and functions. Certain of the arguments advocating that position are likewise fallacious in that they inevitably appeal to spurious inductivist arguments. Moreover, neuroimaging experiments show that the doing of mathematics activates portions of the brain that are far from the language centers; indeed much of the fMRI evidence arguably shows that mathematical doing activates areas of the brain that even work *in opposition to* the language centers.

We should take another look therefore, at the position on the origins and nature of mathematics that, though spurned by nonmathematicians such as Lakoff and Núñez, commands respect among some of the foremost mathematicians in the world, notably Sir Roger Penrose and the renowned Kurt Gödel. We will return to the epistemological question later, the question as to the processes and structures of mathematical reasoning.

Gödel's famous theorems, the consistency and completeness theorems have been examined in great detail by many philosophers, mathematicians, and logicians. I want to examine these yet again, from an ontological, epistemological, and natural intelligence point of view, in order to highlight the pivotal disagreements many have with what is called Platonistic mathematics.

These issues will be framed within the context of Gödel's disagreements with what is called Hilbertian or Mechanistic mathematics.

Platonic and Hilbertian Mathematics: The Issues

The central issue comes down to a famous fundamental decidability problem. In a commentary on Roger Penrose's *The Emperor's New Mind* (1989), referring to Gödel's 1951 Gibbs lecture (Feferman et al. 1995), Martin Davis (1993) says the following:

(1) "Gödel understood very well what Penrose seems to have missed: that a Platonist position concerning mathematical entities is perfectly consistent with a mechanist view of mind" (1993).

He explains that he thinks it quite likely that Gödel would have agreed with Penrose's judgment that mathematical intuition cannot be the product of an algorithm, but that Gödel did not assert this as a consequence of his famous theorem. Furthermore, (in the same source) Davis says:

(2) "The question of whether mathematical insight[5] is the product of an algorithmic process is an important and difficult question. But Gödel's theorem has nothing to say about it."

I believe it can be fairly quickly shown that Davis' claims in (1) about the "perfect consistency" between Platonistic mathematics and mechanistic (formal) mathematics are entirely wrong and that they entail what Penrose called the "slippery character" of the insight found in Gödel's [second] theorem.

Furthermore, contrary to his claim in (2), that Gödel's theorem has nothing to say about whether mathematical insight is a product of an algorithmic process I believe the non-algorithmic nature of the insight found

---

[5] It should be noted that Penrose uses the terms 'intuition', 'direct awareness' and 'insight' more or less interchangeably.

in Gödel's theorem is betrayed even by Davis' own arguments. Gödel's famous proofs included the following conclusions:

- He showed that it is impossible to give a *meta*-mathematical proof of the consistency of a system comprehensive enough to contain the whole of arithmetic, unless the proof uses rules of inference essentially different from the transformation rules used in deriving theorems within the system.

- He demonstrated a fundamental limitation on any axiomatic method used to develop any *Principia*-like system within which arithmetic can be developed, showing once and for all time that such systems are necessarily incomplete.

The Second Theorem

At the risk of being repetitious, we should go over the Second Theorem again. Gödel's famous Second Theorem shows the incompletability of mathematics. That theorem runs as follows:

"...for any well-defined system of axioms and rules, in particular, the proposition stating their consistency. . .is undemonstrable from these axioms and rules, provided these axioms and rules are consistent and suffice to derive a certain portion[6] of the finitistic arithmetic of integers" (Feferman et al. 1995, p. 308).

The theorem essentially says that nobody can set up a formal system and then consistently state *about* that formal system that he or she perceives (with mathematical certainty) that its axioms and rules are correct, and that they contain all of mathematics. This is so because anyone who says they perceive the correctness of the axioms and rules must *also* claim to perceive their consistency.

But the assertion of the consistency of those axioms is itself *not provable* in that formal system. Hence the person who claims to perceive the truth of something that cannot be proved in the system [*i.e.* the assertion of the consistency of its axioms] has to abandon the claim that the system contains all of mathematics (Feferman et al. 1995, p. 292).

But then we are left with the larger problem of explaining how it is they perceive such truth. How can someone *know* that the axioms are consistent? What is the source of the knowing?

---

[6] Boolos makes clear in his footnote to Gödel's paper that this is Peano's axioms and the rule of definition by ordinary induction, with a logic satisfying the strictest finitistic requirements.

Gödel distinguished between a system of all *true* mathematical propositions, which he called mathematics in the "objective sense," from a system of all *demonstrable* mathematical propositions, which he called mathematics in the "subjective sense." His Second Theorem demonstrated that no axiom system (subjective sense) can possibly fully comprise all true mathematical propositions (objective sense). The reason that no axiom system can do this is because of the indemonstrability of the assertion of consistency of that formal system.

Returning to Davis' comment in (1) above, it appears he is asserting both an *intra*-consistency as well as an *inter*-consistency between the two mathematical systems, Platonistic mathematics (objective) and mechanical [formal/Hilbert] mathematics (subjective). To assert that a Platonist position concerning mathematical entities is, as he says, "perfectly consistent" with a mechanist view of mind, Davis must be claiming to perceive the consistency of the axioms and rules of Platonistic mathematics[7] and to perceive the consistency of the axioms and rules of mechanistic mathematics used to define the mechanical view of the mind.

But the consistency of either mathematical system is not provable in either system. It would also follow that the assertion of consistency between them is likewise not provable in either or both systems together.

Nonetheless, Davis is asserting that it is *true* that Platonistic mathematics is perfectly consistent with a mechanistic view of mind. In sum, *Davis is claiming to have mathematical insight that is not derivable from axioms of either system.*

Contrary to Davis' claim, however, there is no perfect *inter*-consistency between the two systems because, by Gödel's theorem, one system [the formalist] is incomplete while the other is not. The formalist system of mathematics is incomplete because though it may be a well-defined system of correct axioms, consisting of all demonstrable mathematical propositions, it does not contain all *true* mathematical propositions.

The Platonistic or objective system of mathematics, on the other hand, does allow true propositions which are not provable in the system. *It is formalist mathematics, the subjective sense of mathematics (in Gödel's terms), that is incomplete because it is limited to only those propositions which are demonstrable. It cannot state its own consistency because the assertion of its consistency is not demonstrable with its own axioms and rules.*

Objective Platonistic mathematics, on the other hand, is not incomplete by Gödel's theorem precisely because it allows true propositions which are

[7] To the extent that the axioms and rules of Platonistic mathematics can be well-defined.

not provable by the axioms and rules of that system, including the assertion of its own consistency.

In a real sense, Davis has shown himself to be a true Platonist by his own words, though I believe his claim regarding inter-consistency between the two systems is wrong.

The Non-algorithmic Nature of Mathematical Insight

With regard to Davis' comment in (2) above, that Gödel's theorem has nothing to say about whether mathematical insight is the product of an algorithmic process, one must counter that it has a lot to say about it.

In the Gibbs lecture, Gödel introduced a distinction between the First and Second Theorems by noting what he meant by a well-defined system of axioms and rules. He stated:

> ". . .this only means that it must be possible actually to write the axioms down in some precise formalism or, if their number is infinite, a finite procedure for writing them down one after the other must be given. . This requirement for the rules and axioms is equivalent to the requirement that it should be possible to build a finite machine, in the precise sense of a "Turing machine", which will write down all the consequences of the axioms one after the other" (Feferman et al. 1995, p. 9).

Again, Gödel's Second Theorem showed that no well-defined system of correct axioms can comprise all of objective [Platonistic] mathematics because the proposition stating the consistency of the system *is true but not provable in that well-defined system.* Hence it is not included as a proposition in that system.

But he also stated that ". . .as to subjective [formalistic] mathematics, it is not *precluded* that there should exist a finite rule producing all its evident axioms" (Feferman et al. 1995, p. 11). That is, Platonistic mathematics, as mathematics proper, is non-algorithmic in part because it contains true propositions not demonstrable within that system. It contains those true propositions of which we have a non-algorithmic insight but which are not demonstrable.

On the other hand, even though subjective formalist mathematics may have a finite rule [algorithm] producing all its evident axioms, Gödel stated (same source):

> "we with our human understanding could certainly never know it to be such. . .we could never know with mathematical certainty that all propositions it produces are correct."

We would not know this precisely because the consequence concerning the consistency of the axioms would constitute a mathematical insight not derivable from the axioms and rules of the system.

In at least these instances, Gödel has most effectively stated that mathematical insight is not the product of an algorithmic process. He has further demonstrated an inconsistency between formalist and Platonistic mathematics.

Implications of Non-algorithmic Insight to a Science of Intelligence

Based on the above, there are significant ontological, epistemological, and natural intelligence questions to ask. Gödel's consistency theorem has been thoroughly examined many times since 1931 by foremost mathematicians and logicians in the world. None has had any significant objections to his proof.

Assuming that Gödel has in fact shown that mathematical insight in his consistency proof is non-algorithmic, how do we explain it? Clearly, if that non-algorithmic insight is there, that alone shows the inconsistency between Platonistic and mechanistic mathematics, disproving the mechanical view of the mind.

It also has tremendous significance to our concept of our own natural intelligence. That is, our *knowing* that the corpus of formal mathematics is incomplete even while we cannot prove that it is, somehow at least in part resides *in us* rather than in those axioms, theorems, and deductions, a "proof for the record," (Devlin, 1997) that we may try to represent alphanumerically on paper.

There is the additional epistemological question that must be asked. What is the nature of that mathematical *knowing* which permits our minds to transcend all that can be gotten from a precise step by step procedure and all that is written down or somehow represented alphanumerically in a proof?

Other Mathematical Sources of Non-algorithmic Intelligence

In some ways, this non-algorithmic element is also found in other mathematical problems. One such problem is the famous Berry paradox involving the paradox of naming the unnameable. We run into such problems when faced with *immense* and other numbers such as the *googol* and the *googolplex*. Even though immense numbers are finite and countable, they are nonetheless clearly *unnameable* for human beings where "naming" is defined in its constructive sense as a description. In that sense, a name is a linguistic, alphanumeric, or some other symbolic representation or label, of an object.

Though we conceive of and reason perfectly well with immense numbers, there is no way we can alphanumerically represent or linguistically label, that is *constructively name* those numbers. More to the point, they cannot be handled even in principle by the most sophisticated digital or super-digital computer imaginable.[8]

Again, how do we explain the non-algorithmic element of the mind which permits it to "see" or know a truth that cannot be proved and permits it to reason with numbers it cannot even name?

Penrose and Hameroff have turned to an analysis of the behavior of the cytoskeletal microtubule (Langton 1989) with Penrose in particular seeking a quantum theoretical explanation he calls Objective Reduction (Penrose 1994). Since I am addressing Penrose's arguments for an Objective Reduction elsewhere, I will not speak to them here. Though I believe his may be one of the most promising approaches to explaining the physical sources of this element of the mind's *know how*, I also believe there are fallacies in his effort, not the least of which may be the genetic fallacy.

5.2 Problems with Representation Theories Revisited

I have argued elsewhere (Estep 2003) that an explanation of how the mind handles such large numbers and gödelian type proofs rests on a more comprehensive theory of *knowing* than heretofore found in the philosophical literature.

That theory of knowing must be built upon principles of an emergent natural realism which fundamentally rejects much of the current nominalism one finds pervading philosophy and the sciences concerned with mind, brain, and cognition.

Though there are varieties of nominalism, it is essentially the ontological and epistemological view which explicitly or implicitly holds: (a) there are no real objects [abstract and/or spatiotemporal]; (b) if there are objects, they do not exist independently of our experience or know*ledge*[9] of them; and (c) there are no universals, such as properties and relations which exist independently of the language with which we describe them.

With respect to theories of knowing, this view boils down to various kinds of representation theories in which knowledge *is* knowledge *representation*. Though nominalism fundamentally holds that there must be

---

[8] For an interesting discussion involving these numbers and their relation to an emergent theory of mind, see Scott 1995.

[9] The concept *knowing* as a different and distinct cognitive relation from the concept *knowledge* is not recognized by nominalists.

a language interface between our cognitive powers and the world, in its stark form the view is that there is *only* that interface. A language interface is some kind of alphanumeric code or symbol system.

Paraphrasing Frege, nominalists consistently confuse a symbol *of* or *about* some thing with the *thing* it is a symbol of. One consequence of this has been to relegate concepts such as *immediate awareness* and a more primitive *nonrepresentational* concept of *naming* to the conceptual trash bin or to stipulatively redefine them as noncognitive.

Though we have witnessed recent efforts to correct this with an emphasis upon enaction, bodily movement, and visual thought processes, from the standpoint of scientific inquiry, it appears we are still unwilling or unable to recognize the price to be paid.

Naming, Indexes, Classification, Sets, Kinds and Types

The concept *naming* has both a constructive and non-constructive sense. The constructive sense is the one with which we are all familiar. Essentially, it is a definite description of an object, including the use of proper names such as "Mary" or "John." This is the language use of the term. However, there is the non-constructive sense which is not tied to any alphanumeric, linguistic, or other symbolic representation or function, though it is clearly a cognitive, *indexical* function. As we earlier noted, an index functions to point to an object.

In this non-constructive sense, "naming" refers to a primitive immediate awareness relation of a subject to an object in which one "points" to [that is primitively indexes] an object. Such a relation does not require an interface.

Indexicality presents problems to anyone who would map natural intelligence to computers or even theoretical structures. As we also earlier noted, most theoretical efforts to map the mind and its cognitive functions either to a computer or to theoretical structures intended to explain it rely on meaning representation languages to do so.

Representations are sometimes viewed as analogous to "impressions" with causal, *not* cognitive links to one's environment, and that the mind makes inferences from these representations.[10] Among other problems, this view is subject to the same criticisms we raised earlier against the bogus process of abstraction and the use of a spurious sense of induction.

Certain meaning representation languages include, for example, Zadeh's PRUF (1977, 1978), a meaning representation language which uses fuzzy

---

[10] See Hilary Putnam's criticism of this view in "The Dewey Lectures 1994: Sense, Nonsense, and the Senses: An Inquiry into the Powers of the Human Mind," *The Journal of Philosophy*, Vol. XCI, Number 9, September, 1994.

set theory and Test Score Semantics (TSS) for translating natural language expressions into PRUF. In PRUF, an entity generally is viewed as having the effect of inducing elastic constraints on a set of objects or relations in a universe of discourse.

The meaning of an entity is identified with those constraints. However, meaning representation languages have been shown to be insufficient to characterize the semantics of cognitively functioning indexicals found in everyday natural language and in what human beings seem to naturally *know how* to do. Meaning representation languages fail in part because of recurring indexicality and because knowing *how* is not reducible to knowledge *that.*

The occurrence of indexicality in natural languages and in common everyday *knowing how* is a primitive relation between subject(s) and object(s) in which the objects are unique. The term "uniqueness" means that these objects are not members of classes *in that primitive relation* between subject and object. As such, they are *sui generis* objects.

But what is a unique as compared to a class object and how is that related to representation theories? We should again clarify what a representation is. A representation is a classification by use of an alphanumeric symbol or set of these denoting some thing or set of things. It is a proposition *about* or *of* something, a *that*-clause, which *is a classification.* The concept of classification includes *kinds* and *types*. Kinds and types are categories of classifications. And classes are categories.

Fundamentally, categories, classes, sets, kinds, and types[11] involve *extension.* That means they are all quantitative in nature. They all fall into the broader category *extended,* thus the use of the universal and existential quantifier in an alphanumeric calculus or language [representation] about them. A representation just is an alphanumeric code or language of some kind, usually unary, binary, denary [for computer programs].

In computer programs these are used to represent declarative sentences (knowledge *that, i.e.* "that" clauses), where the word "sentence" here has a technical meaning[12] required for computer programs.

---

[11] Generally speaking, nominalists reject the existence of classes and concepts. However, deferring to kinds and types does not avoid the issue. The accepted hierarchical order of these concepts is: category, class, kind, type. Sometimes 'set' is used interchangeably with 'class'. They all belong to a broader category, *extended* or *quantitative*. The latter term is used because of the use of universal and existential quantifiers to bind variables substituted for terms. A class is the extension of a concept, such as the class of extended objects itself.

[12] The term 'sentence' here means a mathematical sentence made up of variables (*x*,y,z), quantifiers ($\forall$ for 'all' and $\exists$ for 'there exists'), connectives (~ for 'not', & for 'and', $\vee$ for 'or', $\rightarrow$ for 'implies'), and mathematical symbols such as '=', '+','-', '×', '0','1', and

The basic unit of these codes or languages is (usually) either 1s, or some combination of (sequences of) 1s and 0s. But what are 1s and 0s? Numbers are classes or sets.[13] Again, what are types and kinds? These are themselves categories of *classifications*. Computational processes are classification processes.

Unlike highly context-sensitive and elastic human and animal intelligence which handles unique objects, machine intelligence only classifies class objects. Even though situation theory (Barwise and Perry 1983) was developed to account for *context-dependent* information not necessarily tied to linguistic representations, it nonetheless relies on representational [class] concepts to provide an interface. This in turn fails to account for non-algorithmic knowing or insight of the kind we see in gödelian type proofs.

One way or another, in nominalist-inspired representational approaches, the appeal is to an interface, to classes, types, kinds, and a principle of similarity or association tying things together which fall under or into them. Regarding situation theory, "information is always taken to be information *about* some situation. . ." (Devlin 1997).

In sum, representation theories can only deal with *class* objects. By definition, these are objects which have already been classified, where they have been sorted[14] according to some *rule of similarity* [with other objects of a class, kind or type]. To sort things by some similarity rule means to sort them in terms of properties or predicates those things have in common with other things of a class or kind.

But the human mind not only sorts objects in its experience in terms of properties or predicates things have in common. The human mind also sometimes selects an object *in spite of* properties that object may have in common with others in a class, kind, or type. It sometimes selects an object because it is unique, it is "unlike any other" in spite of properties it may have in common with others of *any* kind, type, or class.

'>', and parentheses. A sentence is distinguished from a formula in that a sentence has no free variables, they are all bound by quantifiers. 'First-order' means that quantification is over elements of the set and not over the subsets. Gödel coding can be done by assigning distinct positive integers to each of the basic symbols of the language, then to get a gödel code number for a given sentence one takes the product of successive prime powers, 2^{e1}, 3^{e2}, 5^{e3}. . .p_i^{ei}. . .p_n^{en} where p_i is the *i*th prime and e_i is the integer assigned to the symbol occurring at the *i*th place in the sentence.

[13] I am using the term 'class' here as synonymous with 'set'.

[14] I am using the term 'sort' here in the sense of classifying (or assigning a kind or type to) something. Sorting does not have to be tied necessarily to explicit linguistic, alphanumeric, or other signs or forms. Where it is not used in the sense of classifying, the word 'select' is often used.

I have elsewhere (2003) argued that "unlike any other" also cannot be gotten from identifying differences among objects. Identifying differences would still be merely another classification in terms of identity (or sameness) relative to properties or predicates. There is a real sense in which we know *sui generis* objects which are unique, unlike any other, in spite of properties the object has in common with or is different from others of a kind, type, or class.

I have also argued in the same source that it is precisely this kind of *knowing the unique* which is embedded in mathematical *know how* which permits the mind to transcend explicit, alphanumerically formalized step by step procedures, algorithms, as well as deal with numbers it cannot even name.

5.2.1 Classification and the Nature of *Sui Generis* Objects of Immediate Awareness

An understanding of immediate awareness and *sui generis* objects hinges on an understanding of what it means to classify something. This understanding must come by way of both ontological and epistemological analysis of primitive cognitive levels.

Again, classification requires the use of language, some kind of types or tokens of alphanumeric symbols or representations used to denote sets of things. When one classifies some thing, what one is doing is sorting that thing from other things based upon a set of criteria, usually a set of properties or traits, taken to define the members of a class. The very notion of "class" is defined as a set, group, or configuration containing members seen as having certain traits in common (American Heritage 1993).

Thus the very act of classifying or sorting some thing as a member of a class, means that thing is sorted based upon traits it has in common with others. It is sorted based upon a principle of similarity among properties. Thus it is a *class* object to the one doing the selecting or sorting. It is not a *sui generis* object.

A *sui generis* object, on the other hand, is a unique object. It is selected not on the basis of properties or traits in common with others, but in spite of properties or traits it may have in common with other objects. As such, in that relation, a *sui generis* object is not a member of a class. Moreover, *sui generis* objects are not selected based on difference. Sorting a thing by picking out properties that make it different is just another way of sorting by similarity, which is classification.

Understanding what a *sui generis* object is cannot be gotten from reading a dictionary meaning of the term because of the contradictions found in the meaning there. Any standard English dictionary will states that "*sui generis*" means "Being the only example of its kind, unique. [Lat. *Sui generis*: sui, of its own + generis, genitive of genus, kind]" (American Heritage 1993).

But strictly speaking, it is a logical inconsistency or contradiction to claim that something can be "the *only example* of its kind." That is because a kind or class requires that there be more than one thing sharing properties or traits in common by virtue of which they are sorted or classed as "*of a kind* or class." There cannot be a *single* thing that is or has its own class or is its own kind.

There is an obvious tension if not outright contradiction inherent in the English language meaning of "*sui generis*". This is no doubt due to centuries of the influence of nominalism and representationalism, the "usurpation of metaphysics by language" lamented by James (1884).

But strictly speaking, *sui generis* objects in an immediate awareness relation with a subject do not belong to any kind or class. They are selected by a subject precisely because they are unique and are of no kind or class in that immediate relation with the subject.

Because *sui generis* objects in the immediate awareness relation are not class objects, they are not linguistic objects either. As such, they cannot be reduced in any way to objects of *knowledge by description*. The immediate awareness knowing of such objects is *knowing the unique*. Immediate awareness is called "non-classificatory" precisely because the object in the Subject-Object immediate awareness relation is *sui generis*. It is an entirely unique object to the subject in that relation.

Some cognitive scientists and researchers object to the use of the word "object" to refer to these. Their objections rest largely on the claim that these objects do not exist in the material world, hence they are not "real." If they are not found within the structures of the brain somewhere, then they do not exist, so the argument goes.

However, these objects are very real to subjects; they are in immediate relations with subjects and have direct affect upon their intentional and intelligent behavior. Moreover, they are not "figments" of the imagination either, but are in many instances necessary for survival. They are found spanning the entire cognitive domain from the preattentive to the fully attentive, conscious phases of intelligent activity, especially within the structures of *knowing how*.

5.3 Phenomenal Experience and Mathematics

I take functions to be mappings in the mathematical sense as sets of ordered *n*-tuples. But even where we are discussing linguistic grammatical meanings, it is clear that these are kinds of cognitive structure present in thought, hence they cannot *literally* be mathematical functions, though this appears to be a major assumption underlying meaning representation languages such as those referenced above. Meaning representation languages assume language expressions, including indexicals, can be reduced to mathematical functions.

Taken literally, this assumption means the following: In a context C or situation S about a domain U of entities there is always a function i mapping the set T of singular terms (or "infons") used in C or S to the entities in U:

$$i : \langle \mathrm{T}, C \rangle \rightarrow U \tag{5.1}$$

The function i posited is constituted by the syntactical and semantical meanings of T, (and in the case of natural language sentences asserting propositions, the function can be assumed to be constituted of fuzzy or classical truth-values).[15] Each meaning t in T: $C \rightarrow U$. That is, the syntactical-semantical meaning of an expression, including the cognitive content of an expression, is a function that maps contexts [of speech or meaning] into [speech or meaning] contents.

Clearly, functions such as this are operative in veridical thinking, as found in scientific theories. And as Castañeda notes it may also be appropriate for formal semantics or those concerned with issues of veridicality to ignore the cognitive mechanisms, such as primitive indexicals or *non-constructive* naming, that actualize the mappings.

But those concerned with apparently non-algorithmic cognitive activities of the mind, and ultimately the indexicals found even in natural languages themselves, cannot ignore those mechanisms. Otherwise, one makes the mistake of assuming that human intelligence, understanding and knowing consist only in those conditions under which intelligence, understanding and knowing can be *verified.*

For example, even in a natural language context or situation, the function i in the above must be understood in the following way: In a natural language context C, there are factors which we might assume cause the thinker/speaker to think the (one) appropriate functional value in that situation or context.

However, the thinker/speaker does not think *the function itself.* Interpretive functions i are *external* to the cognitive relations or content of thinking (Castañeda 1989, p. 141), though they are obviously needed in

---

[15] For more discussion of assumptions underlying set-theoretical approaches to semantics, see Castañeda, 1989.

theories of the processes of thinking and veridicality. What Castañeda is actually referring to here is the error mentioned earlier, paraphrasing Frege, of mistaking a symbol *of* or *about* something for the thing it is a symbol of.

Demonstrating the Problem with Indexicals

Castañeda presented an interesting demonstration to show that a basic interpretive function *may fail to operate* in illusory linguistic and conceptual experiences. Such experiences are found in such everyday expressions as someone pointing in the distance and saying to a friend, "That tiny dot over there in the distance is Martha's house."

Though a functional argument is available for this sentence, the functional value is missing. It is missing because the properties of a white dot are found in the speaker's *subjective* phenomenal experience; they are not found in "Martha's house."

That function *cannot* be cognitive content and neither can it be the same as linguistic (grammatical) meaning. The fact is, there is no objective proposition that is the external target of that statement, even though the natural language speaker who says it speaks and thinks demonstratively (*i.e.*, indexically). The speaker non-constructively *names* an object.

The statement expresses a kind of discounted *illusory* meaning. The speaker knows that the tiny dot in the distance is not literally Martha's house. Moreover, though the hearer understands perfectly well what the speaker has said, it does not follow from his/her statement that Martha lives in a tiny white dot. The statement is cognitively meaningful even though it includes a discounted illusory meaning. And neither the speaker nor the hearer makes an obvious inference from it.

What is needed to explain non-algorithmic insight found in gödelian type proofs is an emergent natural realist account of indexicality, subject phenomenal experience, including an account of discounted illusory meaning, and immediate awareness.

It also requires retroduction, a form of reason that goes beyond simple deduction and induction. This view rejects the nominalist principle of the necessity for an interface between mind and world. Moreover, it does so without committing the fallacies of an earlier traditional metaphysical realism, or those genetic and other fallacies of contemporary postmodernism.

A fully developed theory of natural realism would provide a classification of primitive indexical relations between Subjects and Objects, where the latter include unique objects. Obviously, it also requires a non-reductionist, emergent theory of mind and knowing in which our perceptual and conceptual powers are continuously and dynamically extended by indexical means.

Retroduction, Reality and Non-algorithmic Insight

Retroduction[16] is a form of inference used to *generate* ideas, hypotheses, not deduce them from prior ideas or statistically verify them. Peirce (1839-1914) originally recognized and named this form of inference and later proposed as a logic of scientific discovery (Hanson 1972). It is a form of reasoning about states of affairs for which one does not yet have an explanation. Peirce also called retroduction Hypothetic Inference and described it as follows:

"When one contemplates a surprising or otherwise perplexing state of things . . .he may formulate it into a judgment. . .he will often finally strike out a hypothesis, or problematical judgment, as a mere possibility, from which he either fully perceives or more or less suspects that the perplexing phenomenon would be a necessary or quite probable consequence. . .That is a retroduction" (Peirce Vol. 8, 1931-1958, pp. 227–231).

Retroduction takes the following form: (1) The surprising phenomenon C is observed; (2) But if A were true, C would be a matter of course; (3) Hence, there is reason to suspect that A is true.

Though retroduction has been criticized as possibly committing the fallacy of affirming the consequent, that criticism is actually misdirected. Strictly speaking that fallacy applies to deductive arguments, not retroductive (or inductive) arguments. The very wording of retroductive form indicates that the conclusion is hypothetical; it is not certain as deductive conclusions are.

In many respects, retroduction can be a kind of rigorous reasoning by analogy. Peirce considered it the most important of the three[17] kinds of logical inference because it was the only one that "opens up new ground."

For example, though far more complicated than the simple retroductive form above, Russell utilized retroductive inference (not in the above sequence) to reason about the scope of all possible human experience and the scope of all possible mathematical facts in order to compare the two. He (1903) reasoned as follows: the number of functions of a real variable [a real number on the real number line] is infinitely greater than the number of moments of time. Even if we could live forever and spent the entire eternity of the universe thinking of a new function every single

[16] Retroduction was also referred to as 'abduction' by Peirce.

[17] The three kinds of inference are deduction, induction, and retroduction.

instant, there would still be an infinite number of functions which we cannot have thought of. There would be an infinite number of facts which cannot enter our experience. Thus the scope of mathematical facts is greater than the scope of possible human experience.

His reasoning demonstrates the limited scope of human experience as compared to the broader scope of mathematical functions of a real variable. That broader scope of mathematical functions contains an infinite number of real objects, functions, whose reality does not depend upon human existence.

Moreover, building on Russell's argument and recalling earlier arguments against the "language thesis," (that numerosity is causally dependent upon language) we may wish to consider the relation that *names* and *descriptions* have to those infinite functions which "we cannot have thought" and which "cannot enter our experience". How do we *name* those functions and how do we describe them? How do we even have a *concept* of those functions which "we cannot have thought" of in the first place?

To make this problem very concrete, we might follow Rucker's (1982) analysis of the Berry Paradox and ask ourselves questions of the following kind: What is the smallest natural number that cannot be described (or named) to a person in words? If we assume there are numbers that cannot be described to a person in a lifetime—or even in the entire life of the universe— and if we assume there is a *least* such number, call it $\mathbf{u}_0$, it appears that we have just described (named) a particular natural number called $\mathbf{u}_0$.

But $\mathbf{u}_0$ is supposed to be the first number that *cannot* be described (named) in words. That is, we're left with an apparent paradox which *points* to the cognition and existence of mental concepts which cannot be named or formalized. Can we nonetheless be said to *know* such objects?

To understand these issues, we must also understand nonlinguistic indexicality, *immediate* awareness, rational forms of inference beyond deduction and induction, and the part all these play in our overall natural intelligence.

5.3.1 Perception and Mathematical Objects

In a famous essay, Gödel drew an analogy between sense perception and the perception of abstract objects of mathematics. Though this analogy has been extensively commented upon, we might take yet another look at what Gödel had to say.

"But, despite their remoteness from sense experience, we do have something like a perception also of the objects of set theory, as is seen from the fact that the axioms force themselves upon us as being true. . .it seems that, as in the case of physical experience, we *form* our ideas also of those objects on the basis of something else which is immediately given. ." (Gödel 1964b).

At least two things are clear in this statement. First, he distinguishes between sense perception and the perception of abstract objects of set theory. Second, as with sense perception, we experience the objects of set theory based upon something immediately given.

His reference to the "immediately given" makes clear his differences with Russell and others regarding reliance upon language-based assumptions. He later explained in more detail what he meant by the "immediately given."

"Evidently, the 'given' underlying mathematics is closely related to the abstract elements contained in our empirical ideas. It by no means follows, however, that the data of this second kind, because they cannot be associated with actions of certain things upon our sense organs, are something purely subjective. . Rather they, too, may represent an aspect of objective reality, but, as opposed to the sensations, their pres*ence in us may be due to another kind of relationship between ourselves and reality"* (1964b, pp. 271-2).

We should certainly ask what Gödel meant by "the 'given' underlying mathematics is closely related to the abstract elements contained in our empirical ideas." Just what those "abstract elements contained in our empirical ideas" might be, he did not say. However, we should look at Gödel's comparison between the formation of ideas of physical objects and the formation of ideas of abstract objects.

We also need to understand his statement that though mathematical intuition need not be conceived as a faculty giving an *immediate* knowledge of mathematical objects, we form our ideas of these objects on the basis of something which *is* immediately given.

That something, however, is not the *sensations* which are the data for our formation of ideas of physical objects. Moreover, he says that by our thinking we cannot create qualitatively new elements but only reproduce and combine those that are given.

Gödel has been criticized for not stating what plays the role of data in the case of mathematics (Feferman et al. 1990). But he *has* stated that it is *not* sensations. Moreover, he gives us an idea of the data he's referring to as follows: That something besides the sensations actually is immediately given follows (independently of mathematics) from the fact that even our ideas referring to physical objects contain constituents qualitatively different from sensations or mere combinations of sensations, for example *the idea of object itself* (Benacerraf, 1964, p. 271).[18]

We can get an idea of the data Gödel is referring to by looking at constituents "qualitatively different from sensations or mere combinations of sensations" such as the idea of *object.* For purposes here, I want to focus upon properties and relations as the given which we use to form our ideas of abstract mathematical objects.

This is consistent with at least part of Russell's approach, though his view and Gödel's on the reality of the objects recognized will differ for reasons which we might briefly explore. This will require that we look at Russell's analyses of the paradoxes of set theory and his proposed solutions, as well as Gödel's arguments against them, specifically against Russell's no-class theory. These have been thoroughly investigated by others, but our efforts will be to clarify the issues and remaining questions relative to the recognition of objects of set theory.

5.3.2 The Reality of Sets and Concepts

We may note that Gödel's realism extended to both *sets* (as characterized in axiomatic set theory) and to concepts. Concepts are the properties and relations of things existing independently of our definitions and constructions. Thus these properties and relations are real and exist whether we can name them or not. Parallel with the concept "object" is Gödel's concept of "set" which he explicates as follows (Benacerraf 1964, p. 262n):

"The operation 'set of *x*'s'. . . can only be paraphrased by other expressions involving again the concept of set such as: 'multitude of *x*'s', 'combination of any number of *x*'s', 'part of the totality of *x*'s', where a 'multitude' ('combination', 'part') is conceived as something which exists in itself *no matter whether we can define it in a finite number of words (so that random sets are not excluded.*"

The last comment would indicate that for Gödel there are real objects extending beyond our capability to name or formalize them in any language. Thus, any formal system we have is *incomplete,* but we are also left with questions regarding the *limits of naming* objects, including concepts, which we nonetheless can (in some sense) think. We are left with the issue of how we come to know such objects.

[18] In contrast, for Russell it is sense data that are immediately given in the acquaintance relation; an "object" is an *inference* from sense data.

Assuming that sets and concepts are real, how might Gödel explain our *forming* ideas of those objects? Gödel requires a consideration of properties and relations of the object that involves a *totality* of properties and relations to which they belong, as well as a relation between a *subject* and those properties and relations of the object (as well as the totality to which they belong).

That is, he supports what is called an *impredicative* theory. These are theories that permit quantifiers over a universe of sets of things, including (implicitly) the set being defined. Some critics have charged that impredicative theories are circular, but their criticisms have not been supported.

Gödel's support for an impredicative theory is tied to his arguments regarding Russellian paradoxes of set theory and the vicious-circle principle. Russell had shown as a result of analyzing the paradoxes of Cantor's set theory that our notions of "truth", "concept", "being", and "class" are self-contradictory. He concluded that the problem was the axiom assuming that for every propositional function there exists a class of objects satisfying it, or that every propositional function exists as a separate entity. "A separate entity" means something separable from the argument (that is, that propositional functions are abstracted from given propositions).

The question is if we reject the existence of a class or concept in general, how do we determine under what further hypotheses such entities exist? Russell's strategy was the "no-class" theory in which classes or concepts *never exist as real objects*.

Thus, sentences containing these terms are meaningful only if they can be interpreted as "a manner of speaking" about other things (Gödel, 1964a, p. 217). To some extent, this was Russell "throwing in the towel" on mathematical realism and adopting a nominalism that Gödel held to be largely destructive of mathematics.

In response, Gödel set forth several arguments against Russell's vicious circle principle. He argued that neither the formalism of classical mathematics nor the formalism of Russell's own *Principia Mathematica* satisfy the vicious circle principle (in its first form).

The axioms in the formalism of classical mathematics imply the existence of real numbers definable in the formalism only by reference to all real numbers.

"And since classical mathematics can be built upon the basis of *Principia* (including the axiom of reducibility), it follows that even *Principia* (in the first edition) does not satisfy the vicious circle principle. . ." (Gödel 1964a, p. 210).

Moreover, one can deny that reference to a totality implies reference to *all* single elements of it. That is, one can deny that "all" means the same thing as an *infinite logical conjunction.* Though there may be problems with this, Gödel suggests that if one takes "all" to mean analyticity, necessity, or demonstrability, the circularity of impredicative definitions disappears.

But even if "all" *does* mean an infinite conjunction, he argues that the vicious circle principle applies only to entities we construct. In that case, there must be a description [definition] of the construction which does not refer to a totality to which the object defined belongs because the *construction* of a thing cannot be based on a totality of things to which the thing *to be constructed* itself belongs.

"If, however, it is a question of objects that exist independently of our constructions, there is nothing in the least absurd in the existence of totalities containing members, which can be described (that is uniquely characterized) only by reference to this totality" (Gödel 1964a).

Moreover, he argues that this would not contradict the second form of the vicious circle principle since one cannot say that an object described by reference to a totality "involves" this totality, although the description does. Additionally, it would not contradict the third form of the vicious circle principle "if 'presuppose' means 'presuppose for the existence' not 'for the knowability'".

Gödel concludes from his arguments that Russell's vicious circle principle, on which his "no-class" theory depends, applies only to the constructivist or *nominalist* position regarding the objects of mathematics and logic, particularly toward propositions, classes, and notions. 'Notion' is understood as a symbol together with a rule for translating sentences containing the symbol into sentences not containing it, such that a separate object denoted by the symbol "appears as a mere fiction".[19]

Contrary to Russell's no-class theory and the efforts to render impredicative definitions impossible, leading to talk of classes as "mere fictions," Gödel's critique of Russell's arguments show that classes and concepts may indeed be conceived as real objects. The term "real" here

[19] Gödel states that the concept of "notion" may appear to involve one in an infinite regress. However, he states that this does not preclude the possibility of maintaining this [Russell's] viewpoint for all the more abstract notions, such as those of the second and higher types, or for all notions except the primitive terms.

means existing objectively and independently of persons, and objectively and independently of language.

On Gödel's realistic account, a recognition of a mathematical object, say a set, would seem to require first of all a perception (or thinking) of elements and relations—a multiplicity—which is perceived (or thought) as a unity. In some sense, this would also appear to be entirely consistent with Cantor's definition of a set as ". . .a Many which allows itself to be thought of as a One" (Cantor 1932).

Cantor later defined this in the following way: "By a 'set' we mean any gathering into a whole M of distinct perceptual or mental objects *m* (which are called the 'elements' of M)" (Cantor 1932, p. 282). For both Gödel and Cantor, sets exist whether anyone ever perceives or thinks them. The set M, consisting of $\aleph_{\theta+\omega}$ ordinals, exists whether humans can perceive or think M or not. Thus, a set is the form of a *possible* thought [of a multiplicity as a unity].

It would also seem that *forming* abstract concepts of mathematics, say forming membership relations among sets, requires *knowing how* to perform a set operation, in addition to *knowing* a form of a possible thought [of a multiplicity as a unity]. Thus, forming abstract concepts of mathematics includes many levels and kinds of knowing.

5.3.3 Intersubjective Requirements of Mathematical Thought

In spite of or because of the continuing debate between mathematical realists and postmodernists on the origin and nature of mathematical thought and ideas, the ideas of mathematics must meet rigorous standards to be accepted by mathematicians themselves as well as the rest of the world. Regardless of the origins of mathematical ideas, their formal expression and use must meet agreed upon standards that can be repeatedly tested and subjected to intersubjective scrutiny.

Given its foundation in logic (at least in part) and analytic nature of all of mathematics, there are consistency as well as coherence requirements that any mathematical idea must meet. Moreover, the pragmatic requirement as well must be met.

That pragmatic requirement has consistently been met over centuries when mathematical ideas, theorems, concepts are objectively and repeatedly tested, both in abstract problems as well as in applied empirical problems in demanding fields such as physics, chemistry, biology, and engineering sciences, among others. These ideas, concepts, theorems, and principles not only get tested out in mathematics proper. As noted, they also get tested out in other disciplines as well.

5.4 Summary

We investigated the relation between awareness and universals by closely analyzing arguments regarding the origins and nature of mathematical ideas and thought in this chapter. We surveyed a number of opposing positions.

As we saw, Lakoff and Núñez argue that the body shapes the development and content of mathematics. They sought to find the explanation for the origins of mathematics in their notion of "embodied mathematics."

Among other things, the authors' questionable use of "conceptual metaphor" at the level of the sensorimotor system seems to beg many questions. Unlike the literary definition of "metaphor" the authors claim that conceptual metaphor is a basic cognitive function and that it is by means of this function that mathematical ideas arise.

We raised a number of questions that Lakoff and Núñez either do not address at all or do not adequately address with factual evidence. We questioned what "conceptual metaphor" as a process or mechanism amounts to. We also questioned whether or not there is an empirical basis in neurological science to support that such a process at the level of the sensorimotor system exists. Though the authors say that it functions to "project embodied reasoning" of the sensorimotor system to abstract reasoning, this must itself be a metaphorical use of the term "reason." The sensorimotor system does not reason in any understandable sense of that word.

Additionally, we questioned what the notion of "projection" amounts to. The authors do not adequately explain this and there appears to be no empirical evidence in the neurological sciences to support the authors' assumption that conceptual metaphor as a function at the level of the sensorimotor system exists. Moreover, even if it does exist, their concept of metaphor *already entails* the explanation they seek.

Among other things, their arguments are characterized by a fundamental genetic fallacy and wholesale fallacious biological reductionism, along with a great deal of ambiguity and equivocation. It is a relativistic postmodern view born of nominalism and conceptualism.

We also assessed the language causal argument which holds that the genetic language centers in the brain are the causal determinants of mathematical thought. This is also a biological reductionist and determinist argument. The evidence from many sources, however, does not support this position.

Indeed, neuroimaging experiments show that the doing of mathematics activates portions of the brain that are far from the language centers; indeed much of the fMRI evidence arguably shows that mathematical doing activates areas of the brain that even work *in opposition to* the language centers. Certain of the arguments advocating that position are likewise fallacious in that they inevitably appeal to spurious abstractionist or inductivist arguments that beg the question at issue.

We also assessed arguments regarding neural substrates underlying mathematical doing, in particular those parts of the brain activated during kinds of deductive reasoning as in theorem-proving. A number of fMRI studies have shown that it is the visuo-spatial parts of the brain that are activated, not those having to do with language processing. Evidence suggests that deductive reasoning is based on spatial representations and processes, corroborating the mental model theory of reasoning.

We then assessed lengthy realist arguments for mathematical realism, drawing upon complex arguments from Bertrand Russell as well as Kurt Gödel, among other notable mathematicians.

Though Gödel's theorems have been and continue to be analyzed among the world's foremost mathematicians and other scholars, his proof with the Second Theorem showed that no well-defined system of correct axioms can comprise all of objective [Platonistic] mathematics because the proposition stating the consistency of the system i*s true but not provable* in that well-defined system. Hence it is not included as a proposition in that system.

This proof has tremendous significance to our concept of our own natural intelligence. Among other things, we can raise questions regarding our *knowing* that the corpus of formal mathematics is incomplete even while we cannot prove that it is. If Gödel's proof is correct, that knowing proves a kind of non-algorithmic intelligence that somehow resides *in us* rather than in those axioms, theorems, and deductions that we may try to represent alphanumerically on paper.

These results, of Gödel's proofs as well as supported arguments as to their significance, at once support a realist theory of mathematical thought and ideas (as opposed to postmodern biological reductionist theory) and disproves the language causal argument as well.

However, we also argued that the origins of such ideas are not necessarily the most scientifically significant thing about them. We argued that whatever the origins of mathematical ideas, their formal expression must meet agreed upon objective standards that can be repeatedly tested and subjected to intersubjective scrutiny. They must "test out" in the real world. Mathematical ideas are the basis for all our knowledge, and all the most advanced knowledge-generating research in the world, providing forms for our laws and law-like statements.

I argued as well that an explanation of how the mind handles gödelian type proofs and large numbers that appear to defy our ability to even name them rests on a more comprehensive theory of *knowing* than found in the philosophical literature.

That theory of knowing must be built upon principles of an emergent natural realism which fundamentally rejects much of the current nominalism and biological reductionism one finds pervading philosophy and the sciences concerned with mind, brain, cognition, and intelligence. Among other things, it must incorporate the concept of non-algorithmic insight in the mind's handling of *sui generis* objects as well gödelian-type proofs. It must also include nonlinguistic indexicality, *immediate* awareness, rational forms of inference beyond deduction and induction, and explanations of the part all these play in our overall natural intelligence.

6 Intelligence as Self-Organizing Emerging Complexity

Prevailing theories of natural intelligence focus upon where intelligence comes from and what it is made of instead of what it does. Above, we reviewed various arguments that general intelligence as well as particular kinds of intelligence, such as mathematical doing, is found in neural language centers of the brain. We showed those arguments are not supported either by the evidence or logic.

6.1 Categories of Natural Intelligence

Substantial evidence in neuroscience, intelligence research, and fields such as kinesthetics demonstrates that there are at least three major categories or kinds of natural intelligence: verbal (referred to as *knowledge that*), *knowing how,* and immediate awareness. Each of these activates different neural substrates of the brain. In some cases, there are overlapping activated substrates. But they are not all reducible to language centers nor are they reducible to one kind of intelligence.

The intelligence of *knowing how* is self-organizing and emergent from structures spanning many levels of major parts of both the central nervous system and many levels of cognitive functions. At least certain of those structures, preattentive, I have grouped together with others and called "immediate awareness." Preattentive structures are below the threshold of attention processes yet are cognitive. These structures arise from across the sensory and sensorimotor systems, though they have largely been empirically demonstrated in the visual system.

Immediate awareness makes knowing how possible. Together, they are the most basic kinds of intelligence, underlying everything else in our rational behavior, including verbal intelligence.

What natural intelligence is, from a biological point of view, is less significant than what it does. Living intelligent systems are made up of enormous numbers of cells that interact with one another, network together, moving, growing, reproducing and dying. This is true as well of those cognitive components at all levels of the natural intelligence system. These large numbers of elements of intelligence, of different kinds of knowing, are also networked together, moving, growing, reproducing, and dying.

A large body of neurological, psychological, physiological and epistemological evidence suggests that intelligence in all forms has to be understood as a function of such numerous interacting systems on many levels. Thus, a key ingredient of any theory of intelligence is an account of how those multiple component processes can be integrated and how large-scale coherence can emerge within distributed self-organizing activity patterns.

Moreover, there are highly complicated dynamic feedback systems both within the organism on many levels and interacting with the environment that make possible, among other things, the smoothly timed and executed tasks that human beings do every day. This includes the intelligence of a mathematician who immediately sees complicated but elegant proof patterns which he or she uses to tackle an enormously difficult theorem. It also includes a young child learning to recite a poem for the first time; or a young recruit learning how to fire an M16A2 on target.

The major mistake prevailing single-capacity theories of intelligence make is to assume that the intelligence observed and experienced at the level of the entire human is caused by single (or a few) genes or neural clusters found in certain areas of the brain, such as the language centers. The pursuit of simple, direct chains of causality, however, has led to a narrow and fallacious theory of intelligence premised upon classical self-fulfilling assumptions.

In this chapter we will be concerned to set forth our approach to theory of self-organizing emerging natural intelligence by taking advantage of the best theory models. We will also look at some of the problems associated with devising theoretical means to measure and analyze the dynamics and information transmission occurring in natural intelligence networks.

6.2 Self-Organization and Pattern Formation

Self-organization refers to kinds of pattern-formation processes found in both physical and biological systems. Patterns in self-organizing systems emerge at global levels from large numbers of interactions among lower level components of those systems. Additionally, just to be technically precise, the rules or algorithms that characterize those lower level interactions are executed based on local information in a self-organizing system without reference to global patterns (Camazine et al. 2001).

Patterns of self-organization emerge from the internal dynamic processes of a self-organizing system; they are not imposed on the system from any external source. That is, a self-organizing system is just that: it organizes itself without any instructions from outside itself.

Our concern here is with self-organization in living things, specifically self-organization in natural intelligence. We should also briefly compare properties of self-organization in physical (inorganic, mechanical) systems with those in biological systems. Self-organization in physical systems involves large numbers of components that are inorganic. For example, the grains of sand in a desert or chemicals in a reaction experiment. Compared to biological components, these inorganic components are relatively simple. They obey physical laws and their large number of interactions produces predictable deterministic patterns.

Biological systems also obey the laws of physics, but the mechanisms of self-organization in living things involve a much greater level of complexity precisely because the interacting components are living. Moreover, the rules governing the interactions of large numbers of living components also differ from physical systems in part because they are they are influenced by genetically controlled properties not found in physical systems (Camazine et al. 2001).

Though incomplete in crucial respects, the difference between mechanical physical self-organizing systems and biological self-organizing systems is driven home with the following:

> ". . the subunits in biological systems acquire information about the local properties of the system and behave according to particular genetic programs that have been subjected to natural selection. This adds an extra dimension to self-organization in biological systems, because in these systems selection can finely tune the rules of interaction" (Camazine et al. 2001, p. 13).

Fine tuning the rules of interaction is not an option available to self-organizing physical systems. However, this description is incomplete at best. In biological systems interactions between components are minimally based upon information transfer as signals or cues. As noted earlier, signals are stimuli shaped by natural selection specifically to convey information, while cues are stimuli that convey information only incidentally.

However, complex social environments may be intentionally designed with cues to deliberately, not incidentally, convey information. Again, teaching is one example of this. Moreover, with sufficient genetic engineering, signals originally shaped by natural selection can also be altered. They can be reshaped by human intervention. Because human beings are also biological systems that are self-aware, they can and do use their intelligence to alter the rules of self-organization themselves.

Emergence

The defining characteristic of self-organizing systems such as natural intelligence, is that the organization emerges from multiple (in some cases immense numbers of) interactions among their components.

As with any organism, we must view natural intelligence, *knowing,* as a large population of simpler components which through time works upwards, synthetically constructing larger aggregates of rule-*governed* or rule-*bound* objects.

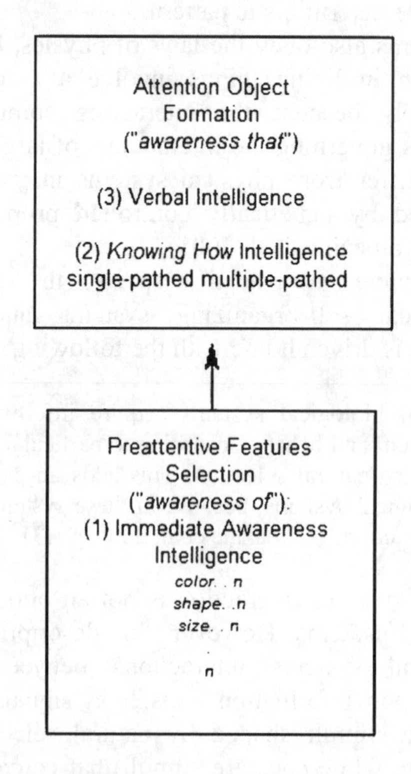

Fig. 6.1. Levels of Awareness and Kinds of Intelligence

These objects interact *nonlinearly* with one another and with their environment in support of the overall life-like dynamics and emergent patterns and qualities of a natural intelligence system. The classical top-down linear science approach to intelligence will entirely miss these properties. On the other hand, the self-organizing organismic theory model approach, which we will address later, is a bottom-up, highly distributed and massively parallel nonlinear view of knowing, of natural intelligence.

On the most primitive levels of natural intelligence, the interactions between the components generate emergent properties. The most primitive level of cognition is the preattentive phase of feature selection. At that phase of cognition, the living intelligent system selects primitive features such as color, shape, size prior to forming conjunctions of those features into objects. This is the level of "awareness of" as depicted in the above graph. Recall also that though these are examples of feature selection in the visual system, preattentive feature selection occurs across other sense modalities.

The interactions among these features form the next level of cognitive structure in the natural intelligence system. Those system component interactions, at the level of "awareness that" in turn form the components and their interactions for the next higher level of cognitive organization, with different emergent properties. This process of generating levels of emergent properties based on interactions of components proceeds to higher levels of intelligence in turn, with each level manifesting its own self-organizing properties and emerging properties and their interactions.

6.2.1 Interactive Systems and Self-Organization

We earlier rejected classical reductionist explanations of natural intelligence. These arguments hold that understanding of a phenomenon can be gotten from analysis of its parts. Specifically, it is the view that descriptions of higher level processes and structures, such as intelligence, can only be explained in terms of descriptions of some lower level processes and structures.

The classical reductionist approach to natural intelligence, in which a single "cause" is identified to account for higher level phenomena, is opposed to self-organizing emergent explanations in part because of the radical reorientation required of the scientific methodological approach necessary. An emergent view of natural intelligence necessitates finding formal ways to handle very large numbers of components of data sets at all levels of the inquiry. From a purely mathematical point of view, these large numbers themselves can present challenging problems for any researcher.

For example, even when intelligent organisms such as human beings seem to be doing relatively simple things, their interactions add up to unmanageable complexity very quickly.

Complexity

In general, complexity of a system can be defined as the amount of information necessary to describe it (Cohen and Stewart 1994). More specifically, complexity is the size of a problem measured by some natural number, *n*. Algorithmically, the number of steps an algorithm needs for particular problems (of a class of problems) would be some natural number *N* which depends upon *n*, the size of the problem (Penrose 1989, p. 141).

A theory of natural intelligence has at least two kinds of complexity of a natural intelligence system to address. One is the complexity of the system itself; the other is the complexity of its development or growth in intelligence in an environment. Also adding to measures of complexity are the complicated positive and negative feedback mechanisms one finds in natural intelligence systems as well. These complicated causal loops result in nonlinearities and can lead to high levels of unpredictability.

We can see how the amount of complexity of data sets consisting of certain numbers of components can rise very quickly with the number of interactions among those components. The number of ways pairs of components can interact is, let us say, half the square of the number of components. If you have 10 components, that means there are 50 interactions between them; if you have 100, then there are 500 interactions, and so on. The following table gives one an idea of how quickly the interactions increase relative to the increase in the number of components.

Table 6.1. Sample Interactive Components Matrix

Number of Components	Number of Interactions
10	50
100	500
1,000	500,000
10,000	50,000,000
100,000	5,000,000,000

To give a real-world example of the problems, we could cite research in *knowing how* where typical problems revolve around smoothness and timing in the execution of a given task. Keep in mind that smoothness and timing are clear indicators of knowing how to do something. Depending upon the task, these often must be taught. One must come to know how to perform such that smoothness and timing are exhibited in one's doing. Such problems are involved in performing kinds of surgery, performing rescue operations that involve fast movement such as running and quickly accessing casualties who may be trapped in various kinds of accidents; or military recruits learning just about any kind of basic infantry task.

All these examples involve highly complex physical movements in combination with highly complex mental operations related to immediate planning and execution involving large numbers of variables. Research into any facet of such tasks involves many factors. Taking only one of those as an example, gait analysis, we can demonstrate the highly complex nature of the inquiry involved.

For example, gait data sets consist of many kinds of variables, including kinematic, kinetic, electromyographic, metabolic, and anthropometric, to name just a few (Chau 2001). The data sets are high-dimensional sets, hence statistical analysis methods become almost useless beyond five variables. Moreover, conducting research observations of such whole body performances are limited because human observation and interpretation of performance events are limited to three-dimensions.

Additionally, classical scientific reductionist approaches such as factor analysis are based on linear assumptions while the phenomenon itself is nonlinear. Multivariate statistical methods are useful only to the extent that the data is reducible to very few dimensions, and principal components analysis is useful only to reveal linear relationships among variables.

High-dimensionality of nonlinear phenomena requires alternative methods. Because of the very large number of components involved and complicated connections between them, research in natural intelligence systems involving *knowing how* inevitably requires computational and formal methods to process massive amounts of data.

A number of highly effective mathematical methods have been devised to generate general principles governing the system under study, even if we do not know everything about all the components involved. Nonetheless, we can still make useful predictions about system behavior, given certain known conditions.

Natural intelligence systems, like other living systems, behave in highly distributed, massively parallel and extraordinarily integrated fashion. Complex feedback systems and damping are the norm even for common intelligent actions. Network models are necessary to increase our understanding of such systems and to overcome narrow perspectives of single or limited causal chains.

6.3 Mechanism and Organicism Revisited

In Chapter 1 we looked at two broad categories of theory models underlying and governing most intelligence research. We sorted out the single-capacity and sociological-mechanistic effects models governing much of that research. Both of these models are mechanistic but in different ways.

In the context of the discussion of self-organization and emergence, we need to revisit those models and introduce even more differences between those and organismic models, models of living things.

6.3.1 Organized Simplicity and Unorganized Complexity

We earlier noted that a mechanistic point of view is one in which states of affairs are represented like a machine, where "machine" is defined as an object consisting of parts ordered to function in predetermined ways to bring about certain specified effects. The parts of a machine have *fixed* actions with the actions specific to a certain kind of machine resulting from a combination of its parts. The effects of a machine are additive.

The mechanistic point of view can be either statistical or non-statistical. A non-statistical point of view characterizes or generates representations of *organized simplicity*. This point of view is limited to representations of systems of a few parts because of the computational complexity problems which result with many parts. This was represented earlier in the single-capacity mechanistic effects model.

That is, the *organized simplicity* point of view and associated methods work extremely well with systems which can be split into isolable causal chains. These are causal chains in which there are connections or relations between two or just a few variables. To determine the effects of such a system, what is needed is an equation for each part in isolation, one for each combination of parts, and one for the context (Steiner 1988).

Thus, for a system of two parts, only four equations are required, but for one of ten parts, the number of equations increases to 1,035. The growth in the number of equations is a result of the possible combinations of parts. For n parts, there are 2^n combinations. For a machine of 20 parts, there are 2^{20} or over a million combinations, thus the complexity of a system of organized simplicity which results is exponential (Steiner, 1988).

A statistical mechanistic point of view, as depicted in the sociological mechanistic effects model in Chapter 1, and as found in some cognitive science and intelligence theories (Gärdenfors, 1988), generates representations of *unorganized complexity*.

Ultimately, the underlying principle in systems of unorganized complexity is the second law of thermodynamics, with the probability of the system tending toward maximum entropy. This is based in part on viewing the components of the system as discrete, atomistic, linear, additive, and in isolation. It is based on such a view without taking into account the interrelations of the components and functions of the entire intelligence state of affairs, including the directiveness, self-organization, and emergent properties found in human knowing.

Taking their model from classical physics, rather than accounting for each combination or interaction of parts, the statistical mechanistic point of view addresses *average* combinations or interactions. This is precisely the approach taken by Gärdenfors (1988) in his model of intelligence dynamics and its applications.

The major problem with this approach, however, is that viewed as a whole, the system must be very large to secure accuracy (Steiner 1988). The relative error of average values is of the order $n/\sqrt{n}$. A system of 20 parts would be too large for *non-statistical* mechanistic treatment, and such a system would be too small for the *statistical* approach because the error would be too great, an error rate of approximately one in five.

Thus far we have the following characterizing principles for the mechanistic point of view:

- States of affairs represented as a machine, that is they are represented as an object in which parts are combined or ordered to function in predetermined ways to bring about specific effects. One-way causal chains.
- Determining factors are non-alterable parts combined.
- Effects of a machine: linear and additive [summative].
- The entire state of affairs is not a determining factor.
- May be either statistical or non-statistical [discrete mathematics].
- If non-statistical, it represents *organized simplicity*; if statistical, it represents *unorganized complexity*.
- Limited to one- or two-variable problems; *not multivariable.*

There are other properties as well. The emphasis in these models is upon homeostasis, the maintenance of an equilibrium or "balance" between a system, its components, and whatever is taken to be outside the system (whatever is *not* the system). This includes whatever is contradictory of or inconsistent with the system. Such systems are not characterized by growth or evolution, but by entropy, a tendency toward disorder.

Whether explicitly identified or not, this theoretical focus on intelligence or belief revision is for purposes of maintaining *consistency,* or to rid one's "doxastic (belief) system" of unwanted structural tensions of whatever sort. A kind of mental hygiene approach is a clear principle of the mechanistic point of view as applied to natural intelligence systems.

6.3.2 Organized Complexity

The organismic point of view is one in which states of affairs are represented like a living thing. This point of view stresses the organization and interactions among the parts. It does not stress the properties of the parts in isolation from one another. In general, an organism is viewed as a structured whole in which the content and form of its parts are determined by the function of the whole.

In an organism, the parts do not have non-alterable natures and fixed actions. The parts *act interdependently* to maintain the function of the whole organism. The parts of an organism are not simply combined and then it is determined what the whole is to be. Rather, the content and form of the parts change relative to a whole. The emphasis in an organismic state of affairs is on the whole taken as determining its parts, thus generating representations of organized complexity. As such, the effects of the whole are not linear, isolated and additive. Rather, they are *configurational,* they are organized to be interactive, continuous, and transactional (Steiner 1988; Jen 1990).

Research directed to organized complexity of any kind, including natural intelligence systems would necessarily require concepts, models, and mathematical *cum* logical tools sufficient to represent multivariable complex dynamic configurations of complex organization, of the interaction of large (immense, infinite) numbers of variables or components. Logical and mathematical concepts and tools appropriate for nonlinear, continuous, complex dynamic systems are needed.

Organisms differ from machines in many ways. Among those differences is the fact that though both can exhibit homeostatic equilibrium, only an organism exhibits heterostasis. Examples of homeostasis in a living organism would include maintenance processes such as thermoregulation in warm-blooded animals. In cool weather, for example, certain centers of the brain are stimulated to turn on heat-producing mechanisms of a normally healthy body, and body temperature is monitored back to the center so that temperature is maintained at a constant level (so long as the temperature of the environment of the system does not exceed a certain level). Similar mechanisms exist throughout the body to maintain constant physicochemical variables.

But living organisms also exhibit a tendency toward increased differentiation, organization, tensions, and other *improbable* states. That is, living organisms exhibit *growth*, which machines do not. To fully understand the concept "heterostasis" and the limitations on homeostasis, we must explore the nature of open systems in contrast to closed systems. This will enable us to better understand the inadequacies of machine models, which are closed, as models for theorizing about natural intelligence systems.

While demonstrating the differences between open and closed systems, we will also explore the mathematical tools necessary to characterize phenomena of open living systems. To summarize what we have said thus far with respect to the organismic point of view, we have set forth the following characterizing principles:

- States of affairs represented as an organism are represented as a structured whole determined by the function of the whole.
- *Alterable* parts of an organism act interdependently to maintain the function of the whole organism generating organized complexity.
- The emphasis is on the entire organism as a determining factor of effects not parts determining effects. The organismic system is *multivariable*. It is *not* limited to one- or two-variable problems as most causal, discrete, linear model problems are.
- The system exhibits *both* homeostasis and heterostasis.
- Characterizations of organismic systems require continuous, dynamic, nonlinear mathematics.
- Organismic systems are characterized by *negative* entropy [negentropy], tendencies toward highly improbable states.

As earlier noted, mathematically, emergent phenomena are *nonlinear*. That means that they are phenomena that do not obey the mathematical superposition principle (Saaty and Bram 1964). For example, where φ is an operator, it is said to be linear if the effect of operating on the sum of two entities, for example functions, is equal to the sum of the effects of operating on them separately: $\varphi(f + g) = \varphi(f) + \varphi(g)$.[1] Where this equivalence does not hold, the operator is nonlinear.

[1] This is the distributive law.

Thus in place of correlation functions and linearizations, which are appropriate to some degree for deterministic systems, there is a need to turn to nonlinear mathematics and information-theoretic measures, measures of order and randomness, to understand the behaviors of these systems.

Causality

Machine models usually emphasize static single, one-way causal chains in which the sum of causal determinants is equal to causal effects. However, living things are characterized by dynamic complex positive and negative feedback loops, with nonlinear multiple causal determinants, effects, and multiple directional causality.

For these and other reasons, the machine model view of causality is not sufficient for scientific analysis and research into living systems, especially natural intelligence systems.

Causality is a property of certain kinds of connections between things. Analysts of networks use graph theory to help sort out those connections and possible causal relationships between their components.

We can initially sort out inventories of causes (determinants) of certain effects (resultants) that comprise a concatenation that does not contain bidirectional connections or chaining. This is depicted in the following:

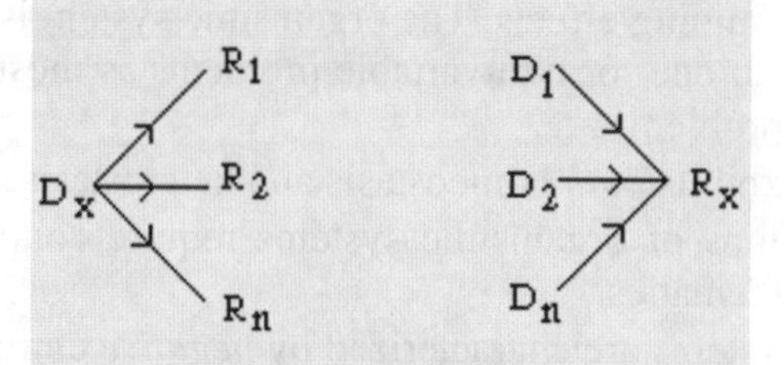

Fig. 6.2. Single and Multiple Determinants and Resultants

Reductionist models assume that causality acts deterministically upward through any material hierarchy, from atoms to molecules, cells to organs, eventually to the level of the entire organism. However, it is clear that there is downward causation as well (P.B. Anderson, et al. 2000; Scott 2002), as evidenced in intentional as well as unintentional human intervention on many levels in the genetic development and growth of living species, including ourselves.

Weak downward causation[2] assumes that molecular components of living organisms are governed by nonlinear dynamics with basins of attraction.

"Death, in this formulation is but another of the attractors shared by the interacting molecules, and a physician's job is to keep the molecules of a patient within the basin of the living state" (Scott 2002, p. 302).

However, living systems are characterized by complicated feedback information systems including both positive and negative feedback. Though mechanistic-minded reductionists may prefer to focus upon negative feedback because of its obvious stabilizing features, self-organizing systems use positive feedback. Above, I cited thermoregulation of body temperature in warm-blooded animals as an example of negative feedback. If external temperatures rise above or drop below certain thresholds, physiological responses by the animal will act to stabilize body temperature by bringing it back into a certain range.

In contrast, positive feedback tends to reinforce change in a system in the same direction as initial information indicates. This can result in growth or heterostasis. On the other hand, it left uncontrolled, it can have a snowballing effect.

Positive feedback has an amplifying effect that can prove destructive. But in living organisms, one finds both positive and negative feedback, with negative feedback tending to keep positive feedback under control. These mechanisms for information input, toput, output, and feedback create self-organizing emergent structures and patterns in systems involving large numbers of components.

6.4 Nonlinear Theory Models Approach to Natural Intelligence

The domain of natural intelligence is *multi-* and *high dimensional.* The scope of intelligence is extended far beyond the single capacity theory to reflect the facts of natural intelligence found in human and animal experience. An extended domain of natural intelligence includes many kinds knowing and their components that extend far beyond the traditional view of intelligence as language-based propositional knowing. These kinds act on other kinds to produce even newer kinds of intelligence.

[2] Three kinds of downward causation have been sorted, including *strong, weak,* and *medium,* though strong downward causation is generally rejected.

Because of the self-organizing dynamical and emergent nature of the elements of the domain of natural intelligence, we take a nonlinear theoretical model approach to characterize that dynamics. This approach has a number of advantages over the classical science approach:

1. Nonlinear theory models approach to natural intelligence permits a more complete and exhaustive classification of kinds of intelligence; it does not limit the scope of cognition to only one kind or category of knowing, verbal (linguistic) intelligence.
2. Nonlinear theory models approach incorporates the strengths of classical statistical and nonstatistical models and methods, permitting characterizations of organized simplicity and unorganized complexity, but goes beyond these to permit characterizations of *organized complexity.*
3. Nonlinear theory models approach is not reductionist as are classical models; it incorporates reductions at levels where those are appropriate but extends beyond reductionist methods to include retroductive models of hypothesis formation, description and explanation of emergent phenomena.

The theory models approach to inquiry is a methodology for seeking out theory as a *source* of wanted theory. It is a methodology for forming the theory sought out into a model or point of view from which the wanted theory can arise. As explained by Steiner and Maccia (Maccia and Maccia 1966; Steiner 1988) the theory models approach is a retroductive approach to devising concepts and hypotheses. As such, it is not reductionist.

A reductionist approach would mean that the theory one is seeking is equivalent to the source theory. On this approach, one assumes that concepts and hypotheses are already given or ready-made in the source theory.

However, not all concepts and hypotheses are ready-made. With emerging new facts and phenomena, ideas must be rationally devised. A deductive approach would mean that the wanted theory is derivable from the source theory, that is, one would search for concepts and hypotheses from which concepts and hypotheses can be derived. This again assumes that ideas are already made and waiting to be derived.

On the other hand, as explained by Steiner (1988), the theory models approach, because it incorporates retroductive forms of inference, is based upon analogy. An example is the use of digraph theory to theorize about influence relations. On the basis of an analogy between the relations of points and the relations of persons, concepts and hypotheses about influence relations can be devised.

More specifically, this same analogy can be used to form concepts and hypotheses about natural intelligence relations between persons as knowers (or cognitive "centers") and object(s) of those relations. The dynamics of natural intelligence can be gotten by means of the theory models approach with the use of digraph theory.

The nature of configuration complexes is central to understanding the nature of organized complexity. Moreover, an understanding of these is necessary to see the differences between the mathematics of summative complexes and the mathematics of configuration complexes.

In general, the differences are found in discrete linear mathematical approaches and continuous *nonlinear* mathematical approaches, with much of our current understanding of the nature of complexity founded upon discrete mathematics. In general, differences between discrete and continuous theoretical approaches look something like the following (Blum, 1989).

Table 6.2. Complexity Theory: Discrete and Continuous

	Discrete	Continuous
Underlying Mathematics	Logic, Combinatorics, Number Theory	Analysis, Differential Equations
Foundations	Firm	Weak
Formal Models	Turing Models, RAMs (all equivalent)	Various (incomparable)
Analysis of Algorithms	Systematic	Ad hoc
Complexity Theory	Highly Developed	Not Developed

As earlier noted the view of intelligence as a living thing requires a mathematical and scientific approach very different from the classical top-down, verbal rule-governed, logic-based, linear and additive approaches. Those approaches are largely anti-theory and anti-concept formation; are either reductive or, as indicated by the primary value attached to data collection and verification procedures, or inductive.

Given the enormous variety and range of intelligence experience, a theory of natural intelligence must incorporate the major mechanisms by which intelligent beings, both human and animal, acquire and act upon information.

What is needed in a new scientific approach to intelligence is an integrated mathematical *configuration* model permitting characterizations of *organized complexity*. Again, these characterizations are based on nonlinear assumptions. Such characterizations can be provided by an integrated organismic theory model, formed from the integration of *set* theory, *information* theory *graph* theory and *general dynamical system* theory. This theory model will be referred to with the acronym "SIGGS" throughout the remaining chapters.

6.4.1 The SIGGS Theory Model

SIGGS is a formal theory model originally devised in the 1960's by Steiner (Maccia) and Maccia (1966). Since that time, it has been greatly extended by Thompson (2005) and Frick (1983, 1990). In particular, Thompson has developed a consistent nomenclature and definitions of the properties of the model. In general, those will be followed here, though the explication here will tend to be less formal and more intuitive in order to more fully explicate the properties of the theory model.

Primitive terms for SIGGS, as corrected and extended by Thompson include *set, element/component/object, contained in, ordered pair, universe of discourse, characterization, occurrences, parameters, connection, relation, affect relation* (for Steiner Maccia information system), *event*, and *sequence*.

The mathematics of set theory provided the means to formally define "system" as a group of at least two components with at least one affect relation and with information (Maccia & Maccia, 1966; Steiner, 1988). In essence, a system is an ordered pair consisting of an object set *S*, and a relation set, *R*. Set theory is used *to give meaning to* the definition of "system" in that a group of at least two components becomes a *set* of at least two elements which form a sequence.

In addition to formally defining "system" set theory is of course necessary to demarcate the scope of the natural intelligence universe based upon logic and facts found within the domain of animal and human experience. As we demonstrated in an earlier chapter, with set theory, we more precisely carved the problem space of natural intelligence.

Also, set theory allows the quantification of a complex organization as a whole. Information theory allows the quantification of action; graph theory permits the quantification of structure; and general systems theory permits an *organismic* perspective to treat complexities of *configurations* of a whole.

Relations between components (elements, constituents) of a system, affect relations, are also given meaning through digraph theory. Digraph theory is in turn based upon set theory. Digraph theory is mathematical theory which characterizes between pairs of points lines which can be directed. The group of a system becomes a set of points and an affect relation becomes a set of directed lines. Digraph theory fully captures and extends beyond Russell's (1903, 1984) notion of sense or direction of a relation.

Within a system interpreted as a group of at least two components with at least one affect relation and with information, the set theoretic notion of "function" permits quantification of influence or affect relations in that an affect relation is a mapping of the group into itself.

With the use of information theory, information is a *qualifier* of affect relations. It becomes a characterization of occurrences at categories in a classification. These occurrences may be of system components or system affect relations or both. Because a classification is a set of categories, set theory is also basic to information theory.

Properties of a system are subsets which are sorted out from the set of all systems, the power set, because they have conditions on them over and above the conditions which make them a system. For this reason, properties of a system are not part of the definition of "system". Though we will clarify this later, a subset of systems with an environment consists of those systems from the set of all systems which have an added condition of a *negasystem* [given meaning by the set-theoretic notion of "complement"].

Not all systems necessarily have an environment, but all systems do have a negasystem. Use of set theory is explicit in the conditions regarding sizeness and homeomorphismness. The set theoretic characterization, cardinality, is explicit in the condition sizeness, while homeomorphic mapping is explicit in homeomorphismness.

The set characterization "complement" marks off the system from its surroundings, the negasystem (or "not-system"). Within whatever universe of discourse is selected, components which do not belong to the system are the negasystem. For example, the system could refer to the formal intelligence conditions constituting qualitative knowing obtaining of a person or persons while the negasystem could refer to all else which does not belong to that set of conditions.

Because the function from one set into another is constituted by an association of elements in one set with those in another, the association between intelligence conditions in one person with those conditions in another may be mapped. Similar mappings may be obtained by selecting a person as the system and a group as the negasystem or a machine as the system and the person might constitute the negasystem. Where intelligence properties of a person, P_1 affect (influence) those properties of another, P_2, the formal intelligence conditions obtaining of P_2 are a function of the intelligence conditions obtaining of P_1.

More technically, the SIGGS theory model consists of a group of related terms, such that some of the terms are primitive or undefined and others are defined. Primitive terms are obviously required to prevent circularity. All defined terms are defined through the primitive terms or through defined terms which have already been defined by means of primitive terms. The terms constituting the SIGGS theory model are characterizations with respect to a system in general and not with respect to only one kind of system, for example a biological system. The theory models a group of related characterizations about a general system or a general system descriptive theory.

6.4.2 Information Theory

The advantage of utilizing SIGGS to devise theory of self-organizing emerging natural intelligence is that it permits representations of *organized complexity*. This is due in part to the fact that it is based upon assumptions of non-linearity and complex interactions among the parts which make up the system.

It also includes information theoretic measures which are not provided on the standard uncertainty increasing (unorganized complexity) and uncertainty maintaining (organized simplicity) theory models. Obviously, information theory is necessary to adequately characterize how organisms, including humans, acquire and act upon information. The term "information" used here is not identical to the ordinary language use which often refers to content or meaning. It is used here as defined within communication theory in a selective sense in terms of uncertainty.

The concept "information" can be sorted into at least two or possibly more senses[3] depending upon whether there are alternatives in the characterization. There is information in its *selective* sense, when there are alternatives available; and information in its *non-selective* sense, when there are no alternatives available.

Where "information" is defined in terms of alternatives available for selection, the more alternatives, the less specific a (decision) situation is, hence the greater the amount of information. When there is only one alternative possible, the situation is fully specific hence there is no information (from the point of view of the selective sense of "information").

Following Maccia and Maccia's (1966) example, in the characterization "C_6H_6 is the formula for benzene", there are no alternatives. The characterization is fully specific, hence from a *non-selective* point of view the characterization is information.

[3] Klir (1989) has expanded beyond these two senses of "information."

However, from a *selective* point of view the characterization is not information since there are no alternatives. There is no uncertainty. In the following characterization, there is uncertainty: "penicillin is related to recovery from streptococcus infection or it is not so related". There is uncertainty because not all occurrences can be characterized by means of one category of the two possible.

The above characterization of information is an extension beyond Hartley's (1928) original formulation, which was concerned to set up a quantitative measure whereby the capacities of various systems (in electrical communication) to transmit information may be compared.

Looking upon an organism as an intelligence system, either in isolation or in the context of other organisms, information theory provides a way to give meaning to the categorization of components of the system, connections of components to one another, connections between the system and its environment, and the uncertainty of occurrences at those categories.

Information is a characterization of occurrences by means of categories. For example, biological information would be a characterization of occurrences at categories with respect to living organisms. Additionally, characterizations of occurrences are sometimes themselves made into other characterizations. An example given by Maccia and Maccia (1966, p. 9) is in telegraphy in which a message characterized in terms of categories of letters is made into a characterization in terms of categories of dots, dashes, and spaces.

Klir's analysis (1989) has shown that several characterizations of information are possible, depending upon the mathematical theory assumed. In the following characterization of the SIGGS theory model, information will be taken in its *selective* sense, within a communication context, and with a frequency interpretation of probability.

The frequency interpretation of the concept of probability is used in SIGGS. Probability theory is mathematical theory which characterizes frequencies of occurrences with respect to classifications, that is sets of categories. An occurrence is at a category in a classification if it is assigned to that category. The probability that a given occurrence can be assigned represents the ratio of the frequency of occurrences at that category to the frequency of all occurrences at every category in the classification.

The incorporation of information theory in SIGGS permits the characterization of the *transmission* of natural intelligence components. As information theory permits the characterization of occurrences at categories, we are concerned with occurrences at intelligence categories of *signed* expressions of intelligence that is symbolic, iconic, and enactive. Thus, with SIGGS, we have the following:

Where T = measure of shared information:
- P_1, P_2 = systems;
- H = basic information function
- $H(P_1)$ = amount of information in one system's action
- $H(P_2)$ = amount of information in another system's action
- $H(P_1,P_2)$ = amount of information in both systems' action taken jointly then,
- $T(P_1, P_2) = H(P_1) + H(P_2) - H(P_1, P_2)$

Information-Theoretic Extensions of Simple Feedback Model

SIGGS theory model provides information measures which extend the usual cybernetic models, as seen in the simple feedback model in Chapter 1. On the usual cybernetic model, input is what the system takes in, but measures of what is available to the system from the negasystem and how that relates to the input remains unknown. In addition, output on the cybernetic model is what is available from the system, but what the negasystem gets and how that relates to the output remains unknown. SIGGS distinguishes the following "feed" functions and "put" properties. The "feed" transition functions move components from one subsystem to another, while the "put" properties identify the component partitioned sets (Thompson 2005).

- FI, Feedin is transmission of toput from a negasystem to input in a system. It is information from the system's surroundings to the system.
- FO, Feedout is transmission of fromput from a system to output in a negasystem. It is information from the system to its surroundings.
- FT, Feed-through is transmission of toput from a negasystem through a system to output of a negasystem.
- FB, Feedback is transmission of fromput from a system through a negasystem to input of a system. It is information from the system through its surroundings and back to the system.

Feedthrough is feedback with respect to the negasystem. The property of a system having an environment is referred to as "toputness" on the SIGGS theory model. As with the system, a negasystem can also have properties. For example, as a system can have the condition of selective information *on a negasystem* available for its [the system's] selection, so too a negasystem can have the condition of selective information *on a system* available for its [the negasystem's] selection.

- TP, Toputness is negasystem components for which system input control qualifiers are "true." It is the system's environment or selective information on a negasystem which is available to a system for selection.
- IP, Input is resulting transmission of toput components; systm components for which system input control qualifiers of toput components are "true."
- FP, Fromput is system components for which negasystem fromput control qualifers are "true."
- OP, Output is resulting transmission of fromput components; negasystem components for which negasystem output-control qualifiers of fromput components are "true."
- SP, Storeput is system input components for which system fromput control qualifiers are "false."

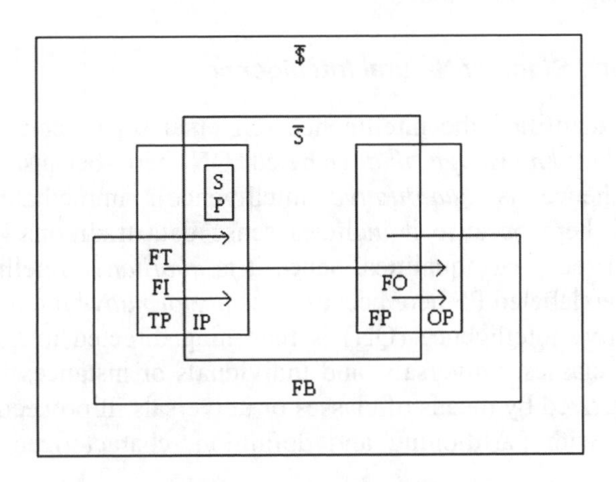

Fig. 6.3. Graph of SIGGS Theory Model[4]

Every system has information in the sense that occurrences of its components or affect relations or both can be classified according to categories. Here, we are concerned with categories of natural intelligence.

The added condition of uncertainty of occurrences at categories is necessary to develop information properties on the system and its

[4] This is an earlier version taken from Maccia & Maccia 1966. The outer rectangle is the negasystem; the inner centered rectangle is the system.

environment or *negasystem*. The concept "information" can be mathematically defined in our theory model in terms of uncertainty of occurrences at natural intelligence categories.

Though we have sorted three distinct pathways of information transfer that organisms use, including signals, cues and clues, a fully developed theory would require rigorous clarification and classification of the categories of occurrences of that information. This classification must span the entire intelligence repertoire of intentional behavior, including interacting components of verbal (by linguistic means), visual and sensorimotor categories.

6.5 SIGGS Applied to Natural Intelligence Systems

Given the set, information, graph, and general systems theoretic properties of SIGGS, we will outline the application of that theory model as a model for theorizing about natural intelligence. Where appropriate, we will extend concepts of that model.

Elements and Signs of Natural Intelligence

Above, we identified the intelligence universal set as consisting of the subsets verbal *knowledge that* (labeled QN here because it involves extension, hence is *quantitative* intelligence); immediate awareness (labeled QL here because it includes sense data traditionally known as primary and secondary qualities, hence it is *qualitative* intelligence); and *knowing how* (labeled PF here because it is *performative* intelligence).

Quantitative intelligence (QN) is reasoning directed to knowledge or knowing of classes, universals, and individuals or instances which are in turn characterized by means of classes or universals. It proceeds largely by class logic, with partitioning and definition, characterized by abstract inference.

Qualitative intelligence (QL), on the other hand, is reasoning directed to individuated qualities or attributes given in experience that *uniquely* mark off a person, thing, or universal. Such reasoning does not proceed by classification of shared properties or attributes, but in part by *ostensive* indexical *non-logical* operators such as "none-other". "None-other" is not the logical operator negation of "other" as "non-other" or "not-other", but is an operator *which separates one from all others*. Far more research is necessary to isolate those mechanisms by which immediate awareness, qualitative knowing, proceeds across the entire spectrum of human and animal sensory systems.

Performative intelligence (PF) proceeds in actual doings that exhibit or otherwise discloses patterns of smoothness and timing. This kind of intelligence cannot be reduced to class structures of quantitative reason because, in part, performative knowing has qualitative knowing structures embedded within it. Qualitative intelligence characterizes human intelligence, but in certain crucial respects does not characterize artificial intelligence.

Elsewhere (Estep 2003) I identified the universe of signs of intelligence consisting of the symbolic (SYM), iconic (IK), and enactive (EN) subsets. We should clarify that we are defining the intelligence universe so as to focus upon *signed* knowing, that is public expressions, representations, or artifacts of knowing. These are exhibited or disclosed as patterns of intelligence.

The *context* of knowing is defined in our theory in terms of the *position* that is the artifacts (public expressions, *signed* expressions) for *coming to know*. The intelligence universe is different from the intelligence universal set which includes the subsets: subject, object, content, and context of knowing. Our concern here is to provide at least the outlines of a general theory and some of the formal and contingent relations obtaining of the intelligence universe, so as to later relate these to computability (decidability) and complexity theory.

In essence, we are concerned with two sets, the set of intelligence categories of knowing K = {QN, QL, PF}, and the set of categories of signs of knowing, S = {SYM, IK, EN}. From their Cartesian product, we obtain nine ordered pairs:

K × S = {(QN, SYM), (QN, IK), (QN, EN), (QL, SYM), (QL, IK), (QL, EN), (PF, SYM), (PF, IK), (PF,EN)}

We could include in the universal intelligence set not only the categories of knowing but also the categories of signs of knowing. With this, we have the following intelligence set:

Intelligence Set E = {QN, QL, PF, SYM, IK, EN}. We can then form the power set, $\wp$ of E denoted by $\wp$(E), which is the set of all subsets of E, that is:

$\wp$(E) = {∅, E, {QN}, {QL}, {PF}, {SYM}, {IK}, {EN}, {QN, QL}, {QN, PF}, {QN,SYM}, {QN, IK}, {QN,EN}, . . . {n_{64} }}.

The formation of our power set $\wp$(E) yields 64 elements, the possible classes of intelligence subsets. We are primarily concerned with setting forth at least the outlines of a hypothetical axiomatic material system with theoretical ordering by means of set theory, information theory, and digraph theory.

6.5.1 The Use of Digraph Theory to Characterize Intelligence Relations

We earlier noted that digraph theory is mathematical theory which characterizes between pairs of points lines which can be directed.[5] Directed graphs consist of ordered pairs. The links in a directed graph are called arcs. Directed graphs or digraphs can be used to represent non-symmetric relations like "is the mother of" or "is hostile to." Undirected graphs[6] consist of *unordered* pairs. They are used for relations which are necessarily symmetric, such as "is the sibling of" or "conspires with."

The lines are used to indicate relations; the concept "relation" refers to function. Digraph theory can be applied to analyze groups of a system. The group of a system can be interpreted as a set of points (vertices or nodes) and an affect relation can be interpreted as a set of directed lines.

The use of graph theory in the SIGGS theory model permits the characterization of relations as intelligence affect relations between intelligence systems or between their components.

For example, persons or machines (natural or artificial intelligence systems) can be represented on directed graphs as (lettered) points and a communication (intelligence) relation between any two systems as a directed line segment (line segment and arrow) connecting the two systems.

Where there is an arrow or arrows between two points, there is a directed connection or pairing. Where there is a line without an arrow, a directed connection is assumed in one or the other or both directions.

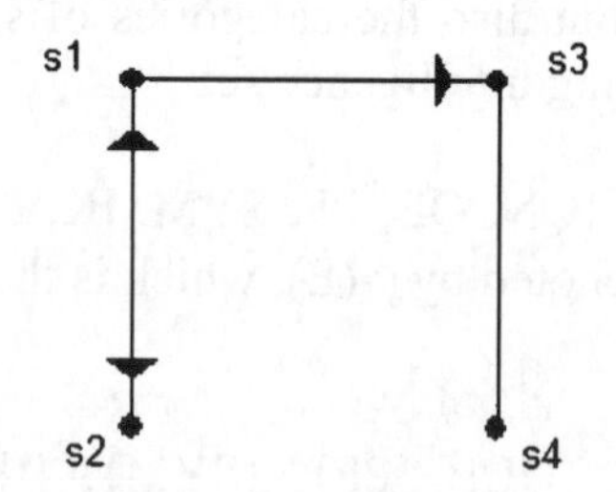

Fig. 6.4. Directed Graph

[5] The prefix "di" indicates that digraphs are graphs with directed lines.
[6] These are also known as graphs.

In the above graph, there is a directed connection or pairing between s1 and s2, s1 and s3, s2 and s1, and s3 and s1. Where there is only one arrow between two points the directed connection is direct. This is found in the connection between s1 and s2; s2 and s1; and s1 and s3. The graph is not completely connected, but is disconnected.

In non-valued graphs, nodes are either connected or not. Either Mary is the mother of Jill or she is not. In valued graphs, the lines have values attached to represent characteristics of the relationships, such as strength, duration, capacity, flow.

Reflexive graphs allow a node to have a tie to itself. These are useful when analyzing collectives or large groups such as cities. Nodes might represent banks with ties representing money transfers between them; a given bank will have ties to itself.

A graph is *connected* if there exists a path from every node to every other. A maximal connected subgraph is called a *component*. A *maximal subgraph* is a subgraph that satisfies some specified property (such as being connected) and to which no node can be added without violating the property.

A digraph that satisfies the connectedness definition given above (*i.e.* there exists a path from every node to every other) is called *strongly connected*. That is, for any pair of nodes *a* and *b*, there exists both a path from *a* to *b* and from *b* to *a*. A digraph is *unilaterally connected* if between every unordered pair of nodes there is at least one path that connects them. That is, for any pair of nodes *a* and *b* (and *a b*), there exists a path from either *a* to *b* or from *b* to *a*. A digraph whose underlying graph is connected is called *weakly connected*.

A maximal strongly connected subgraph is a *strong component*. A maximal weakly connected subgraph is a *weak component*. A maximal unilaterally connected subgraph is a *unilateral component*.

Moreover, digraph properties of a system result when certain conditions are placed on the system's affect *relations* or its *group*. For example, complete connection, strength, unilateralness, weakness, and disconnection exemplify digraph properties of a system arising from conditions on its affect relations. The properties of passive dependency, active dependency, independency, and interdependency also exemplify digraph properties of a system due to conditions on the *group*.

With digraph theory, an intelligence system group becomes a set of points; system affect relations become sets of directed lines; and digraph properties of a system result when certain condition are placed on its affect relations or its group.

Given the large number of connections and relations among components at local levels in self-organizing systems, emerging in kinds of global dynamics, digraph theory, with dynamical system theory, provides powerful mathematical tools to characterize those connections and relations. These tools are necessary to describe the integrated behavior of intelligence systems coordinating the actions of many, possibly an immense number, of components.

In addition to the above properties of SIGGS, two additional parameters are necessary to define a general natural intelligence system. These include a linearly ordered time set, *T*, and a system state-transition function, σ (Thompson 2005).

- Linearly ordered time set, $T =_{df}$ linearly ordered set. This set is required to give the natural intelligence system a dynamic property which may be mapped over the reals. This set assigns an appropriate time that an event occurs. Without this set there would be no order or sequence to the events of a system.
- System state transition function, $\sigma =_{df}$ the function that maps a current system state onto a subsequent system state. This is required in order to alter the state of a system. Whereas *T* moves objects about the system, σ changes the state of the system due to new affect relations defined by the move.

Because our theory is largely about human intelligence phenomena, we cannot have a purely formal theory. It concerns contingent relations in reality. This presents some problems for checking on the coherence of the theory, but given its partial formalization a check on logical consistency is nonetheless possible. This is so because there are deductive links to be checked out.

Ordering through digraphs also gives us the advantage of presenting a theory which expresses relations which are contingent and also recursive and asymmetrical. That advantage is the use of path analytic techniques to check out correspondence of the relations expressed in the theory to those in reality. Path analysis is a procedure for estimating the path coefficients from correlational data using regression techniques (Steiner, 1988).

In our theory, we are interested in properties of *relations* (specifically affect relations) on natural intelligence sets, the properties of reflexivity, symmetry, and transitivity; properties of *functions* of intelligence sets, such as surjection, injection, and bijection; as well as *graph-theoretic* properties on intelligence groups, such as *connectedness*.

For the latter, we are concerned with *contingent* relations rather than pure formal relations between the sets, thus we will focus upon *directed graphs* in which the relations are recursive and asymmetrical.

Moreover, we will later be concerned with the computability of those relations that is functions, on natural intelligence sets. We are interested in formalisms which establish that we can map the objects of one intelligence set onto the objects of another.

Social Network Theory and Patterns of Intelligence

Social network theory and analysis, based upon digraph theory can be helpful to demonstrate properties of SIGGS on certain levels. This section is intended solely for illustration.

Nodes and edges in digraph theory and analysis can consist of many things. Here, we are concerned with natural intelligence networks. Recall that the structure of our inquiry (where this is limited to human natural intelligence) permits at least two major categories: Analysis of intelligence relations between a given person or persons and other person or persons; and between person or persons and objects in the world.

Thus the population of nodes (the object set) can consist of persons and objects, while the edges consist of relations, specifically affect relations between persons and objects. This is the relation set.

We also assume that boundaries on the population of nodes (persons) are imposed on the persons themselves, so this is not necessarily a naturally occurring cluster or network such as a neighborhood group. It is a network of persons brought together by shared objectives. There are many levels of analysis of such networks, and we will assume that some individuals in the network, which we will call Network A (assuming only one network for simplicity that is not connected with other networks) have not had any face-to-face contact with others in the network, whereas others have. However, they are all bound together nonetheless by a larger organization or network of networks, which we will call Network B, sharing the same objectives.

Such a network as Network A is unlike a family because (a) individuals in the network are not related to one another; and (b) one assumes that most if not all family members have met one another face to face at some point in time.

Just as individuals who are nested in work relations, this particular Network A is institutionalized largely by those shared objectives. Moreover, each individual in Network A is part of a personal family, lives in a neighborhood, and is nested as well within a work environment. This individual may or may not be nested in each of these with other members of the original Network A. In effect, the original Network A is a network nested within a network that is nested within a network, and so on.

A digraph or social network analysis of such a network, from the point of view of natural intelligence, is faced with a multi-modal hierarchical structure nested design. It requires that the analyst focus on multiple levels of analysis simultaneously to ascertain how any given individual is embedded within that total structure.

It will generally be the case that any given individual member of the original Network A will be connected by many different kinds of ties and relations to any number of other individuals who may or may not be part of the original Network A or Network B.

Indeed, from the point of view of one who may be conducting an inquiry into Network A, it will be those relations of a given member that will initially become known prior to the researcher learning of the existence of the Network A. Like picking up single strands of thread to follow back to their original source, a researcher may follow a single strand to many other interconnected strands and networks before coming upon the network that ties so many together. *It is the relation set that is often found before the object set is discovered.*

If we are intelligence researchers wanting to collect data about certain kinds of relations among persons, we can take any number of approaches. Effectively, we will want to look at a universe of possible relations. Our research question and background assumptions may indicate the kinds of relations most significant that we may decide to start with. If we are searching for intelligence related to either a criminal or terrorist threat, for example, we may know from data mining that a cluster of electronic transfers of funds has taken place between networks of individuals who may otherwise be beyond suspicion.

Thus, we might want to sort out both material and informational relations to look at, starting with any nodes (persons) whose names may have popped up in data mining. Material things in general are usually located at or near a given node (person or group of persons) at a given time. Movements of money, people, and things can take place between nodes (persons). Informational relations, however, can be shared by more than one person at a time.

Intelligence can be transferred between persons and groups just as material and information[7] can. If researchers are focused upon intelligence related to criminal or terrorist activities, they must use a prior set of indicators or patterns by which to analyze the dynamics of any network. That is obviously beyond the scope of this work; our intention is to use

[7] This sense of information is neither identical nor equivalent to the concept of intelligence as we are using it here, though it is related.

indicators of kinds of natural intelligence by which to analyze a hypothetical network.

Fundamental Properties of Networks: Density and Connectedness

We are especially interested in fundamental properties of networks and how to use those to analyze specific network configurations on all levels of a natural intelligence system. Specifically, network analysts are concerned to discover *how the network structure emerges from the micro-relations between individual parts*. The usefulness of network methods to map such multi-modal hierarchical relations is the core of its power.

Keep in mind that digraph theory, or graph theory in general, is a mathematical theory that can be used on the most micro-levels of analysis of the intelligence of a given organism to the most macro-levels of intelligence of groups of organisms such as human beings.

Thus, digraphs are used in the theory and analysis of neural networks in the brain as well as in computer architectures, social networks, and in the theory and analysis of natural intelligence in general.

There are two fundamental properties of digraph theory or network analysis in general that are of crucial concern in understanding theory and analysis of any network. These are *density* and *connectedness*.

Through axiomatization and digraphing, because they are both ways of ordering explanatory theoretical sentences, we have evidence of completeness. Any gaps in the theory will be shown because missing deductive links will be apparent in axiomatization and missing connections will be apparent in the case of digraphs. With respect to digraphs, presented as path diagrams meeting the requirements for path analysis [connections must be asymmetrical], the density and connectedness of the digraph indicate whether connections are missing.

Density is the number of direct connections over the number of possible connections. Density is given by the following equation:

$$D = DC/N(N\text{-}1) \tag{6.1}$$

Where "D" stands for density
"DC" stands for number of direct connections
"*N*" stands for the number of nodes

Density cannot fall below some minimal value because obviously less than *N*-1 direct connections results in some nodes not being connected.

Connectedness is the number of direct and indirect connections over the number of possible connections, given by the following equation:

$$C = DC + IC/N(N\text{-}1) \tag{6.2}$$

Where "IC" stands for the number of indirect connections.

Our theory is generally concerned with real and possible intelligence elements obtaining of natural or artificial systems. This includes a concern for the characterization of the association between the set of intelligence properties obtaining of one system and the intelligence properties obtaining of another. An example would include interactive systems or sets, with any number of elements, where the intelligence properties of one system affect those of another.

That is, where A and B denote the individual sets A intelligence properties and B intelligence properties respectively, if to each element in set A there is assigned a unique element in set B, then such an assignment is called a function: f: A → B

The set of intelligence properties obtaining of A is the *domain* of f and the set of intelligence properties obtaining of B is the *co-domain* of f. If a ∈ A, then the element in B which is assigned to **a** is the *image* of **a**.

For example, let f assign to verbal quantitative knowledge (QN) or *knowledge that* about each discipline its formal intelligence classification, such as analytic-*a priori*; synthetic-*a posteriori*; *a priori*-synthetic. Here the domain of f is the set of intelligence conditions constituting quantitative knowledge about each discipline. The co-domain of f is the list of possible intelligence categories which make up the logical and evidential criteria for propositional assertions. For example, the image of quantitative knowledge about mathematics is analytic-*a priori.*

The *range* of f: A→ B is f(A) is a subset of B. Applied to A and B above, f(A) = {analytic-*a priori;* synthetic-*a posteriori*; *a priori*-synthetic}. Obviously f (A) ⊂ B. That is, quantitative knowledge about disciplines and the elements in B which appear as the image of at least one element in A are a subset of B.

In our theory, we also want to focus upon the sets (QN), (QL), and (PF) between which certain elements are related in some way. For example, we may want to focus upon the set of ordered pairs of intelligence conditions of a certain category satisfying a certain equation. This set would be a subset of A × B and would contain all points of the graph of the equation. The subset is a relation from A to B.

Partial Order on the Intelligence Set

A *partial order* on a set A is a relation [sometimes denoted by the symbol $\leq$] such that (i) $a \leq a$ for all $a \in$ A; (ii) $a \leq b$ and $b \leq a \Rightarrow a = b$; (iii) $a \leq b$ and $b \leq c \Rightarrow a \leq c$. That is, a relation partially orders a set if and only if it is reflexive, transitive, and antisymmetric.

For purposes of illustration, we will cut down on our power set, limiting it to three elements. Thus, assume that S = {QN, QL, PK,}. The power set of S we will denote by $\wp$(S), the set of all subsets of S. That is,

$$\wp\,(S) = \{\varnothing, S, \{QN\}, \{QL\}, \{PK\}, \{QN, QL\}, \{QN, PK\}, \{QL, PK\}\}.$$

A partial order on the set $\wp$(S) is set inclusion ($\leq$ is replaced by $\subseteq$). If we have T, P, Q, R $\in \wp$(S), the following properties obtain: (i) $T \subseteq T$; (ii) $P \subseteq Q$ and $Q \subseteq P$ imply that $P = Q$; (iii) $P \subseteq Q$ and $Q \subseteq R$ imply $P \subseteq R$.

Where we substitute the natural numbers $\boldsymbol{N}$ = 1, 2, 3, 4, 5, 6, for some of the above subsets of the power set, the partial order is graphically presented below where the arrows indicate set inclusion. (Not all admissible lines are drawn in the figure).

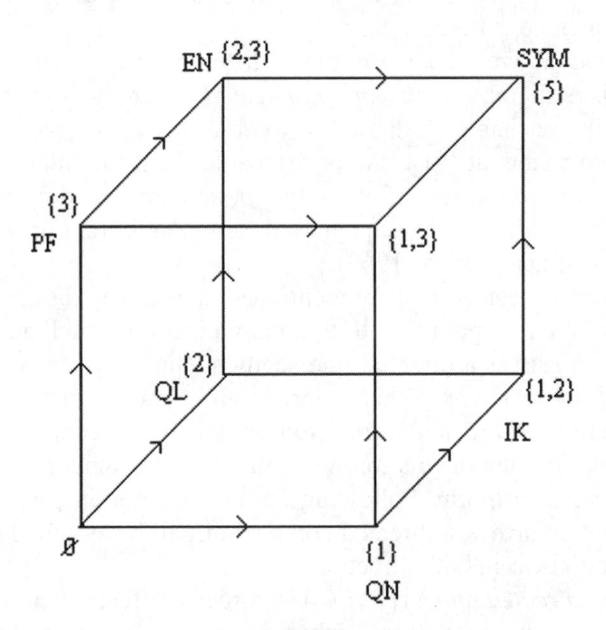

Fig. 6.5. Partial Order on Natural Intelligence Set

In the above graph, $\{3\} \rightarrow \{2, 3\}$ means that $\{3\} \subset \{2, 3\}$. Also, $\{3\} \rightarrow \{1, 2, 3\}$ is implied in $\{3\} \rightarrow \{2, 3\} \rightarrow \{1, 2, 3\}$. Some pairs of elements in the power set will not satisfy any of the relations of $\subseteq$, $=$, $\supseteq$. As an example, $\{1, 3\} \not\subset \{2, 3\}$.

A set A and a partial order $\leq$ together form a pair (A, $\leq$) called a partially ordered set. Where $a \leq b$, a is said to be smaller than or precede b and b is greater than or follows a. Where $a \neq b$ then $a \leq b$ is written as $a < b$ (with $a, b \in \mathbf{R}$). A partially ordered set (A, $\leq$) is totally ordered if, for every $a, b \in$ A, either $a \leq b$ or $b \leq a$.

As earlier noted, digraph theory is mathematical theory which characterizes between pairs of points lines which can be directed. A digraph consists of a finite collection of points, $P_1, P_2, P_3, \ldots P_n$, together with a described subset of the set of all ordered pairs of points which are directed lines. If D is a digraph, the graph obtained from D by removing the arrows is called the "underlying graph" of D.

Again, a digraph D is said to be *connected* (or weakly connected) if it cannot be expressed as the union of two disjoint digraphs. This is equivalent to saying that the underlying graph of D is a connected graph. A *subdigraph* of a digraph is a subset of points and directed lines of the digraph which constitutes a digraph.

As noted above, there are three different ways a digraph may be connected. A digraph is *strongly connected* or strong if every two points are mutually reachable. A digraph is *unilaterally connected* or unilateral if for any two points at least one is reachable from the other. A digraph is *weakly connected* or weak if every two points are joined by what is called a *semipath*, an alternating sequence of points and arcs (or directed lines) which are distinct (Harary, 1969).

Intelligence sets A and B mentioned above can be represented on graphs as lettered points and a communication (intelligence) relation between two sets as a directed line segment (line segment with an arrow) connecting the two sets. Where there is an arrow or arrows between two points, there is a *directed connection* or *pairing.* A connection of one or more points or components to one or more other points or components is called an *affect relation.* Following earlier assumptions, where there is a line without an arrow, a directed connection will be assumed in one or the other directions or in both directions.

A *direct directed affect relation* is a directed affect relation in which the channel (line) is through no other point. An *indirect directed affect relation* is a directed affect relation in which the channel (line) is through other components (points). A completely connected intelligence affect system is, by definition, not possible since such a system would have complete connectedness if and only if all its intelligence affect relations were direct directed ones, that is direct channels from and to each intelligence component.

Where there is only one arrow between two points, the directed connection is *direct*. Where there is no line between a given point and other points, there is no connection or pairing with any of the other points.

A *directed walk* in a digraph is an alternating sequence of points and arcs (or lines). A *closed walk* has the same first and last points, and a *spanning walk* contains all the points. A *path* is a walk with all points (hence all arcs) distinct, except the first and the last.

A digraph is *unilateral* if and only if for every pair of distinct points, P_x and P_y, there exists a directed path from P_x to P_y or (in the exclusive sense of "or") from P_y to P_x. It follows that all strong digraphs are unilateral however it is not the case that all unilateral digraphs are strong.

A digraph is *disconnected* if and only if the points constitute two disjoint digraphs, that is with no directed line joining any point in one digraph to any point in the other.

Digraphs, hence, fall into four mutually exclusive subsets with respect to connectedness:

C_0 = disconnected digraphs
C_1 = weakly connected digraphs
C_2 = unilateral digraphs
C_3 = strongly connected digraphs

The use of digraph theory added to the SIGGS theory model permits a high level of precision in theory and analysis of properties of systems, such as natural intelligence systems. Any intelligence network would have complete connections if and only if all its affect relations were direct directed ones, direct channels from and to each component of the network.

Theory and analysis of the transmission of intelligence between nodes (persons) in an intelligence network requires the use of information theory. This will permit us to characterize occurrences at categories.

6.5.2 Information-Theoretic Measures on Natural Intelligence Systems

We are interested in measures of information (uncertainty) in an intelligence system in general, and measures of information in the *transmission* of intelligence between persons or groups, either within or between systems, though we will tend to focus primarily upon the former.

Propositional knowledge *that* (QN) is relatively non-problematic to measure as long as we have publicly expressed linguistic or alphanumeric propositions which represent a system's intelligence *that*. But we will see that the above approach to *non*-propositional and *nonlinguistically* expressed or disclosed intelligence, immediate awareness and *knowing how,* (PF) and (QL) presents unique problems when we look at elements of kinds of intelligence thought to be uniquely human. This will require viewing these uniquely human kinds of intelligence from a purely syntactical perspective that is from the perspective of the amount of information resulting from actions or performances at natural intelligence categories.

Information-Theoretic (Uncertainty) Measures of Intelligence

Again, we assume that human and machine intelligence systems and transmission networks are not complete or "all knowing," hence there *must* be *uncertainty* of occurrences at the categories where we interpret that uncertainty as explicated in terms of probability distribution. The existence of alternatives for the occurrence of any intelligence (for example qualitative intelligence) component indicates the selective sense of information, which can be measured and such measures of the *transmission* of information shared between two or more systems can be calculated with the Shannon (1949) formula.

The basic information function is designated by "H.". By summing over the amount of information associated with each selection, weighted by the probability that the selection will occur, the value of H can be obtained. To state this more precisely, H(C) is the average uncertainty per occurrence with reference to the classification C. *It is the average number of decisions needed to associate any one occurrence with some category c_i in C,* with the provision that the decisions are appropriate; it is a function of the probability measures in C:

$$H(C) = -\sum_{i=1}^{n} p(c_i) \log_2 1/p(c_i) \tag{6.3}$$

A measure of joint uncertainty would be:

$$H(C_{IJ}) = \sum_{i=1}^{m} \sum_{j=1}^{n} p(c_i, c'_j) \log_2 1/p(ci\, c'_j) \tag{6.4}$$

The measure for conditional uncertainty would be:

$$H\,(C_I\ C_J) = \sum_{i=1}^{m} \sum_{j=1}^{n} p(c_i\,c'_j)\,\log_2 1/p(ci\ c'_j) \tag{6.5}$$

The three H measures are related as follows:

$$H(C_I) + H(C_J\ C_I) = H(C_{IJ}) \tag{6.6}$$

The T measure is the amount of shared information:

$$TIC_I,C_JP = H(C_i) + H(C_J) - H(C_{IJ}) \tag{6.7}$$

Measures of Uncertainty and Intelligence Categories of Occurrences

To address the extension of the above information-theoretic measures on intelligence sets, and to evaluate our own use of Shannon's probability measure on those same sets, we will continue with a geometric approach to the intelligence universe.

We are interested in the limits of probability theory to capture the information of an intelligence system in general and the transmission of information in natural intelligence systems in particular. Moreover, we are concerned to set out the limits of any of the measures to address the nature of information-theoretic measures of intelligence.

The information base from which affect relations are determined (which also determines the intelligence network or system in general and the components of that network) may be well-defined or it may not be. As Thompson (2005) points out, we may only have an idea or a guess as to what components are actually contained in the information base. Even though we may only have a working hypothesis as to what components are contained within it, whatever guess or idea we are working with will produce the system components. Thus, the definition of the information base is crucial, especially when referring to criminal or terrorist systems. Integrating data mining technologies within the larger theory of natural intelligence makes possible real-time predictions.

Given the ambiguities surrounding definitions of the information base, we should give consideration to alternatives to classical crisp measures of information. Though this is not the place to explore all the limitations on probability theory to capture information transmission of a natural intelligence system, we should at least outline certain of the problems.

Klir et al. (1989; Klir and Yuan, 1995; Klir and Wierman, 1999) has made evident certain of the limitations on probability theory for dealing with various situations under uncertainty. First, it is clear that Dempster-Shafer theory is more general than probability theory with respect to the bodies of evidence that can be conceptualized by either of the two theories. The additivity axiom of probability theory is replaced in the Dempster-Shafer theory with the weaker superadditivity and subadditivity axioms. A consequence of this replacement is that it is not required in the Dempster-Shafer theory that degrees of belief be allocated to singletons of the universal set. When they are, Dempster-Shafer assumes the mathematical structure of probability theory.

With respect to evidence, possibility theory and probability theory are virtually disjoint sets. As Klir notes (1989), the only body of evidence they share consists of one focal element that is a singleton, that is the case of total certainty (which means we are referring to the *nonselective* sense of information).

Moreover, they involve different types of uncertainty. Probability theory captures uncertainty in terms of dissonance, measured by Shannon (1948) entropy. In possibility theory, uncertainty is expressed primarily in terms of nonspecificity measured by function V, with some amount of potential conflict, measured by function C. Also, since possibility distributions are in a one-to-one correspondence with fuzzy sets, it is also meaningful, according to Klir, to characterize possibility distributions by their degrees of fuzziness.

In essence, the two theories are complementary in the applications of their formalisms. Probability theory characterizes the number of occurrences expected at a given category in a classification at a given time. Possibility theory characterizes the number of occurrences that can occur at that same category at any one time.

Moreover, possibility theory is less computationally sensitive to error in the assessment of possibility degrees than probability theory is in the assessment of probability degrees. This is because possibility theory employs the operators *maximum* and *minimum* and the error does not accumulate when we operate on possibility distributions. No matter how many times we operate with them, the error cannot exceed the largest error in the assessment of possibilities. On the other hand, in probability theory, where the basic operations are *sum* and *product*, the error increases with the number of operations performed.

With respect to fuzzy set theory, the membership grade functions are in the range [0, 1]. This is a generalization of classical set theory which includes the range {0, 1} of characteristic functions. For standard fuzzy set operators, the law of excluded middle and the law of contradiction are violated. Fuzzy set theory and probability theory cannot be directly compared because fuzzy set theory does not involve the power set $\wp(X)$ or a σ-field of selected subsets of X. Both the latter are necessary in the formalization of probability theory.

Given the increasing complexity of self-organizing emerging systems, there has been greater acceptance in the second half of the twentieth century of alternative mathematical formalizations of uncertainty. Until recently, probability theory was the only available theory for formalizing uncertainty. Alternatives allowing more refinement in measures of uncertainty, especially of cognitive processes have become more accepted. Among these alternatives are those Klir has elaborated upon as noted above.

Information-Theoretic Measures of the Universal Intelligence Set

Setting aside issues related to probability interpretation of information, the information-theoretic measures incorporated within SIGGS provides means for categorizing major components of intelligence networks. Occurrences of kinds of intelligence behavior, in terms of the major categories of natural intelligence we sorted above (QN, QL, PF) can be viewed as selective information.

The probable occurrence of instances in each of these categories can be determined, especially within the context of the above theory of signs, including symbolic (SYM), iconic (IK), and performative (EN). Patterns of natural intelligence occurring within these categories can be determined with the use of information-theoretic measures provided by SIGGS.

Moreover, the array of patterns of interactive natural intelligence behavior can also be characterized with the information-theoretic measures. That is, flow of natural intelligence occurrences according to categories of intelligence and signs of intelligence from person(s) to other person(s) and groups can be determined. Measures of TP, toput, IP, input, with measures of commonality, T, may be obtained.

6.6 From a Symbol-based View to a Geometric View of Natural Intelligence

The nonlinear theory model approach as illustrated with SIGGS permits reorienting the perspective on intelligence from a classical top-down *symbol*-based, single-capacity view to a *geometric*-based performance of intelligent agents in a fitness landscape. By turning to a phase space of intelligence possibilities, utilizing a broad array of variables from major categories of intelligence, the very dynamics of a natural intelligence agent can be evaluated as that agent works his or her way through the landscape. As an additional tool, Boolean networks can provide well-developed concepts and methods to increase our knowledge and awareness of the dynamic nature of natural intelligence systems.

6.6.1 Boolean Networks

The properties of Boolean networks have been thoroughly studied and techniques have been developed for determining the dynamical properties of specific kinds of networks (Forrest and Miller 1991). Thus they are of tremendous value in studying the fundamental properties of highly complex dynamical systems which may have very large, even immense, numbers of interrelated components.

The dominant dynamical behaviors of Boolean networks ffound in complex, dynamical systems, are characterized by properties such as *state cycles, frozen components* [in which components of the network are "locked" into either the on or off position], and *stability to perturbation.* These properties in turn are dependent upon the *structural* properties of the network, that is, *N* (*network*), *K* (*variables*), and number of Boolean functions for each element.

Even though most natural intelligence phenomena manifest themselves over continuous domains, it is sometimes useful to initially simplify the data and functions. A Boolean network N (I, F) is defined by a set of nodes corresponding to intelligence components, $I = \{x1, \ldots xn\}$ and a list of Boolean functions $F = (f1, \ldots fn)$.

The state of an intelligence component is determined by the values of other intelligence components at a given time, *t*, by means of Boolean logic functions. Generally, if an intelligence component is regulated by *K* variables, the number of possible combinations of their presence or absence is 2^K. For each of these states, the response may be 0 or 1. Thus the number of possible Boolean functions of *K* variables is $(2^2)^K$.

Boolean logic uses the basic statements AND, OR, and NOT. Using these and a series of Boolean expressions, the final output would be one TRUE or FALSE statement:

If A is true AND B is true, then (A AND B) is true
If A is true AND B is false, then (A AND B) is false
If A is true OR B is false, then (A OR B) is true
If A is false OR B is false, then (A OR B) is false

The state of each component is functionally related to the states of other components using these logic rules.

The list of Boolean functions F represents the rules of regulatory interactions between the components. Any intelligence component transforms its inputs (regulatory factors that bind to it) into an output, which is the state or expression of the component itself.

Combining gates (switches) and inputs enables us to work out how all the combinations affect each other. For example, with 2 intelligence components and 2 possible states, there are 4 combinations. The table below will give the reader some idea of the increasingly very large numbers and interactions that result in combination.

Table 6.3. Boolean Network Combinations

Switches	Number of Networks
1	$2^2 = 4$
2	$4^4 = 256$
3	$8^8 = 1.7 \times 10^7$

Given their core properties, switching Boolean networks are important for an adequate theory of *complex but ordered systems,* such as the interacting neurons in a neural network or interacting primitive relations in a natural intelligence network.

This is so because these networks facilitate handling the large number of elements involved as well as their connections, their interrelations. Boolean network models are also important because one can trace the behavior of populations of very large numbers of interacting elements making up different network configurations, with varying parameters, across landscapes known as Poincaré maps or spaces of possibilities, so as to evaluate their properties.

Again, there are highly complicated and complex negative and positive feedback connections in natural intelligence. These can become even more complex with input functions that are called "canalizing." This is where a single input in a fixed state is sufficient to force the output to a fixed state, regardless of the state of any other input. A lot of such functions can force portions of the network to a fixed state, breaking it up into active and passive structures.

The simplest pattern behavior is a state cycle consisting of a single pattern of 1s and 0s, a cycle of length 1. The length of the cycle of a network could be the total number of states in state space, in which case the system will repeat every pattern it is capable of displaying. In a 3 relation system, the length of the cycle is 8 possible states (*i.e.* 2^3). A Boolean network can have many state cycles that is attractors (families of trajectories), each one with a collection of trajectories flowing into it that is its basin of attraction.

Because the network operation is determined only by the logic of the gates and their input connections (in terms of 1s or 0s), we can apply it (map it) to any real life situation that has a similar decision structure and interconnectivity.

Moreover, we can devise a specific network that reflects what we know about an actual situation or set of events, such as in natural intelligence or in complex social interactions. If we have properly factored in the variables such that we have the correct configuration, we can study the network to predict behavior.

Random Boolean Networks

Under conditions of incomplete knowledge such that we do not know all the variables and do not know the correct configurations, then we can randomly initialize the values. Given any number of gates (*N*) inputs to each (the connections, *K*), random networks will show some behavior. Generally, the behavior can be roughly categorized as follows:

- The network can rapidly approach a fixed state, a point attractor. In these systems it doesn't matter from which initial state we begin, the network always converges to the same point.
- The network can cycle through a chaotic sequence affecting all the components, rarely giving the same combined output state.
- The network can settle down to a short sequence affecting only a few of the components. Many may be fixed in 0 or 1 states and the remainder cycle through a succession of repeating states.

Discrete Digital and Continuous Analogue Domains

Moreover, the entire scope of natural intelligence phenomena is both discrete and continuous. That is, it consists of discrete problems definable over the integers or any numbers that can be effectively encoded into the integers, and it also consists of continuous problems defined over the real numbers. Discrete problems assume that all underlying spaces are countable. Continuous problems are defined over the real numbers, in which the underlying spaces are not countable.

Though we started the discussion of Boolean networks based on simple binary logic, which would no doubt work for most discrete problems, it is possible to extend these network models to include intermediate real values between 1 and 0.

Smoothly timed movements in the intelligence of knowing how and immediate awareness, for example, require fuzzy values in place of binary values. Generally, fuzzy valued networks will also exhibit attractor dynamics and self-organizing criticality, when the system develops or evolves to a critical point and then maintains itself at that point.

Because *knowing how* and immediate awareness depend upon structures and connections of the sensory and sensorimotor systems, a research focus permitting intermediate real values is crucial to those facets of natural intelligence. Recall, however, that this will not work with certain kinds of indexical operators in human language.

Taking only one rather complicated example where such a research focus will largely work, the kinds of stealth movement tactics developed during any military basic infantry training demonstrate the necessity for these and the total inadequacy of models based upon binary logic.

Disciplined moving, touching, and even apparently simple tasks such as knowing how to hold a weapon and squeeze a trigger with just the right amount of pressure do not come naturally and must be learned. Top-down, digital linear models used to research such learning inevitably fail because of the lack of context-dependent sensitivity that self-organizing, emergent and continuous models can provide.

The geometric orientation to the study of natural intelligence can also be greatly effected with the use of high-speed computers, particularly for the study of *knowing how* and immediate awareness. Methods for doing this will be more closely assessed in the next chapter.

This requires the use of highly parallel distributed processing and connectionist models capable of simulating rule-*bound* know*ing*, as opposed to limiting simulations to rule-*governed, symbol*-based know*ledge*. Such computer-based research programs have already proven highly useful in the study of gait analysis (Simon 2004), part of the intelligence of *knowing how*.

For a study of a natural intelligence system, where the computer is used as an instrument permitting experimentation of a kind in a hypothetical universe, what is needed is an approach to computation permitting a focus upon *on-going, dynamic* interactive knowing behavior—rather than a focus upon final results. That is, what is needed is a computational architecture of an intelligent system permitting a natural method of knowing behavior *generation*.

This natural method must reflect the distributed and parallel structures of human knowing systems. These natural intelligence systems include a hierarchy of the above mentioned populations of simple components constructing aggregates of simple rule-governed or rule-bound objects interacting nonlinearly with one another and with their environment to produce emergent structures of intelligence.

At each level, the primitives must be identified and rules governing or bounding their behavior under conditions at that level must be specified. Primitive knowing behavior, our species of knowing how in conjunction with immediate awareness, must be organized in the architecture of the artificial system similarly with their natural counterparts. From this organization, emergent properties of knowing, of natural intelligence, arise.

These considerations also apply to an understanding of the relation between natural and artificial intelligence. The classical approach to artificial intelligence has been premised upon a discrete, top-down view of natural intelligence as symbol-based, rule-governed *knowledge that.*

That approach, the usual Artificial Intelligence (AI) approach, is a serial processing strategy, with problems defined over the natural numbers, integers, rationals or domains encodable in the integers, requiring a great deal of elaborate programming and *prior* know*ledge* engineering. It is an approach built upon a centralized control structure with access to large sets of *predefined* data structures, operating with algorithms defined by mathematical formulas and discrete procedures.

We will more thoroughly address these issues in the following pages.

6.7 Summary

The major objective of this chapter was to emphasize the crucial importance of focusing on what natural intelligence does as opposed to where it comes from. We set forth our approach to theory of natural intelligence by examining properties of self-organization while also looking at some of the problems associated with devising theoretical means to measure and analyze the dynamics and information transmission occurring in natural intelligence networks.

We focused upon natural intelligence as a large population of simpler components which through time works upwards, synthetically constructing larger aggregates of rule-*governed* or rule-*bound* objects. We noted that a theory of natural intelligence has at least two kinds of complexity to address: the complexity of a natural intelligence system itself; the other is the complexity of its development or growth in intelligence in an environment. Also adding to measures of complexity are the complicated positive and negative feedback mechanisms one finds in natural intelligence systems as well. These complicated causal loops result in nonlinearities and can lead to high levels of unpredictability.

We also addressed the problems associated with any kind of research directed to organized complexity of any kind. Such research requires concepts, models, and mathematical *cum* logical tools sufficient to represent multivariable complex dynamic configurations of complex organization, of the interaction of large (possibly immense, infinite) numbers of variables or components.

Seeking those concepts, models, and mathematical tools we turned to SIGGS, a formal theory model originally devised in the 1960's by Steiner (Maccia) and Maccia (1966) and extended by Thompson (2005) and Frick (1983, 1990).

The mathematics of set theory provide the means to formally define "system" as a group of at least two components with at least one affect relation and with information (Maccia & Maccia, 1966; Steiner, 1988). In essence, a system is an ordered pair consisting of an object set S, and a relation set, R. Set theory is used to give meaning to the definition of "system" in that a group of at least two components becomes a *set* of at least two elements which form a sequence.

In addition to formally defining "system" set theory is of course necessary to demarcate the scope of the natural intelligence universe based upon logic and facts found within the domain of animal and human experience. As we demonstrated in an earlier chapter, with set theory, we more precisely carved the problem space of natural intelligence.

Also, set theory allows the quantification of a complex organization as a whole. Information theory allows the quantification of action; graph theory permits the quantification of structure; and general systems theory permits an *organismic* perspective to treat complexities of *configurations* of a whole.

The advantage of utilizing SIGGS to devise theory of self-organizing emerging natural intelligence is that it permits representations of *organized complexity*. This is due in part to the fact that it is based upon assumptions of non-linearity and complex interactions among the parts which make up the system.

It also includes information theoretic measures which are not provided on the standard uncertainty increasing (unorganized complexity) and uncertainty maintaining (organized simplicity) theory models. Obviously, information theory is necessary to adequately characterize how organisms, including humans, acquire and act upon information.

Looking upon an organism as an intelligence system, either in isolation or in the context of other organisms, information theory provides a way to give meaning to the categorization of components of the system, connections of components to one another, connections between the system and its environment, and the uncertainty of occurrences at intelligence categories.

The added condition of uncertainty of occurrences at categories is necessary to develop information properties on the system and its environment or *negasystem*. The concept "information" is mathematically defined in our theory model in terms of uncertainty of occurrences at natural intelligence categories.

Though we have sorted three distinct pathways of information transfer that organisms use, including signals, cues and clues, a fully developed theory would require rigorous clarification and classification of the categories of occurrences of that information. This classification must span the entire intelligence repertoire of intentional behavior, including interacting components of verbal (by linguistic means), visual and sensorimotor categories.

We identified the intelligence universal set as consisting of the subsets verbal *knowledge that* (labeled QN because it involves extension, hence is *quantitative* intelligence); immediate awareness (labeled QL because it includes sense data traditionally known as primary and secondary qualities, hence it is *qualitative* intelligence); and *knowing how* (labeled PF because it is *performative* intelligence).

We used concepts of digraph theory such as group and directed lines, to interpret the object set and relation set of an intelligence system as a set of points and system affect relations as sets of directed lines. We demonstrated that digraph properties of a system result when certain condition are placed on its affect relations or its group.

We were especially interested in fundamental properties of networks and how to use those to analyze specific network configurations on all levels of a natural intelligence system. Specifically, we were concerned to discover *how the network structure emerges from the micro-relations between individual parts*. The usefulness of network methods to map such multi-modal hierarchical relations in intelligence systems is the core of its power.

We demonstrated that the use of digraph theory added to the SIGGS theory model permits a high level of precision in theory and analysis of properties of intelligence systems.

We also used information theory to demonstrate information-theoretic (uncertainty) measures on an intelligence network. The existence of alternatives for the occurrence of any intelligence (for example QN, quantitative intelligence) component indicates the selective sense of information. This sense of information can be measured in an intelligence system and such measures of the *transmission* of information shared between two or more components of systems (or two or more systems) can be calculated with the Shannon (1949) formula.

We also noted that the information base from which affect relations of a group are determined (which also determines the intelligence network or system in general as well as the components of that network) may be well-defined or it may not be. A researcher may only have an idea or a guess as to what components are actually contained in the information base. The definition of the information base is crucial, especially when referring to criminal or terrorist systems. Thus, given the ambiguities surrounding definition of the information base, we determined that consideration to alternatives to crisp measures of information, including formalisms of the Dempster-Shafer theory of evidence; imprecise probabilities; and possibility theory should be explored and possibly included. Moreover, integrating data mining technologies within the larger theory of natural intelligence utilizing SIGGS makes possible real-time predictions.

In addition to the SIGGS theory model, we also noted the value of Boolean networks as an added tool of the model. The nonlinear theory model approach permits reorienting the perspective on intelligence from a classical top-down *symbol*-based, single-capacity view to a *geometric*-based performance of intelligent agents in a fitness landscape.

Properties of random Boolean networks in analysis and theory of self-organizing and emergent configurations of natural intelligence were explored in an earlier publication (Estep, 2003). By turning to a phase space of intelligence possibilities, utilizing a broad array of variables from major categories of intelligence, the very dynamics of a natural intelligence agent can be evaluated as that agent works his or her way through the landscape. Boolean networks can provide well-developed concepts and methods to increase our knowledge and understanding of the dynamics of natural intelligence systems.

[illegible] Boolean [illegible] and [illegible]

7 Mapping Natural Intelligence to Machine Space

A recurring theme throughout has been that classical science approaches to theory of intelligence are not at all adequate to address the full scope and depth of natural intelligence actually observed in human and animal experience. The classical approach underlies the single-capacity theory which holds that natural intelligence of both humans and animals reflects the unfolding of genetic structures in the brain already determined at birth. In part, the failure of these approaches is due to methods and concepts that are overwhelmingly discrete, serial, top-down and linear.

Likewise, the classical approach to *artificial* intelligence as well is premised upon a discrete, serial, top-down view of intelligence as symbol-based, rule-governed (algorithmic) *knowledge that*. And in spite of an enormous increase in our understanding of biological systems, brought about in part by growth in the neurosciences, these basic structural approaches to artificial intelligence continue. In general, this is a continuation of the fundamental conceptual and methodological split between mechanistic and organismic approaches to intelligence.

The usual Artificial Intelligence (AI) approach is a serial processing strategy, with problems defined over the natural numbers, integers, rationals, or domains encodable in the integers, requiring a great deal of elaborate programming and prior *knowledge that* (QN) engineering.

It is an approach built upon a centralized control structure with access to large sets of predefined data structures, operating with algorithms defined by mathematical formulas and discrete procedures. Similar to their counterparts in the IQ industry, it is a program premised upon and still seeking to build intelligence systems that bear little resemblance to actual intelligence found in the natural world.

This chapter will address certain facets of this problem in light of the broader theory of natural intelligence we generally laid out in earlier chapters. We will explore a host of other problems as well, comparing the classical artificial intelligence approach to more recent neural network (connectionist) approaches. Though recent years have witnessed burgeoning interest in and development of more biologically-inspired architectures, certain of the same problems identified earlier persist. Many of these problems rest on an inadequate theory of natural intelligence.

7.1 Classical Architectures for Natural Intelligence

In Albus (1981, 1991), "intelligence" is defined as "that which produces successful behavior," resulting from natural selection. More specifically, it is "the ability of a system to act appropriately in an uncertain environment, where appropriate action is that which increases the probability of success, and success is the achievement of behavioral subgoals that support the system's ultimate goal".

Albus' theory is intended to be a theory of intelligence in general, not limited to that of lower animals. His definition is one that is largely shared in the artificial intelligence research and development community. Moreover, the natural or artificial system's ultimate goal and criteria of success are defined externally to the intelligent system in the following way:

"For an intelligent machine system, the goals and success criteria are typically defined by designers, programmers, and operators. For intelligent biological creatures, the ultimate goal is gene propagation, and success criteria are defined by the processes of natural selection" (Albus 1991).

One would hardly dispute that the goals and success criteria for intelligent machines are defined by designers, programmers, and operators. With respect to intelligent biological systems, on the other hand, Albus' characterization of the goals and success criteria are more appropriately a characterization of the intelligence of infra-animals rather than humans.

Similar to the single-capacity intelligence theories (of human intelligence), it is a genetically driven, top-down model that fails to capture the broader scope of intelligence actually found in animal and human experience.

Indeed, it may be distorting the concept "intelligence" altogether to collapse it to what are normally regarded as instinctual or involuntary activities such as gene propagation and to focus exclusively upon what is given with genetic inheritance. Roaches and rats are examples of successful gene propagation which readily come to mind. They are clearly examples of successful natural selection where the goal is gene propagation.

One might assume that learning, where this is not limited to a change in behavior which persists, but in the sense of *coming to know,* might be an important facet of specifically human intelligence.

But for Albus this is not the case. We learn that in his definition of "intelligence", learning in the sense of *coming to know* is not required at all. However, learning may be required only to become more intelligent as a result of experience. Thus he defines "learning" as consolidating short-term memory into long-term memory, and exhibiting altered behavior because of what was remembered (Albus 1991, p. 474).

In other words, "learning" as at least one major mechanism of system dynamics is limited to infra-animal learning and is largely reactive. Any kind of envisionment and conceptualization in light of knowledge of the past and consideration of future possibilities, are not included in learning. *Coming to know* is not considered at all.

7.1.1 Learning, Knowledge, Knowing and Intelligence

According to Albus, intelligence is solely a product of and a mechanism for biological advantage. It is a product and mechanism for generating biologically advantageous (gene propagating) behavior. On such a view, hence, knowledge, knowing, belief, and learning (as well as coming to know) will be justified only to the extent that they have survival value in the process of genetic natural selection. And we will see that in general that is precisely the view Albus takes in his design of an architecture for intelligent systems in general.

Of course, this approach to intelligence is not new having been tried by or is implicit in the work of certain philosophers, for example, Quine (1969). It is also explicitly apparent or implicit in the formal learning [inductive] theories found in Holland (1975; Holland et al. 1986) and other theorists of *computational* learning. Thus, before proceeding with Albus, we might review certain pivotal assumptions and arguments of such a position as it has been set forth elsewhere.

We saw earlier that in Quine's naturalist theory that conditioned or stimulatory response is held to be a determinant of claims to know. Quine's position on theory of know*ledge* is parallel to Albus' view, holding that knowledge, mind and meaning are to be explained and modeled solely in terms of single causal chains of natural science.

In Quine's case, that means that knowledge, mind and meaning are reduced to sensory stimulation and linguistic reports about those sensory stimulations. Knowledge is made up of linguistic reports, verbal or written, of our sensory observations, and analytical hypotheses. Observation reports or sentences are clearly the most important, according to Quine, since they are what we learn to understand first. *Knowing how* is not considered at all, and neither is *immediate* awareness. Nonlinguistic knowing of any kind is excluded altogether.

Moreover, the only method of discovery or learning allowed for by Quine and also Albus (when he speaks of knowing) is inductive inference. And as earlier noted of Quine's concepts of learning and induction, found also in Albus' theory, they are concepts fraught with logical inconsistencies and confusion. Just as we saw earlier in Quine's notion of learning, Albus has confused two entirely different senses of the concept "induction" which must be distinguished from one another.

Of course, as earlier noted, there is a valid sense of induction as statistical form of argument in which an inference is made from some members of a class of things to the whole class. There is also a mistaken notion of induction *as process*. This is a sense of induction which is taken to be a process of reasoning in which one derives "theory" or an hypothesis from (sense) data.

Often, one finds the notion of derivation interpreted as the process of "abstraction," which essentially begs the question. Processes of abstraction assume the principle they seek to establish. As earlier noted, this spurious sense of induction is traceable back to Francis Bacon (Anderson 1960), who presented induction as a way of discovering truth. This mistaken notion is built into Albus' proposed architecture of intelligence.

Keeping in mind certain of the basic arguments for and earlier criticisms of a theory of knowing reduced to single causal chains of natural science, and the explanation of knowledge, knowing, belief, and learning (as well as *coming to know*) in terms of natural selection, we should look closely at an architecture of intelligent systems based on such arguments combined with a spurious sense of induction.

For Albus there are four major system elements of intelligence: *sensory processing, world modeling, behavior generation,* and *value judgment.* Inputs to and outputs from intelligent systems are by means of sensors and actuators. Machine actuators are motors, pistons, valves, solenoids, and transducers; natural actuators are muscles and glands. Actuators move, exert forces, position arms, legs, hands, and eyes. They generate forces to point sensors, excite transducers, move manipulators, handle tools, and the like.

Sensors provide input to a sensory processing system. Such input may be provided by sensors which include visual brightness, color sensors, tactile, force, torque, position detectors, velocity, vibration, acoustic, range, smell, taste, pressure, temperature measuring devices. We will largely focus upon Albus' sensory processing system.

Albus' notion of the sensory processing system is based on quantification principles cited above, and entails the same problems we saw with Quine's arguments for sensory stimulations as the basis for knowledge, (observation) sentences and meaning. For example, Albus states:

> "Sensory processing algorithms integrate similarities and differences between observations and expectations over time and space so as to detect events and recognize features, objects, and relationships in the world" (1991, p. 476).

Thus sensory processing in Albus' theory depends on an equivalent of Quine's mysterious appeal to an "inborn propensity to find one stimulation qualitatively more akin to a second than to a third." That is, it depends upon an "inborn sense of similarity," which in turn depends upon recognition of patterns.

The equivalent to this in Albus' theory is an "internal world model," which generates "expectations" with which the sensory observations are compared in a sensory processing (pattern recognizing) system element.

But problems with all this in both Quine's and Albus' theories are immediately evident. Recognition of patterns *already entails* a notion of sense of similarity. Moreover, a sense of similarity entails recognition of patterns. In other words, both Quine's and Albus' explanation of sensory processing begs the very question or problem at hand.

Moreover, their accounts of sensory processing in turn depend upon what we have earlier argued is the spurious sense of induction. That is, in a human system, generalizations or expectations are reached by a process of purported reasoning from discrete sensory data, based on the equally spurious "inborn propensity" or "inborn sense of similarity."

In a computer, these same generalizations and expectations are held to be reached by the same purported inductive sensory reasoning process comprised of algorithms which integrate similarities and differences between observations and expectations in order to detect and recognize aspects of reality.

To continue our discussion, we must introduce certain technical concepts and methods which we will use in our analyses.

Vectors, States, and Trajectories

A vector is an ordered set or list of variables. In the design of a pattern "recognizing" intelligent machine, one must describe many variables and characterize many simultaneous multivariant computations. Thus, vector notation is one way to do this. An ordered set or list of variables defines a vector as follows:

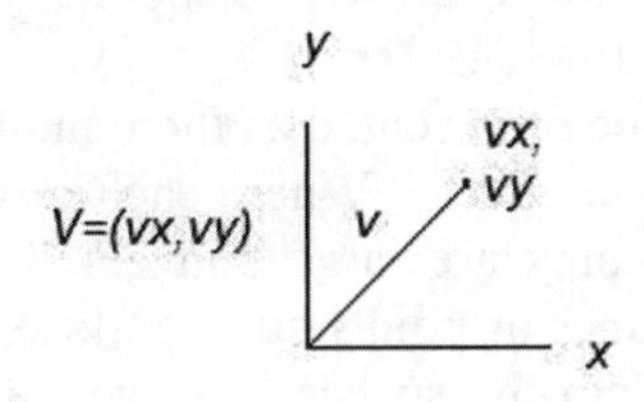

Fig. 7.1. Vector

Components of a vector can be coordinates of a point as in the above graph, (vx, vy) corresponding to the tip of the vector. A vector space is defined as the locus of all pairs of components that can exist. Vectors can have two or more components. A vector with two components is a *surface*; three components define a *volume*; four or more components defines a *hyperspace.*

Vectors can specify a state which is an ordered set of variables. For example, the epistemic state of a person or machine might be characterized by the following state vector: Ep = (ep_1, ep_2, ep_3,), where:

Ep_1 : QN intelligence
Ep_2 : QL intelligence
Ep_3 : PF intelligence

The state vector is in a space consisting of all possible combinations of values of the variables in the ordered set (ep_1, ep_2, ep_3), defining the space S_{ep}. Every point in that space corresponds to a unique epistemic (knowing) condition, and the entire space corresponds to all possible epistemic (knowing) conditions.

In terms of obtaining of a given knower, each variable is time-dependent. Thus we can add one more variable, time (t) to our state vector, Ep = (ep_1, ep_2, ep_3, t). That is, through time, the point defined by Ep will move through four-dimensional space.

Figure 7.2 shows the locus of the point traced by Ep as it defines a trajectory, T_{Ep}. For our purposes, we have added the variable, time (t) to our state vector. Through time, the vector will move along the time axis. Since we take each of the other variables as time-dependent, the trajectory will not be a straight line parallel to the time axis, thus the trajectory will be some curve.

It is important for our purposes in the analysis of intelligent systems and the later discussion that many things (though we will argue not proper names) can be represented as vectors. For example, gestures and motions which "point" can be represented by a trajectory. Pictures can be represented as two-dimensional arrays of points, each with its own brightness and hue (Albus, 1981, 1991).

Each point would then be represented as three numbers corres-ponding to color brightness (red, blue, green). Where each of the numbers is zero, the color is black; where they are large and equal, the color is white. Where there is a large number of points spaced closely together such that the human eye cannot discern the spaces, the eye cannot distinguish that array of a closely spaced large number of points from a real object.

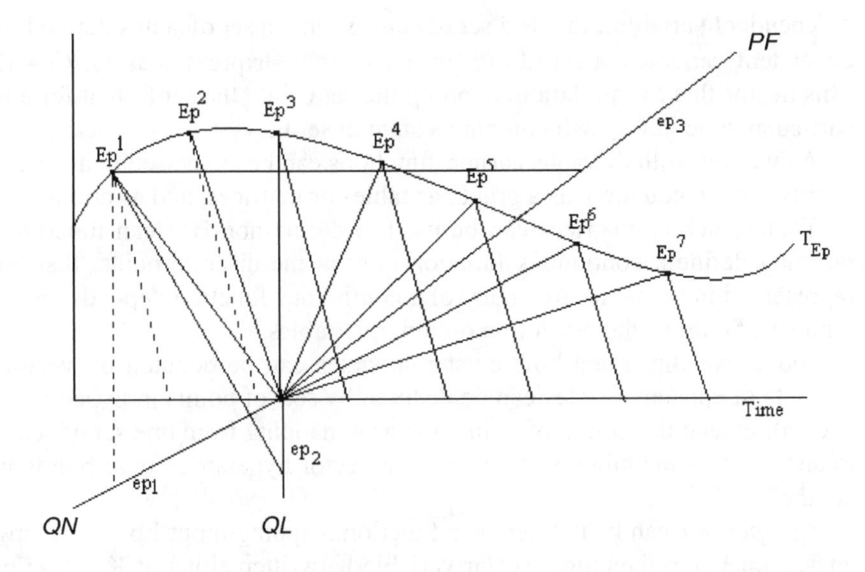

Fig. 7.2. Trajectory of Kinds of Intelligence

Moreover, sounds, musical notes and chords can be represented as vectors, as well as symbols. Any ordered set of binary digits corresponds to components of a binary vector. Significant for the characterization of natural language discourse is the fact that symbols can be represented with continuously variable components.

Thus if the underlying hyperspace is continuous, each point corresponding to some symbol has a neighborhood of points around it which are closer to it than any other symbol's points. Utilizing an example of points in hyperspace corresponding to the symbols *a* and *e*, Albus (1981, pp. 107-108f) discusses how we might view the points in such a neighborhood.

Functions and Operators

Within the context of vectors, a *function* is a mapping of points in one hyperspace onto points in another. In mathematics generally, a function is a relationship between symbols which can sometimes be in a one-to-one correspondence with physical variables [keeping in mind the ambiguity of the term "variable"]. Where the relationship is in a one-to-one correspondence with physical variables, there is sometimes an asymmetry, sense or direction in the relation.

For example, we can map a set of states [that is, a state vector, or all combinations of values of variables in an ordered set] defined by independent variables, that is a set of *causes*, onto a set of states defined by dependent variables, a set of *effects*. This can be expressed as: f: C→ E. This means that f is a relation mapping the set C into the set E, that for any particular state in C, f will compute a state in set E.

As we saw with digraphs earlier, functions can be expressed in a variety of ways: as an equation, as a graph, as tables or matrices, and as circuits.

Tables such as matrices can be used to define non-Boolean functions, but they define a continuous function only at the discrete points that are represented in the table. Accuracy of a continuous function depends on the number of entries, the resolution on input variables.

Above, we illustrated how epistemic states can be denoted by vectors and sets of epistemic states can be denoted by sets of points in hyperspace. We will extend the notion of a function as a mapping from one set of states to another to a mapping of points in one vector hyperspace onto points in another.

An *operator* can be defined as a function mapping input Ep = (ep_1, ep_2, ep_3, . . .ep_n) onto the output scalar variable K, written either as K = H (Ep) or as K = H (ep_1, ep_2, ep_3, . . .ep_n). The functional operator is often indicated by engineers as a circuit or "black box", which exhibits input-output processes.

Where we have a set of operators, h_1, h_2, h_3, . . . h_n operating on an input vector, Ep, as in the following figure below, we then have a mapping H : Ep → K (alternatively, K = H (Ep)) where the operator H = (h_1, h_2, h_3, . . . h_n) maps every input vector Ep into an output vector K. Ep is a vector or point in input space. The information function H is a mapping from input space onto output space.

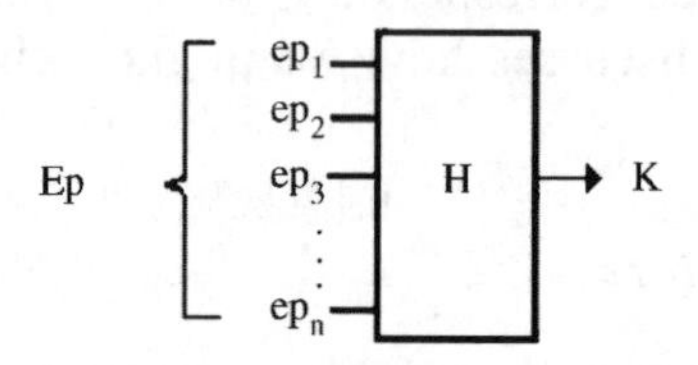

Fig. 7.3. H Function Mapping Input Vector into Output K

In the above Figure, H is a function mapping input vector Ep into scalar variable K. H denotes information functions which we discussed in the last chapter.

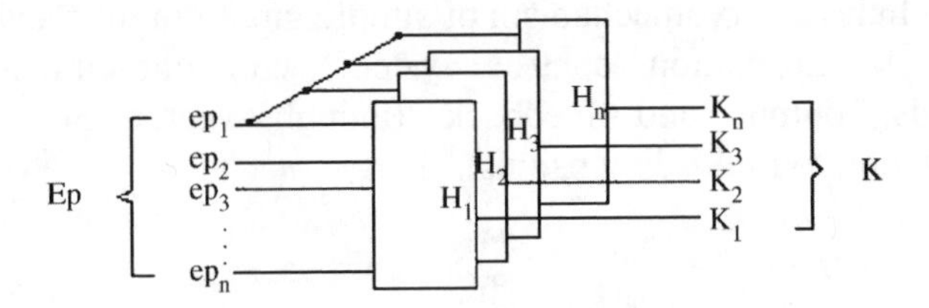

Fig. 7.4. H as Function Mapping Input Vectors Ep into Ouput K

In the above Figure, the set of functions H = (h_1, h_2, h_3, . . . h_n) maps input vector Ep into output vector K. For the sake of simplicity, we will limit our discussion to a single Ep which will map into one and only one K. As we stated above, variables in Ep may be time-dependent, hence Ep will trace a trajectory T_{Ep} through an input space. The H function will map each point Ep on T_{Ep} into a point K on a trajectory T_K in output space.

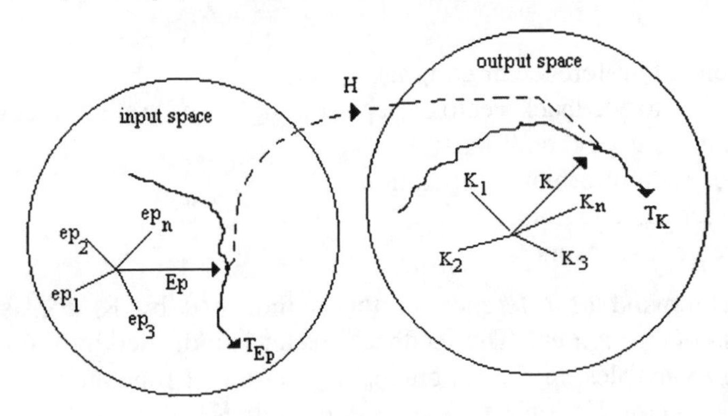

Fig. 7.5. H Maps T_{Ep} Into T_K

In this Figure, H maps every input vector Ep in input space into an output vector K in output space. Thus, H maps T_{Ep} into T_K.

7.1.2 Goal-seeking Intentional Behavior

The proposed structure of goal-seeking and hierarchical control systems on Albus' model is top-down and reactive. It incorporates the usual engineering representations of the structure of control systems for sensory-interactive, goal-directed behavior, where each level of the sensory processing hierarchy includes pattern "recognizers" (classifiers) in the process.

He includes a servomechanism or simple input-output-feedback (simple cybernetic) information control system with one-dimensional input commands, output and feedback (though vector notation permits generalization to multi-dimensions).

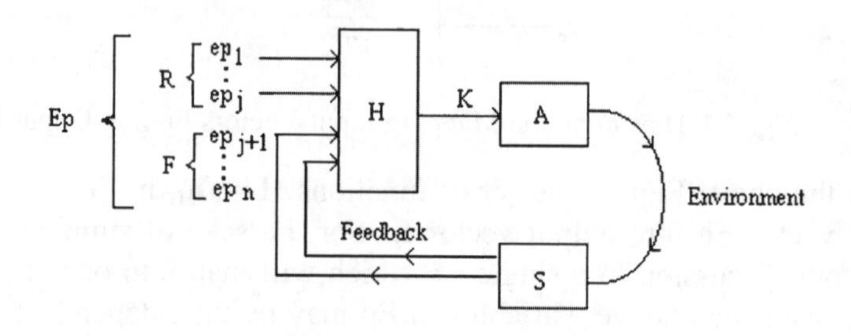

Fig. 7.6. Input-Output Graph

Where:
"R" refers to reference or command vectors;
"F" refers to feedback vectors; "ep_1" through "ep_n" refers to epistemological (intelligence) conditions;
"H" refers to information operations;
"A" refers to actuators;
"S" refers to sensors

The command or reference vector is indicated by R, consisting of variables ep_1 through ep_j. The feedback vector is indicated by F consisting of sensory variables ep_{j+1} through ep_n. Again, the H [operators] computes an output vector K consisting of K_1 through K_k, driving the actuators affecting the environment.

In its usual representation, a command or reference vector *R* establishes a setpoint and over time the feedback vector F varies, thus establishing an *input* trajectory T_{Ep}. The H function or operator computes an output vector K for each input thus producing *output* trajectory T_K.

The *R* [reference or command] vector can be a symbol or string of symbols signifying a goal, task, or plan, and can be viewed as a decomposition into a sequence of subcommands, subreferences, subgoals. Then the output string K_1, K_2, . . .K_n will represent a sequence of subtasks, subgoals or subplans.

In many engineering designs, problems with stability, speed and such are handled with the H functions, provided they are correctly formulated and defined over the space traversed by the input trajectory [Ep input]. If so, then error between the command or reference input [R] and the ouput vector [K] is null and the ouput trajectory T_K will drive the actuators to achieve a goal, perform a task, and so on.

Hierarchical Control

Throughout we have emphasized that *coming to know* and *knowing* are not identical or equivalent to *coming to behave.* That is, knowing and *learning* are not identical nor are they equivalent. Knowing includes knowing *that*, knowing *how*, and immediate awareness. Knowing *that* and what programmers refer to as procedural knowledge are claimed to be embedded in computer programs or wired into their circuitry, though their notion of procedural knowledge (as in Albus' theory) is not identical to *knowing how.*

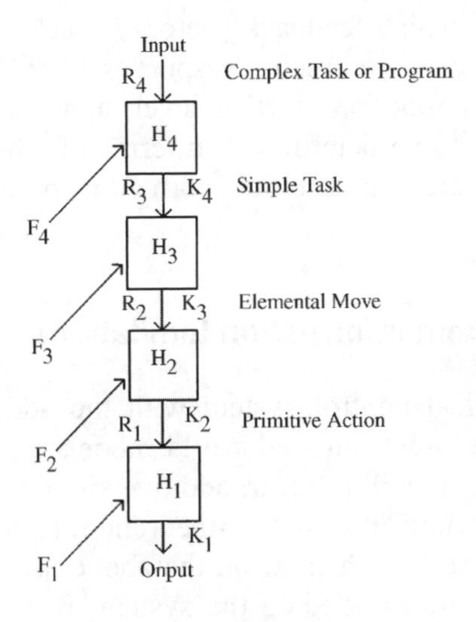

Fig. 7.7. Hierarchical Control System

It is not terribly significant for our purposes whether or not computers can be said to *know* as opposed to behave *as though* they know (for example, meeting some Turing test requirement). I will leave such issues to others.

In Figure 7.7, the highest level input is a symbolic vector denoting a complex task, and there are R vectors at each level of the hierarchy. Information feedback at the highest level includes data identifying a problem or work space and thus decides output vectors K_4 and simple task commands or references R_3 [and their order].

Part of the output K_4 becomes part of the input vector R_3 at the next lower level. R_3 is of course a symbolic vector identifying simple commands or references [with necessary modifications]. This process of information transmission continues at each level of the hierarchy with appropriate changes in the information categories.

Feedback information at F_3, for example, might include identification of position of elements necessary to the performance of a task, as well as state sequencing information from output vector K_3. Feedback information at F_2 might be information from sensors identifying position errors between objects to be grasped, in addition to other sequencing information. The operators H function continuously at each level, producing outputs each instant of time. However, feedback loops at lower levels will obviously be faster than at higher levels.

Response to an environment by such a control mechanism depends in part on the extent to which feedback vectors at each level contain sensory information from that environment. Responses to the feedback vectors in turn depend upon H functions at that level, and success or failure of a performance is held to be determined in terms of whether H functions at each level are sufficient to provide correct mappings to maintain the output.

7.1.3 Control System Information Limitations

The above hierarchical control system with the addition of a complex sensory processing system, internal world models and multiple levels of feedback, is in structure still a linear, additive simple cybernetic model. It is reactive and has serious information-theoretic limitations.

As we noted in the last chapter, on the above usual cybernetic model, *input* is information transmitted to the system. In the above Figure this corresponds to Ep, consisting of R and F vectors fed into the operation H system [space]. *Output* is the information transmitted from the system, corresponding to K vectors in the same Figure. Finally, *feedback*, F vectors is information transmitted from the system through the surroundings and back to the system.

We earlier noted that the simple cybernetic model obviously provides a powerful theoretical perspective for *homeostatic* systems that is *self-preserving operations*, the maintenance of system goals in equilibrium with an environment. It is a theoretical perspective of a reactive, deterministic linear system.

Transmission of information to and from the system [the hierarchy of subsystems within the larger overall system] is directed toward maintenance of specified goals. The major assumption on this model is that information fed back into the system is both necessary and sufficient for the regulation of the system's operations. That is, success or failure of the system's goal-seeking performance is determined by whether or not the H functions provide correct mappings so as to maintain the output trajectory within a range of successful performance in spite of disturbances or other information [uncertainties] in the environment.

However, the above cybernetic model does not provide a perspective for systems characterized by growth, hierarchical differentiation, emergent properties, evolution, and change. In essence, these are systems that have interdependent, transactional relations with that which is outside the system.

This lack of a heterostatic perspective is due in part to the information limitations on the above model. We can see this by looking again at the concepts "input", "output", and "feedback".

In the above model [and amply illustrated in Albus' control system], input is information transmitted to the system. It is what the system takes in. However, information that is *available to* the system from all that is *outside* the system [from the negasystem] and *how that relates to the input* remains unknown. We should perhaps emphasize that the concept "information" here is used in its technical sense as a selection on alternatives, the logarithm [to base 2] of the number of available choices (Shannon and Weaver 1949).

For example, if there are 16 alternative messages among which to choose or decide, then $16 = 2^4$ and $\log_2 16 = 4$, hence the situation is characterized by 4 bits of information. *If we do not know what the alternatives are available to a system for selection from all that is outside the system, then we really do not know how many bits of information actually characterize the situation.* We have an information limitation on the system.

Moreover, on the above model, output is information transmitted from the system. It is information *available from* the system. However, what the negasystem *actually gets* and how that relates to the output remains unknown. Again, we do not know how many bits of information actually characterize the situation. Additionally, feedback on the model relates the output to the input.

There are at least two information measures missing here in addition to the ones cited above: (1) a measure of the selective information from a *negasystem through a system to* a negasystem; and [given (1)], (2) a measure of selective information from a *system through a negasystem to* a system. This reveals tremendous gaps in selective information on this model.

Obviously, an Achilles' heel of the above model is the extent to which variations in an environment are ignored. There is no recognition in the model of information measures sufficient to handle the interdependencies, interactions and transactions with an environment.

In sum, Albus' model of an intelligence system completely misses fundamental necessary conditions of natural intelligence. It completely omits any context-interdependency and sensitivity that are necessary to any natural intelligence system. Moreover, it omits the very goal-seeking and intentionality it was purportedly designed to include, and is instead an essentially reactive system.

7.2 Biologically-Inspired Architectures: VLSI

Top-down, logic based programming such as Albus' model has long been and still is a paradigm for knowledge representation systems. But it is inadequate to address simulations of natural intelligence actually found in human and animal experience. Neural networks offer the best approaches to date, specifically in the form of self-organizing feature maps, though there are major difficulties even with these.

A major difficulty for neural network programs is the fact that most of the current theoretical work done on brain dynamics is still based on the original McCulloch-Pitts (1943) single switch model (Scott 2002; Brooks 1996). In 1996, Brooks emphasized a number of major problems facing artificial intelligence, pinpointing this fact. Along with Hebb's (1949) quantifications of changing synaptic weights, he indicated that "we pretty much have the modern computational model of neurons used by most researchers."

But it is yet unclear whether or not newer models of real neuronal activity have made much of an impact on artificial models of intelligence. An artificial *neural network* is designed to model the way in which the brain performs a particular task or function. The human brain contains around 10 billion neurons. There are many different sorts of neurons, but they all share certain basic properties. The main cell body receives signals from other neurons by means of *dendrites*. The neuron itself builds up a signal inside itself. When it reaches a certain threshold level, it "fires," discharging the signal down its long *axon* and over to other cells through connections at the end of the axon called *synapses*. These are connected to the dendrites of other cells.

Moreover, each neuron has around 10,000 connections on average to other neurons, both incoming by way of dendrites and outgoing by way of axons. The connections can be local with nearby neurons or distant either in adjacent layers of the brain or much farther away.

The neurons in a network are connected together so that the outputs of some feed the inputs of others. The input to the cells arrives in the form of signals down the inputs. In the human brain, the signals take the form of chemical connections between the ends of nodes. The signals build up in the cell and eventually it discharges through the output. We say that the cell fires. Then the cell can start building up signals again.

In the human brain, neurons are connected in extremely complex networks with countless interconnections. An artificial neural net has a much simpler structure, following McCullough-Pitts, when it is simulated on a computer. That model usually has exactly one switch, but will differ in the number of layers and feedback loops.

When brain neuron networks are simulated on a computer, they are designed as adaptive parallel distributed processing machines using massive interconnections of processing units, neurons. They may be structured in several different ways, and the structure or architecture is linked to the learning algorithm characterizing the network's functions Thus, the nature of a neuron in computer control architecture is that of a basic processor in a network of such processors.

Each neuron has a set of connecting links characterized by a weight or strength. An input signal x_i of synapse j connected to neuron k is multiplied by the synaptic weight w_{kj}. (The first subscript refers to the neuron in question and the second subscript refers to the input end of the synapse). Each neuron also has a *summing function* for summing the input signals weighted by the respective synapses of the neuron. Each neuron also has an *activation function* [also referred to as a *squashing* function] which limits the amplitude of the output of the neuron. This function limits the permissible amplitude range of the output signal to some finite value, usually written as the closed unit interval [0,1] or as [-1,1]. An external *threshold function* is also included in the graph. This function has the effect of lowering the net input of the activation function. T

In essence, each neuron sums its inputs with respect to weights, subtracts a threshold and applies an activation function to the result. The activation function, $\varphi(\cdot)$, defines the output of a neuron in terms of the activity level at its input. There are basically three types of activation function, including threshold and piecewise linear, but our concern in part is with *sigmoidal* ("S"-shaped) functions because they more closely characterize actual dynamic behavior of living things.

Some form of nonlinear activation function is necessary in order to try to obtain dynamic, self-organizing mappings central to a mathematical characterization of natural intelligence. It is especially important to realize that the self-organizing patterns of connections among the neurons of these networks is such that they are capable of modification *as a function of experience*. That is, neural networks can learn, and to some degree, they may be said to be able to come to know.

7.2.1 Neuromorphic Architectures

The basic principle behind neuromorphic computational systems is that building such systems true to the representations and architectures used by neurobiology will permit the production of capable, autonomous robots while also suggesting testable hypotheses as to how biological systems accomplish sensory and motor tasks (Higgins 2001).

An assumption underlying this principle is that bottom-up, multi-layered and massively parallel processing found throughout the biological world can be implemented in neuromorphic VLSI (very large scale integration) hardware. Whether or not this is a realistic assumption is an issue I will not address here, though one should note the extraordinary amount of memory and spatial-temporal frequency space required for the computations (Higgins 2001, p. 237; Scott 2002, p. 297). Whatever the case, large numbers of parallel sensors and redundant actuators are necessary.

The neuromorphic (neural network) approach to sensory system programs and architecture is in contrast to Albus' top-down classical approach. Generally, the various neuromorphic architectures consist of the following:

(a) A *single-layer* feedforward neural network is the simplest form of such networks. It consists of an input layer of source nodes or neurons that projects one-way [that is, it feedsforward] onto an output layer of computation nodes. The "single-layer" refers solely to the output layer of neurons.

(b) *Multilayer* feedforward networks, which are the most common, contain one or more hidden layers of computation nodes called hidden neurons or hidden units. The purpose of the hidden units is to intervene between the external input and the network output. By adding one or more hidden layers, a network is enabled to extract higher-order statistics for a more global perspective in spite of its local connectivity with the extra set of synaptic connections and extra dimension of neural interactions.

The outputs of neurons in each layer are connected to the inputs of the neurons in the layer above. The source nodes of the input layer supply elements of the activation pattern which constitute the input signals applied to the computation nodes of the second layer [that is, the first hidden layer]. The output signals of this second layer are then inputs to the third layer, and so on for the remainder of the network. The output signals of the output or final layer are then the overall response of the network to the activation pattern supplied by the source nodes of the initial input layer.

A network or ensemble is said to be *fully connected* if every node in each layer of the network is connected to every other node in the adjacent forward layer. If some of those communication links or synaptic connections are missing, then the network is *partially connected.*

Each connection from one node to another carries a strength which indicates how important the connection is. Strong connections have more influence on the node they connect into than weaker ones. They contribute more to the firing of the cell. The information carried by the network is stored in the differing strengths of the node connections. The strengths between the nodes are called *weights* in the program and are stored as numbers. Neural networks are being implemented in specially created integrated circuits, but most programmers simulate neural networks using software.

The term "feedforward" means that the connections between one layer and the next only run in one direction. There are connections from layer 1 to layer 2, from layer 2 to layer 3 etc. but no connections in the other direction. The opposite of a feedforward net is called a *recurrent* net, which have feedback connections.

(c) *Recurrent* networks have at least one feedback loop. They may consist of a single layer of neurons with each neuron sending its output signal back to the inputs of all the other neurons, with no self-feedback loops, or they may consist of a multilayer system of neurons with one or more feedback loops. Either single- or multilayer networks may also include self-feedback loops. Self-feedback is the output of a neuron fed back to its own input. Moreover, a recurrent network may have a hidden layer or it may not. As explained by Haykin (1994), the presence of feedback loops in recurrent structures has a profound impact on the learning capability and performance of the network. Feedback loops also involve the use of particular branches composed of unit-delay elements which result in nonlinear dynamical behavior.

For our purposes, we should point out that recurrent networks often have attractor states, which we discussed earlier. This means that signals passing through the recurrent net are fed back and changed until they fall into a repeating pattern, which is then stable (*i.e.* it repeats itself indefinitely as it rattles round the loop). The input signals change until they reach one of these attractor states, and then they remain stable. When using recurrent networks, the goal is to train the weights so that the attractor states are the ones that you want.

(d) *Lattice structures* are one- or many-dimensional arrays of neurons with a corresponding set of source nodes supplying input signals to the array. The dimension of the lattice refers to the number of the dimensions of the space in which the graph lies.

Learning Algorithms

An artificial neural network is a massively parallel distributed processor that has a natural capacity for storing experiential learning and making that available for use. It learns or *comes to know* [in a restricted sense of "know"] by means of algorithms which modify the synaptic weights of the network so as to achieve a learning or knowing goal. Generally, it is possible for neural networks to learn in a variety of ways and under certain specifiable conditions.

(a) *Supervised learning.* This occurs when there is an external "teacher", target or desired response the neural network is designed to achieve. The teacher has knowledge of the environment which is represented as a set of input-output examples to the network. The network, however, does not have knowledge of the environment.

Thus, the teacher or whatever makes up the teaching or desired response signal feed provides the network with the desired or target response, the optimum response to be performed by the neural network. Network parameters are then adjusted by both a training signal and error signal, until the network simulates the correct response. This in essence is referred to as the error-correction method of supervised learning. For our purposes, it is fundamentally flawed precisely because the environment is not in the feedback loop of the network. Moreover, without a "teacher" or teaching signal feed, a network cannot learn new strategies for particular or unique situations not covered by a set of defined examples used to train the network.

Other supervised learning algorithms include the Least Mean Square algorithm, which involves a single neuron [thus, it will not be of concern here], and the Back Propagation algorithm, which involves a multilayered interconnection of neurons. With this algorithm, error terms are back-propagated throughout a network layer by layer until it reaches the correct value.

(b) *Reinforcement learning*. This is learning of an input-output mapping by a process of trial and error for the purpose of maximizing a performance index, the reinforcement signal (Haykin, 1994). It may be either non-associative or associative reinforcement learning. In the former, the reinforcement is the only input the network receives from its environment; in the latter, the environment provides information in addition to the reinforcement. With respect to associative reinforcement learning, with many kinds of input from the environment, it is necessary to carefully consider an *evaluation* function on the network, a *critic* function, and a *prediction* function.

A supervised learning system is one largely governed by a set of targets or desired responses. It is an instructive feedback system. On the other hand, a reinforcement learning system is one which is directed to improving performance and learning on the basis of any measure whose values can be supplied to the system. It is an evaluative feedback system (Haykin 1994). In a supervised learning network, an external source [a "teacher"] provides direction to the system. In a reinforcement network, the network has to *probe*, that is *explore*, the environment through trial and error and delayed reward searching for directional information.

(c) *Unsupervised learning*. There is no external teacher or critic in this learning process. That is, there are no examples of the function to be learned by the network. Rather, a task-independent measure of the representation that the network must learn is used and the free parameters of the network are optimized relative to that measure. In effect, the network becomes "tuned" to the statistical regularities of the input data and develops the ability to form internal representations for encoding features of the input, creating new classes automatically (Becker 1991; Haykin 1994).

Self-organizing networks perform unsupervised learning. Thus, with respect to possible computer generated characterizations of natural intelligence, the issue actually comes down to mapping some variety of unsupervised multilayer recurrent neural network topology into a machine.

Self-Organizing Feature Map (SOFM)

Basically, a self-organizing feature map is one in which topographic maps are formed of the input patterns. The neurons are placed at nodes of a lattice and become selectively "tuned" to various input patterns (vectors) in the learning process. Over time, the neurons are supposed to become ordered so that a meaningful coordinate system for different input features is created over the lattice. Thus the spatial locations or coordinates of the neurons in the lattice correspond to features of the input patterns.

Self-organizing feature maps were inspired by the way the human brain is actually organized. It is organized such that different sensory inputs are represented by topologically ordered maps on the brain. The sensory and somatosensory-motor areas of the brain are mapped similar to layered "sheets" onto different areas of the brain.

The human brain is organized relative to these different maps such that we are able to make associations between both spatial and temporal information that is streaming simultaneously, on multiple levels, to our sensory and somatosensory-motor systems. Our brains know how to reinforce connections between things that often appear together in those data streams, which make it possible for us to be aware of our environment and to form higher order combinations of things. This in turn permits us to make sense of our experience and act intelligently in the world. The human brain is a natural, self-organizing feature map, building topographic maps onto itself.

Thus artificial self-organizing feature maps used in computers are designed specifically to be like the human brain in that respect. They are designed to be like natural brain "computational maps," sharing the same (or almost the same) functional properties.

The computational maps are performed in Kohonen's Self-Organizing Feature Map (SOFM) by parallel processing arrays that can handle large amounts of information very quickly. They can rapidly sort and process complex input and represent the results in a simple and systematic form.

7.2.2 The Problem of "Brittleness"

In a thought experiment using a version of the Cocktail Party Problem (2003), I sought to determine whether and how any neural network architecture could handle hierarchical streams of data and the formation of higher categories of things. In general, I wanted to determine whether or not and to what extent any of them could handle relatively *ordinary* kinds of natural intelligence demonstrated in human experience virtually everyday. Because of the unsupervised nature of the natural intelligence disclosed in the Cocktail Party Problem, all but the Self-Organizing Feature Map were ruled out.

In general, though biological systems such as us are remarkably adaptive in most environments, including highly cluttered and noisy ones, the most advanced computer architectures available are "brittle" and virtually unable to adapt at all.

Some history: In 1953, Colin Cherry conducted perception experiments at MIT in which subjects were asked to listen to two different messages from speakers at the same time and try to separate them. His work revealed that humans have the ability to separate sounds from a background. He hypothesized and demonstrated that the human ability to do this is based in part on the characteristics of the sounds such as gender of the speaker, direction from which sound is coming, pitch of voice and speaking speed. Human beings have an ability to make objects of interest pop out from a cluttered, noisy background.

Since his pivotal findings, a number of clinical studies have been conducted that are essentially variations on Cherry's original experiment. To date, given the multi-modal nature of the phenomenon, among many other things, the research community still does not have a solution to the Cocktail Party Problem. That problem might be stated as follows: How are human beings able to make objects of interest pop out from a cluttered, noisy background?

The Party

My own version of the Cocktail Party Problem (2003) differed somewhat from the usual presentation. Moreover, my assessment of the problem from a natural intelligence perspective also differed from the usual assessments. Those tend to focus solely upon sensory mechanics from a biological perspective, leaving out the phenomenological first-person perspective entirely, apparently on the assumption that the phenomenological perspective of the person has nothing to do with the person's ability. My version of the Cocktail Party Problem went as follows:

Think of the last time you were at a party or gathering of some kind, or just in a crowded room with, say, five or more people. Everyone is milling around the room simultaneously in conversation on various topics, at various levels of interest (some dull; some highly animated and loud), at various decibel levels, and with music playing in the background as well. You may have been having a conversation with someone, but during your own conversation with the person in front of you, you nonetheless could overhear conversation of someone else in another part of the room. But you weren't paying any real attention to that other person and what they were saying or to whom they were saying it. Again, you really didn't pay any attention to them and what they were saying because you were engaged in your own conversation with someone standing right in front of you.

During your own conversation, however, midst all the other background sounds and conversational noise, the person you overheard in the background *stopped* talking. They stopped talking long enough for you to then take notice that they had stopped talking. You noticed because you no longer heard their voice in the midst of all the other noise and conversation, coming from the direction of the room where you heard their voice, even in the midst of your own conversation with someone else standing right in front of you.

It is the *absence* of that other person's voice that causes you to lean forward and possibly turn your head in the direction of the room where you last heard them speak. You were trying to determine if they were still there and if they were still speaking, but possibly at a lower level than you could hear. You were trying to *track a single voice* in the midst of all the other sounds and noise, listening for more of the same conversation from that same voice, and in the midst of all the other noise, you knew what to expect to hear. You were also trying to explain why you didn't hear them anymore.

These details of the Party demonstrate relatively ordinary abilities of human beings. We *know how* to isolate and excise objects from their background on the basis of many things, including the *timing* of the occurrences of things. We know how to form expectations about what we should look for and what to anticipate or recognize next, without actually attending to or paying attention at all to what we are doing, or to even be aware that we are doing it. We ordinarily do not even *know that* we are doing this.

On lower levels of awareness, immediate awareness, that we do not ordinarily notice or even recognize, we make associations between both spatially and temporally contiguous information that is streaming on multiple levels, to our sensory and somatosensory-motor systems. Our brains know how to reinforce those connections between things that often appear together in its multiple, simultaneously occurring data streams that permit us to be aware of and recognize objects in our environment and in us.

From all those streams of data on many levels, and in many complex interrelations, our brains form higher order combinations of synchronized features that permit us to make sense of our experience and act intelligently in the world.

The massive numbers of highly complex and interrelated neural networks in our brains process whole objects, such as a human face, knowing it is composed of eyes, nose, mouth and so forth, and that they nearly always appear together. Our brains are aware of and recognize higher-level objects, in part because of the similarities of their parts, to form an ascending hierarchy of related objects. The neural networks in our brains reinforce the interconnections among the parts of things, thus increasing our ability to detect and segment images into objects.

The Cocktail Party Problem was intended to demonstrate that our real-world, real-time environment is very "noisy" and cluttered, yet, assuming a healthy organism, human beings *know how* to "track" single, unique objects such as voices in all that noise and clutter without much effort and without awareness that we are even doing so.

Moreover, as individuals, we experience the world around us and in us from our own points of view, with limited and sometimes distorted perceptions of what is "out there" or "in us." We view a world in which there is a great deal that is often hidden from us by other objects, by lighting and darkness, and even by our own wishes and fears. Nonetheless, we are often more successful than not in maneuvering our way in the world carrying out tasks that demonstrate our abilities to handle a great deal of uncertainty and lots of unknowns.

Our natural intelligence, operated in part by neural networks in our brains, manages our lives and tasks we perform every day by awareness and attending, on some level(s), to significant portions of occluded objects, thereby verifying or falsifying their presence in real-time.

The programs and architecture of *our* brains perform intelligent tasks every single day. Take any human being, equipped with the finest natural intelligence architecture in the known universe, place that human being in just about any context such as the Party, and that human will be able to do all these things. Yet the architectures of our most sophisticated intelligent machines are unable to do virtually any of this.

As Brooks noted, biological systems can adapt to new environments. They do not always adapt perfectly because obviously many biological systems die in some environments. Nonetheless, such systems are in general usually able to adapt. However, our most sophisticated computer programs, as well as our most sophisticated architectures, cannot adapt. They are "brittle." A program compiled for one architecture often cannot run on another.

Even lower biological organisms than us know how to adapt, sometimes much better than we can in some environments. For example, some insects know how to embed themselves in larger organisms that they then use as hosts providing food and protection so as to permit them to "run themselves" within that host. Brooks asked, "Can we build a program which can install itself and run itself on an unknown architecture?" "How about a program which can probe an unknown architecture from a known machine and reconfigure a version of itself to run on the unknown machine?"

The abilities to do these things are flexible, plastic and adaptive capabilities found within the biological world. Yet our approaches to build such architectures and programs still mimic the top-down, linear models or our neural network models are not sufficiently real-world or robust to permit such capabilities.

An obvious advantage of the SOFM model is that it is the computational maps designed to perform in general the way the brain performs. But it does not handle hierarchical structures very well, in part due to problems with classifying data and forming categories. This is the classical problem of universals all over again. The formation of universals cannot be accomplished solely by associationist mechanisms.

Moreover, it is not successful in extracting or organizing hierarchical categories from streams of sensory data, such as the categories of primitive relations found in immediate awareness.

Again, such streams as found in the Cocktail Party are often highly "noisy," containing a lot of irrelevant information that must be "selected out" or otherwise dealt with in order to simulate kinds of intelligence, kinds of knowing, found there. However, there are good reasons to question whether or not the classical view of noise is the right one. That classical view has been that noise is something to get rid of for the sake of having crisp categories, boundaries and decisions.

It appears that much *prior* categorization is necessary to get the SOFM to operate well, and it certainly does not simulate how humans are able to "track" single voices and be aware of and track the *absence* of a single voice in a noisy crowd. The necessity for prior categorization, which would not be possible to computationally solve the cocktail party problem anyway, is clearly a major deficit with respect to architecture necessary to simulate natural intelligence.

Moreover, any adequate model to map such real time streams of sensory data must have an architecture that memorizes the synchronicity among those various sensory inputs. It must score each new sensory experience for its similarity to all previous such experiences (using appropriate association principles), which the SOFM may be able to do to some extent, but it must form higher order categories based on those.

Such a model must also automatically recognize higher-level, emergent objects by logging similarities of component parts to form a hierarchy of related objects. It must have a working memory that categorizes associations by similarity of objects.

With a face, for instance, it must be able to self-organize a higher-level face object with the component parts. By reinforcing associations between spatially and temporally contiguous information, it must be able to reinforce those connections between things that appear together and often in its data stream. Due to those reinforced connections, it must be able to form higher-order combinations of synchronously occurring features. These must occur from the topology of the network.

But SOFM is not adequate to do much of any of these things. For that very reason, even if other problems did not exist, it cannot simulate actual natural intelligence.

Noise and Uncertainty

As noted above, another challenge sometimes posed to Artificial Intelligence involves recognition that noise is often not a nuisance but an integral part of understanding intelligence. This is comparable to arguments made by Edelman (2004) regarding neural variability. In both the organization of the brain (at all levels) and in all external environments, there is an extraordinary amount of variation which should not be dismissed as noise.

Though most top-down computer models attempt to select out noise and treat signals as unambiguous, the fact is human beings more often than not "make sense" of noise as well as messages. They do this seamlessly and instantly, without much hesitation in most normal circumstances. Even in circumstances that are not normal, human beings usually know how to very quickly sort through it, discern patterns, and either dismiss it or take steps to appropriately handle it. Yet there has been little research on "noise understanding" in intelligence research in general or in artificial intelligence in particular.

In the above Cocktail Party Problem, for example, we demonstrated the human natural intelligence ability to reason in the context of a great deal of uncertainty and noise. Our analysis showed that there is a great deal of incompleteness in the data stream flows to the central actor in our demonstration, yet that person was able to handle the incompleteness and information uncertainty very quickly and appropriately in that context. Though the research community still does not have a solution to the Party Problem, nonetheless we have at least something of a baseline data set for artificial designs of such noise handling intelligence systems.

Realistic environments require reasoning systems capable of identifying and implementing effective actions in the face of incompleteness in knowledge, knowing, and awareness of the world. What are the mechanisms of our natural intelligence which enable us to *know how* to isolate and excise objects from their background on the basis of many things, including the *timing* of the occurrences of things? What are the mechanisms of intelligence that permit us to make a great deal of sense of cluttered, noisy environments?

The Role of Indexicals in Natural Intelligence

Elsewhere (Estep 1993, 1996, 2003) I have written extensively about multiple signs of intelligence, particularly those that appear to escape our ability to construct theory and adequate methods of analysis to understand them. Indexicals are just such signs of intelligence. Indexicals are indicators used either in language or in nonlinguistic behavior to point to objects or ideas, including obligations.

These include indicators occurring as words such as "this", "that", "now", "I", or as gestures, images, and patterns of action. Examples of such indicators in language are "This person needs my help." "I must now keep my promise." In general, theories of intentionality that include concepts of moral autonomy must also include adequate theory of both linguistic and nonlinguistic indexicals.

Linguistic indexicals such as "this", "that", "I" are also referred to as proper names, along with the usual understanding of proper names such as "Mary" or "John." Indexicals exhibit a network and structure of *thought* contents when we use them to point to items of experience *as* we experience them. These are decidedly first-person mechanisms of natural intelligence.

We earlier addressed immediate awareness and the role nonlinguistic indexicals play at that level of intelligence. I also argued (2003) that Artificial Intelligence (AI) and Artificial Life (AL) approaches cannot handle either indexically-functioning words or gestures, images, and patterns of action. That is primarily because indexicals are highly context-interdependent; but it is also because such indexicals cannot be conflated with mathematical functions. This is largely due to the first-person nature of their occurrence and meaning.

Of course, a prevailing view in the AI community (especially the "strong" AI community) is that everything is a computer. One needs solely to map things to sets of numbers and so long as we have appropriate algorithms to act on those numbers, we can, in principle, compute anything. As also earlier noted, one of the ways AI engineers and others try to enable a computer to behave as speakers of a natural language behave is to design a meaning representation language for that natural language. The natural language is then translated into the meaning representation language which in turn is then used by the computer to generate natural language behavior.

I analyzed some of the problems with indexicals by looking at the use of those meaning representation languages to map linguistic indexicals into computers. My assessment showed that AI and AL approaches to linguistic indexicals fail precisely because those approaches conflate grammatical meaning with mathematical functions.

Again, "function" is taken to be a mapping in the mathematical sense as a set of ordered n-tuples. Mathematical functions can be sorted into several categories. For some finite function, we can think of an ordered pair in *propria persona,* such as counting the number of pencils on my desk.

Other finite or infinite functions we cannot think in *propria persona* but as falling under a property or set of properties which we can then think in *propria persona,* for example f: $A \rightarrow B$ where A is the set of natural numbers N and B is the set of all integers Z.

We have a complex property representing the infinite function f, that is we have a rule for finding pairs of arguments and values of the function. But we are not thinking the function in *propria persona.* As pointed out by Castañeda (1989, in theory of language we must distinguish between the following two kinds of proposition: (a) The quadruple of 3 is a number; (b) F: $N \rightarrow \{y: y = 4 \times x$ and $x \in N\}$ and $F(3)$ is in N. Proposition (a) is a simple relational proposition; proposition (b) is a functional proposition. The two are equivalent but not identical. To think (a) is not to think (b).

If cognitive or natural intelligence content and grammatical meaning are mathematical functions, then they are sets of ordered pairs. But this is not what Mary L. thinks or says when he thinks and says "This is Richard's house," (using the indexical "this") nor "This appears to be Richard's house."

There are even greater obstacles for computer programs, and the underlying conflation of grammatical meaning with mathematical function with sentences containing discounted *illusory* meaning such as "That yellow dot over there on that mountain is Richard's house." For analysis of just such a sentence containing discounted illusory meaning and first-person indexicals, see Castañeda (1989) or Estep (2003). Can *any* computer architecture or program handle indexicals?

7.2.3 Problems with Pattern Recognition and Limits of Classification

In many respects, the foundations of engineering design, as well as other systems inquiry into pattern recognition, are permeated with logical confusions, conflations, and distortions of meaning. Much of this is the result of the inordinate influence of mechanism along with short shrift given to organismic (biological) models. In general, many engineering models of intelligence apparently have not tried to devise theory of intelligence based upon actual natural intelligence systems, either human or animal, but instead started with an underlying machine model upon which to build their theory. This includes pattern recognition systems as well. Instead of closely analyzing and studying actual human recognition, then proceeding to devise artificial models based upon that, they started with a machine model while ignoring facts of actual pattern recognition. Even where biological considerations were included, mechanistic models tended to have more influence on design.

In effect, what is often called "pattern recognition" is in fact discrete feature pattern matching *classification*. Though many systems scientists and engineers might not find a problem with this (Zimmerman 1993) our arguments above should give them reason to pause. Logically, the concept of recognition and the concept of classification are neither identical nor equivalent, even though they are often used in computer science as though they are. This has resulted in much theoretical as well as practical confusion in the theory and development of pattern recognition systems.

Rosenblatt's (1958) original proposal of the *perceptron* as a "model for information storage and organization in the brain" was the original pattern recognition system. He conceived the perceptron as a layered structure in which each layer of single-switch model neurons receive inputs from another layer, performing logical operations on them, then passing the results to the next layer. The weights of the neurons are modified enabling the system to classify patterns. With more inputs and weight modifications, the ability of the model neurons to classify patterns increases.

As Scott (2002) notes, these pattern recognizers were very limited. The maximum number of patterns recognized could not be any more than the number of neural elements from which the system is constructed; and translations, rotations and scalings of a learned pattern are perceived as new patterns. Moreover, the perceptron had difficulties detecting patterns within patterns. All these things are quite easily performed by human beings.

Since Rosenblatt, any current source in pattern recognition studies will state that the fundamental problem in pattern recognition is to *name* the pattern. As stated above, the overwhelming response to the pattern recognition problem by the systems science and engineering community in general is to treat it as a feature classification problem.

As we saw earlier, any spatial pattern such as a painting or picture, a symbol or a sound can be represented as a vector, and any visual or auditory sequence or a string of symbols can be represented as the trajectory of a vector through hyperspace. Thus when a pattern is given a name it is held to be recognized, and all patterns referred to by a given name are in the same class. The latter shows clearly that the concept "name" is *class* name. An instance of a name is an instance of a class.

For illustrative purposes, we might briefly and only partially outline the design for a robot-control system which is modeled on the brain, which includes goal-directed behavior and pattern recognition. This will of course include a sensory processing hierarchy which provides sensory feedback into each level of the behavior-generating hierarchy.

Each level of the sensory-processing hierarchy processes an incoming stream of data and recognizing (classifying) patterns in the process. We will assess exactly what is involved in pattern recognition in light of our discussion above. To accomplish our goals, we will draw heavily upon Albus (1981, 1991), Mizrai (1990), and others.

Assume we have a continuous hyperspace with points corresponding to symbols, and each symbol has a neighborhood of points. For the sake of simplicity, we will limit the symbols to two in number, *a* and *e*. The following is an adaptation of a diagram provided in Albus (1981):

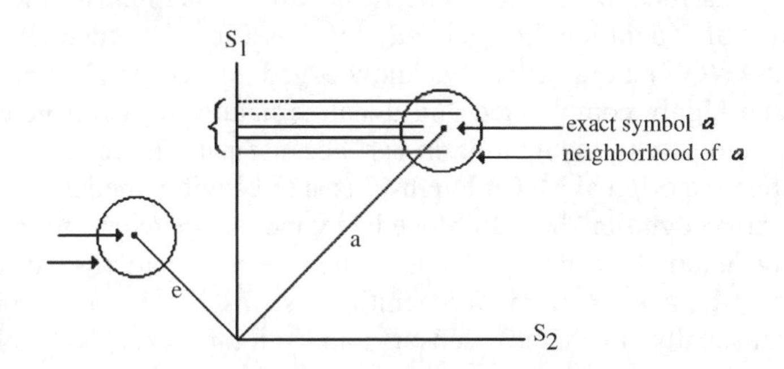

Fig.7.8. Neighborhood of Points

According to Albus (1981, p. 107), we can view the points in the neighborhood in one of two ways: (1) we can view the difference between the neighborhood points and the exact symbol point as deriving from noise on the channel transmitting variables denoting the vector components.

For example, in signal detection theory the detection of a vector within some neighborhood of a symbol vector corresponds to the *recognition* of that symbol against a noisy background; (2) alternatively, we could view the difference between the exact symbol point and neighborhood points as deriving from distortions or variations in the symbol itself.

Albus explains that this makes the best sense if the components of the symbol's vectors are values of attributes or features of the symbol, rather than arbitrary digits. If so, then a neighborhood of points will correspond to a cluster of feature (or attribute) vectors from a symbol set which is not identical, but nearly so.

As an example, the vectors of a printed symbol *e* will vary in each instance of its printing on a piece of paper due to variations in the paper on which they are printed [or variations in the printing apparatus]. And if the feature vectors fall in compact clusters far from the feature vectors of other symbols, the letter *e* will be *recognized* in spite of the fact that no two specimens are exactly alike. He states (1981, p. 108):

"This is a fundamental concept in pattern recognition theory. Hyperspace is partitioned into regions, and the existence of a feature vector in a particular region corresponds to the recognition of a pattern or symbol. By definition, the best set of features is the one that maximizes the separability of pattern vectors. . .it is important to select a set of features that is easily measured and that produces widely separated and compact clusters in feature space."

Again, on this view, the concept "recognition" is reduced to the concept "feature classification", and underlying this reduction is the further reduction of cognition in general to classificatory functions. Only quantitative (QN) intelligence is acknowledged. Moreover, the enormously varied and highly complicated and dynamic nature of neural networks is not built into pattern recognition in part because pattern recognition is still based upon the original McCullough-Pitts single-switch model.

Pattern recognition has developed beyond the perceptron in part by including internal feedback loops which permit patterns to emerge. However, human pattern recognition spans both an immediate (nonlinguistically mediated) sensory and motor sense as well as a linguistically mediated sense. Moreover, the combination of other cognitive, sensory and sensorimotor relations involved in an act of recognition varies with the nature of the object recognized.

Whatever is recognized by a human being is usually immediate. This includes both sensory and motor patterns as well as imagined patterns. Patterned objects may be quantitative (QN); qualitative (QL); or performative (PF). Moreover, patterned objects may not be temporally or spatially present to a subject at all. Examples of these include the very patterns of translations, rotations, and scalings referred to by Scott. These are mathematical patterns that human beings can recognize immediately even if they occur with imagined objects in the mind, not "real" ones.

Additionally, pattern recognition broadly construed must include objects *remembered* as well as mathematical objects (functions) just mentioned as well as (for example) patterns of set theoretical objects. Patterns of objects in sensation are immediately present as are other objects of immediate awareness. These can include immediate objects of attention and even conceptualization, the recognition of ideas.

Any discussion of pattern recognition should also include discussion yet again of the role of linguistic and nonlinguistic indexicals. Patterns of linguistic indexicals such as the use of "this", "that", "now", "I", and others, along with patterns of nonlinguistic indexicals such as gestures and other "pointings" to physical as well as abstract objects and ideas, are woven throughout natural intelligence. These patterns disclose much of the breadth and depth of any natural intelligence system, including the intelligence of human beings. Yet machine pattern recognition to date has little to say about either indexicals or their patterns in intelligence.

Above and elsewhere, I have shown that the *proper name* or indexical relation in acts of recognition is not elliptical for nor a disguised definite description. That is, this relation is not reducible to classification. The *immediate* relation between subject and object actually found in human pattern recognition does not include reflection upon (conscious reasoning about) that object. The object is recognized without any reflection upon its properties or relations.

7.2.4 Kinds of Space: Revisiting the Problem with Universals

Possibly more than anywhere else in his theory of intelligence, it is in his discussion of pattern recognition that two major problems become apparent. One problem is the absence of an account of universals; the other problem is the neglect of perceiver-dependent natural intelligence. In fact, the latter is either completely missing or confused with perceiver-independent action.

To analyze Albus' account of pattern recognition that includes assessment of points, their neighborhoods, and differences between them, we must take into account various kinds of space. Because Albus uses both abstract entities such as points and physical entities (in hyperspace, which cannot be visualized but can be conceptualized) and printed letters (in two-dimensional physical place) to draw the distinctions (1) and (2) above, including reference to properties or attributes of symbols, we must make the following distinctions with respect to the concept "space".

We can divide entities into three classes: (a) those which are not in any place; (b) those entities in one place at one time but never in more than one place; (c) those entities which are in many places at once. Different kinds of space might also include visual, tactile, and physical ("real," Euclidean, or constructed) space, as well as more abstract kinds such as hyperspace.

Russell (1911–1912) illustrates the division with logical relations, human bodies, and general qualities or properties such as colors. Logical relations do not exist anywhere in (physical) space; human bodies can only exist in one place at a time, but general properties or qualities (universals) such as whiteness may be said to be in many places at once.

The latter reference to color must be further clarified. Physicists generally do not include references to secondary qualities such as color in their physical descriptions because they are perceiver-*dependent.*

Above, colors were referenced as numbers corresponding to color brightness. Of course, the numbers represent electromagnetic wave lengths at ängstrom levels. For most human observers, we know how colors are perceived and under what conditions.

But in the above discussion of properties or attributes of the symbols, there is a conflation of talk about perceiver-*independent* properties and perceiver-*dependent* ones. There is a conflation of kinds of space, for example physical or sensible space and hyperspace, and a conflation of *function-argument* (mathematical) talk with *subject-predicate* (natural language) talk. These conflations can and do ultimately lead to distortions and confusions regarding human pattern recognition and natural intelligence in general. Thus, it is best to sort them out as best we can.

Confining ourselves solely to simple physical examples, such as the one given above of a printed character on a piece of paper, we are led to a common sense subject-predicate (natural language) notion of things and their qualities or attributes. In this example, the "thing" is a printed character on a piece of paper that is taken to be a "bundle" of its sensible qualities or properties. Since it is a two-dimensional character printed in ink on a piece of paper, it will have both visual and tactile qualities which are somehow supposed to coexist in one continuous portion of space.

But this common space in which the thing and its qualities are supposed to exist is a scientifically, inferentially derived *constructed* "real" space. The visual and tactile qualities of which an observer is aware *are not in that space.* Those qualities are perceiver-*dependent*, not perceiver-independent. In a sense, the visual and tactile qualities have their own "space."

At this point, we must raise a fundamental conceptual issue regarding Albus' account (2) above of the difference between the exact symbol and its neighborhood. He says that the difference is due to the components (variables) of the symbol itself where these are values of attributes (properties) or features of the symbol.

Thus, a neighborhood of points would correspond to a cluster of feature vectors (from a symbol set) which is not identical, but nearly so. That is, a point [in hyperspace] is taken to represent an exact symbol, (in constructed "real" space) and variables of the symbol are values (numbers represented by points in hyperspace) of attributes (not in constructed real space, but in their own perceiver-dependent space) of the symbol (which is in constructed real space). A neighborhood of points in hyperspace then represents or corresponds to visual, tactile attribute (feature, property) vectors, or list of variables.

The conflation of kinds of space must be sorted out based on a *valid logical, mathematical, empirical, and ontological ground.* Otherwise, inferences based on the above confusions and conflations cannot lead to sound theorizing about pattern recognition.

Costs of Ignoring Phenomenological First-Person Experience

Pattern recognition ultimately refers back to recognition *by someone*. The indexical used to point (either in language or in action) to an object immediately recognized is always in the immediate experience *of someone*.

Patterns immediately present in face recognition are perhaps among the most studied. Computational face recognition is at the forefront of much computer image research. However, existing computer face recognition programs are unable to be deployed in realistic settings, where most face recognition actually takes place, with the normally high variability found in viewing distances, illumination, and sensor "noise."

Yet, all other things being equal, human beings have little trouble immediately recognizing faces under those conditions. It would seem that human recognition capability should provide the model for machine recognition systems. To some extent, there have been efforts to more closely examine those natural intelligence capabilities of human beings when they recognize faces. But to an even greater extent, machine face recognition architectures and algorithms are not based upon the human model.

Sinha et al. (2005) have set forth some findings about actual human face recognition to determine strategies that might be used in the development of machine-based algorithms. A few of those will be mentioned here.

- Human beings can recognize faces in extremely low resolution images. As the authors point out, while the temptation of machine face recognition systems is to increase the amount of detail in face images, high resolution images are not always available. Future research might turn to face recognition with low resolution imagery and more theoretical work done on human recognition to determine how humans are able to recognize faces under extremely low resolution.

- Human ability to tolerate degradations of face images increases with familiarity. Apparently this ability is not the result of some general purpose compensatory processes. Observers' recognition performance with low-quality surveillance video is much better when the individuals pictured are familiar colleagues, rather than those with whom the observers have interacted infrequently.

Moreover, the authors note that this ability is not accrued based on experience in the absence of first-hand familiarity with target individuals; even experienced forensic law enforcement perform poorly unless they are familiar with those target individuals.

The authors pose the following question for research: How does the facial representation and matching strategy used by the visual system change with increasing familiarity, so as to yield greater tolerance to degradations? Clearly, it is with increased first-hand experience *of someone* with an individual, increased familiarity *of someone* with the target individual, which results in greater tolerance to degradation of the target face recognition.

Yet, the authors note, we do not know what aspect of the increased experience with a given individual leads to an increase in the robustness of the encoding permitting face recognition under conditions of degradation. At the very least, the authors propose that researchers draw a distinction between familiar and unfamiliar face recognition.

We might suggest that a thorough analysis and theoretical understanding of immediate awareness intelligence as outlined above and earlier (Estep 2003) would provide the research basis for examining this ability of human beings to tolerate degradations of face images when one is familiar with the person whose face is to be recognized. Immediate awareness is that kind of natural intelligence found within the person and is usually revealed or disclosed in their *knowing how*.

- Facial features are processed holistically. The authors also found that contrary to most machine face recognition algorithms which process features, evidence shows that facial recognition is dependent on "holistic" processes involving an interdependency between featural and configural information. They note that although feature processing is important for facial recognition, their findings suggest that configural processing is at least as important.

Thus conventional machine recognition approaches cannot work (at least by themselves) to simulate natural intelligence face recognition. The actual interdependency of feature and configuration information processing is evidence that conventional and neural network (neuromorphic) programs and architectures are necessary.

7.3 Problems with Complexity

As noted, neuromorphic architectures require large numbers of parallel sensors and like numbers of redundant actuators. This poses complexity problems that are unlikely to go away, especially given the increasing inclusion of computational systems into more devices. This requires more attention to complexity and problems posed by the classical theory.

Classical theories of computation and computational complexity are directed to discrete problems defined over the integers or any numbers that can be effectively encoded into the integers. These theories assume that all underlying spaces are countable hence they cannot address arbitrary sets of real or complex numbers.

From the point of view of the theory of natural intelligence developed above it can be argued that classical discrete computational theories and classical theories of computational complexity are not sufficient to characterize natural intelligence.

Following Blum et al. (1989), we should briefly explore new theoretical attempts to integrate central ideas and concepts from classical discrete theory in a computational approach more appropriate to problems defined over continuous domains. The objective is to briefly provide an overview of the development of theoretical foundations for a theory of computational complexity for such problems defined over continuous domains, for example epistemological problems.

Drawing upon results established elsewhere (Blum et al. 1989), we should look at some comparisons between results over integers with results over the reals and complex numbers. Such comparisons will be instructive by focusing attention on issues regarding basic concepts such as decidability, computability, and complexity.

7.3.1 Decidability

Understanding the notion of decidability/undecidability is intimately connected with the question "What is a computable function?" In general, relative to a universe or domain **U**, a set is decidable if there is some effective procedure for deciding for any given element **u** in **U**, whether or not **u** is in that set. Put differently, in the context of computability, a set is decidable if its characteristic function is a computable function that is if the function defined on **U** has value 1 on the set and 0 off the set.

In the 1930's, the logicians Gödel, Turing, Church, Kleene, *et al.* set forth their own distinct formal notions of effective procedure or algorithm and input-output function on the natural numbers $N = \{0, 1, 2, 3. . .\}$. In each of their efforts, a function *f* from *N* to *N* is defined to be computable if it is the input-output function of a machine (algorithm). A set **S** of natural numbers is decidable if there is some algorithm that, given any natural number *n* will output 1 (yes) if *n* is in **S** and 0 (no) if *n* is not in **S**. Though their formalisms were quite different from one another, the class of computable functions (hence decidable sets) was the same.

However, the resulting classical theory of computation or decidability is a *discrete* theory since the effective procedures are processes over the domain of natural numbers. It can be extended to the integers, the rational numbers, or any domain that can be effectively encoded in *N*.

For example, by gödel coding sentences of a first order language as natural numbers, one can formally ask and answer questions about the set of true sentences of various mathematical sentences.

And Gödel's incompleteness and undecidability theorems showed that, given an arithmetic sentence such as the Goldbach conjecture, there is no effective procedure that will decide whether or not that sentence is a true assertion about the integers.

As Blum (1990) explains, the advent of the digital computer brought an increased interest in the realm of the solvable with the search for efficient algorithms. It was found that a number of problems, for example the Traveling Salesman problem (TSP), while solvable in principle, defied efficient solution.

"Thus, amongst the solvable, there appeared to be yet another rich and natural hierarchy, with the dichotomy of tractability/intractibility, mirroring the earlier dichotomy of decidability/undecidability. Thus, the theory and field of computational complexity was born" (Blum 1990).

That theory of complexity is captured in the NP-completeness results and the basic "P = NP?" question. The latter research has shown the equivalence of thousands of difficult problems from diverse, unrelated areas. Hence, an efficient solution to one can be converted into an efficient solution to another. Some research in the area has been directed to harnessing the complexity of difficult mathematical problems to advantage.

However, the formalisms of classical complexity theory are built upon the models and formalisms of classical computation theory (Blum 1990). Central to this is that complexity [on the classical theory] is measured as a function of word size *L* of a function input or problem instance (where numbers are represented in binary notation and a function input is represented as a finite string of numbers).

The complexity of an algorithm for computing a function, that is solving a problem, is the maximum number of steps (machine operations) for solution over all inputs or problem instances of size L. A function is defined to be *tractable or solvable* in polynomial time if it is computable or solvable by a machine with complexity function bounded above by a polynomial function of L. Otherwise, the function or problem is said to be intractable.

But the classical concepts of computability, decidability, and complexity are all defined over integers or mechanisms encoded into the integers. If the natural intelligence universe is a universe over the reals or complex numbers, then we cannot formulate our decidability/computability and complexity questions about that universe following the classical theory.

There are problems, for example, with the use of the terms "recursive" and "recursively enumerable" as applied to *uncountable* sets. These terms strictly speaking only apply to *countable* sets. Moreover, on the classical theories the construction of a universal machine (algorithm) is based upon an effective encoding of machines by natural numbers. This encoding collapses the sequences of numbers into a single number. Over the integers, this is done with a gödel coding. Such a collapsing destroys the algebraic structure of the underlying spaces.

In the future, we will no doubt see new efforts to redefine measures of complexity, to view a real number as an entity in its own right [rather than as a decimal approximation], and to maintain fundamental mathematical operations, polynomial [or rational] maps and tests as primary rather than reducing all computations to bit operations. With these efforts, the algebraic structure of the underlying space is preserved, that is the algebraic and dynamic structure of the algorithms becomes apparent. These efforts will be shown to provide a more adequate computational approach to epistemic problems over continuous domains.

Computability of Rule-Governed and Rule-Bound Natural Intelligence

Natural intelligence structures (just as physical objects and mathematical structures) can be represented as numbers and symbols in a computer. Programs can manipulate these numbers and other symbols according to algorithms. For physical and mathematical systems, we have scientific and mathematical laws which give us algorithms for determining how systems behave. Thus, when a computer program is run the numbers and symbols are modified by those laws, allowing consequences of the laws to be deduced (or induced; retroduced, depending upon the logical forms governing the modification).

From the point of view of scientific method, it is important to point out that in that sense, as Wolfram (1984) notes, executing a computer program is like performing an experiment, though the objects in a computer experiment follow the laws of the computer program which can be of any consistent form. That is, computation permits experiments of a certain kind in a hypothetical universe. Thus we can arrive at formal characterizations of natural intelligence structures. We can arrive at formal characterizations of natural intelligence.

Natural intelligence generally is a natural learning or coming-to-know phenomenon, though there may be intentional interventions in what is otherwise naturally occurring. Teaching is one such intervention.

Understanding of how humans and animals learn over the past several decades has produced an enormous literature and computational models designed to simulate those processes. Since growth and learning are related, we might briefly review some elementary principles of growth. Thus, we will look to the mathematical bases of conventional models of natural phenomena as bases for growth in intelligence phenomena.

For example, in the simplest case, where we have a system consisting of elements of only one kind, we could have an equation which shows us that the growth of the system is directly proportional to the number of elements present in the system: $\underline{dQ} = a_1Qdt$. Depending upon whether the a_1 is positive or negative, the growth of the system increases or decreases. Q_0 signifies the number of elements at $t = 0$.

This is called the *exponential* law, and it is used to characterize growth in a variety of disciplines such as biological systems. Among other such laws, I consider its use to characterize natural intelligence growth. The exponential law as well as others can be seen to be over-all descriptions or "rules" which govern kinds of natural intelligence behavior.

However, there are interesting problems which arise where the *rule-governedness* of the process of *coming to know* (and reasoning found there) and knowing in general, is correlated with *recursive enumerability,* without taking into account the *rule-boundedness* of certain kinds of dynamic intelligence systems.

That is, where a rule-governed set must have an effective procedure for listing (counting) its members, such a view cannot account for the *rule-bounded* behavior found in the performance of certain kinds of very highly complex cognitive, that is natural intelligence tasks.

In many respects, where all of natural intelligence is correlated with recursive enumerability this correlation amounts to the use of conceptual tools of classical computational theory and formal logical systems to address the nature of and to resolve dynamic, fine-grain intelligence problems. Paraphrasing Mandelbrot (1983) it is similar to attempts to use Euclidean geometry to describe the shape of a cloud, a mountain, or a tree. It cannot work. As he succinctly pointed out, clouds are not spheres, mountains are not cones, and bark is not smooth.

Likewise, the patterns of the natural intelligence universe are irregular and fragmented compared to the precise lines of Euclidean geometry. They exhibit not only a higher degree *but can also exhibit an altogether different level of complexity*.

Possibly the most important facet of this problem to be taken into account is the simple and rather obvious fact that natural intelligence systems *are not complete*. And when they are growing and thriving, kinds of knowing are highly complex dynamical natural intelligence sets.

Recursively Enumerable Natural Intelligence

Rule-governed *knowledge that* (QN) is what we call a recursively enumerable set. That means that it is a set of things we can count. It is a computable set, on the standard digital computer. It is a set of problems or function instances that can be defined over discrete, countable domains.

As such, there are effective algorithms we can use to address those problems. Algorithms are sequential decision procedures to generate an answer to a problem. *Knowledge that* (QN) problems are those requiring "yes" or "no" decisions.

On the other hand, rule-bound know*ing*, which includes *knowing how* (PF), with immediate awareness (QL) embedded within it, requires a different concept or approach to its computability altogether, if it is indeed computable at all. Minimally, *knowing how* requires an approach that generates dynamic self-organizing patterns of interactions among very large numbers of components or elements in the way something is done. That is, a *knowing how* problem requires that we look at the dynamic patterns in the actual doing of something.

Computationally, *knowing how* requires a massively parallel and distributed approach. But any given instance of knowing how to do something may not be entirely computable at all due to immediate awareness embedded within it. Rule-bound knowing behavior may be regular, even predictable behavior to some extent because it is *bound* by a rule.

But "rule-boundedness" simply means that once we input a function[1] into a computer, *we simply have to wait for the computer to show us* how it will map the points onto a real or complex number graph. That is, for example, we have to wait for the computer to show us which points will fall within a fixed circle on that graph, and which ones will fly to infinity. If it is computable, rule-bound knowing would be found in the dynamic

[1] For example, a real or complex number function.

patterns of the components of doing which fall within a fixed circle on a graph.

In a sense, the behavior generated by the computer *is* its own shortest description. It is its own algorithm. In the vocabulary of computation theory, such algorithms are said to be *incompressible*. There are no compressed overall law-like descriptions of the behavior obtained with such algorithms. That is, we do not have a compact overall description or algorithm of what the computer will do *prior* to its generating the behavior (shortest description) or algorithm that it in fact generates of rule-bound knowing.

Such incompressible algorithms may be used to some degree to characterize or simulate rule-bound knowing, but they will fail to do so completely. Rule-bound knowing is made up of those components or elements of a set which include kinds of primitive[2] epistemic [cognitive] relations and their terms constituting *immediate awareness* (QL) in intersection with *knowing how* (PF). But we must beware of mistaking the symbol *of* something for *the thing* symbolized. We must not mistake a representation of something for the something represented. To do so is to fall into a fallacious trap, leading to kinds of fantasy, metaphorical theorizing.

To generate rule-*boundedness* as opposed to rule-*governedness*, we can input arbitrary reals [or complex numbers], iterate certain functions on those numbers, maintaining mathematical operations as primary rather than reducing all computations to bit operations, and simply watch the dynamics of the numbers unfold.

We can use the computer to try to understand the dynamic, self organizing patterns of *knowing how*. But we must not mistake the patterns the computer shows us for the knowing itself. Actual *knowing how* will have immediate awareness embedded within it whereas the simulation will not.

This massively parallel and distributed approach means that it may be possible to characterize and simulate *knowing how* by iterations of mathematical functions, generating discrete as well as continuous dynamical mappings.

But *knowing how* is not rule-governed, where this means we have explicitly formulated the compressible, *overall* algorithm or rule for each step by step procedure of each detail of the knowing behavior to generate

[2] For the sake of readers who may not be familiar with this term in these contexts, the term 'primitive' basically means that "not derived from something else." A primitive object or relation is a basic object or relation that is not based upon anything else. In sound theories, both primitive and defined terms are used. The primitive terms are given meaning through the alternative terms; they are necessary to prevent circularity.

"yes" and "no" answers. "Yes" and "no" answers are responses to a *knowledge that* (QN) problem. They are not answers to a *knowing how* (PF) or immediate awareness (QL) problem.

If we want to know if someone knows how to do something, we ask them, tell them, or otherwise direct them to do it. We then watch to see if the patterns, timing, and context-sensitivity of their doing show that they know how. In a sense, in the computational approach to a *knowing how* problem, the computer takes over and proceeds to show us what it can and will do, what behavior it will generate. And we have to simply wait to see what it does (Wolfram 1984).[3] The shortest way to predict or understand what a *knowing how* system will do is to watch what it does.

This rule-bound (but not rule-governed) behavior was never recognized by the traditional Artificial Intelligence (AI) research program because of its adherence to the top-down, sequential rather than parallel approach to computing. It was also never recognized because classical AI research was (and still is to a large extent) directed to a different set of problems formed as questions requiring "yes" or "no" responses. In essence, classical AI is directed to *knowledge that* problems that are clearly rule-governed.

On the other hand, it may even be the case that the rules of *rule-bound* behavior *are not* formalizable or capable of being made explicit in a sequential algorithm.

Classical computer researchers assumed that if behavior is regular [in some sense] then it is *rule-governed* and is therefore computable on the standard Von Neumann computer. They assumed that all that was needed is enough *knowledge that* engineering and writing enough explicit rules. They assumed *knowing how* is reducible to *knowledge that,* an assumption earlier proved false by Ryle. This explains their emphasis on logic and knowledge-based information processing in classical AI and its subsequent failures with commonsense *know how* and understanding of human beings that Dreyfus recognized.

The futility of such AI efforts is now well recognized and documented, even by early proponents of AI, though there remains a pervasive assumption and misconception that all knowing can somehow be reduced

[3] Wolfram states the distinction in terms of *computational reducibility* and *computational irreducibility*. In computationally *irreducible* systems, general mathematical formulas [algorithms] that describe the overall behavior of such systems are not known and it is possible no such formulas can ever be found. For such systems, we can only turn to explicit simulation of the behavior of that set in a computer. Computationally irreducible systems are not sets of computable problems that can be solved [with "yes" and "no" responses] in a finite time with definite algorithms. Thus, as Wolfram points out, there are questions we can ask about the behavior of such systems that cannot be answered by any finite mathematical or computational process. Such questions are *undecidable.*

to and represented as *knowledge that*. One sees the latter assumption in certain of Penrose's comments (1994) but also in efforts by Lenat (1995) to reduce what he calls "commonsense knowledge" to *knowledge that* statements even though such commonsense knowing entails *knowing how*.

Confusions about the reducibility of *knowing how* (PF) to *knowledge that* (QN) are still pervasive in the artificial intelligence research community. However, the top down sequential approach is not appropriate to a knowing how (PF) problem. These problems are not statable in sentences requiring "yes" or "no" responses. There is no exhaustive list of sentences about performing surgery, running, or firing an M16A2 on target which will represent the immediate awareness and *knowing how* of one who performs surgery, runs, or fires an M16A2 on target. Though there may be elements of commonsense know*ing* which can be so represented, such representations will not exhaust the category of commonsense knowing because there is that commonsense *know how* of which Dreyfus spoke.

The concept "immediate awareness" is a set of primitive relations of *knowing* made evident in what human beings know how to do. It is *nonpropositional* awareness that includes primitive relations of *touching* and *moving* as well as other dynamic primitive relations of *knowing how*. Fully developed, the extended theory of immediate awareness and the theory of natural intelligence in general are within a broader theory of *sign* relations, not limited to alphanumeric *symbolic* relations that continue to dominate current theories of mind and intelligence.

7.4 Summary

The objective of this chapter has been to generally compare and assess the classical and neural network (neuromorphic) approaches to mapping natural intelligence to machine space. The classical approach is obviously built upon a centralized control structure with access to large sets of predefined data structures, operating with algorithms defined by mathematical formulas and discrete procedures. It is a program premised upon and seeking to build intelligence systems that bear little resemblance to actual intelligence found in the natural world.

We examined Albus' (1981, 1991) proposed intelligence system, an essentially classical computer architecture. Though he defined "intelligence" as "that which produces successful behavior," resulting from natural selection, we noted deficiencies of this as it relates to human behavior. Moreover, his proposed structure of goal-seeking and hierarchical control systems is top-down and reactive. It incorporates the usual engineering representations of the structure of control systems for sensory-interactive, goal-directed behavior, where each level of the sensory processing hierarchy includes pattern "recognizers" (classifiers) in the process.

Knowing *that* and what programmers refer to as procedural knowledge are claimed to be embedded in computer programs or wired into their circuitry, though their notion of procedural knowledge (as in Albus' theory) is not identical to *knowing how*. He includes (QN) quantitative intelligence, but omits performative (PF) and immediate awareness (QL) intelligence. His hierarchical control system with the addition of a complex sensory processing system, internal world models and multiple levels of feedback, is in structure still a linear, additive simple cybernetic model. It is reactive and has serious information-theoretic limitations. It does not provide a perspective for systems characterized by growth, hierarchical differentiation, emergent properties, evolution, and change.

This lack of a heterostatic perspective is due in part to the information limitations on the model. We can see this by looking again at his concepts "input", "output", and "feedback". Input is information transmitted to the system. It is what the system takes in. However, information that is *available to* the system from all that is *outside* the system [from the negasystem] and *how that relates to the input* remains unknown. Moreover, on Albus' model, output is information transmitted from the system. It is information *available from* the system. However, what the negasystem *actually gets* and how that relates to the output remains unknown. Again, we do not know how many bits of information actually characterize the situation. Additionally, feedback on the model relates the output to the input.

There are at least two information measures missing here in addition to the ones cited above: (1) a measure of the selective information from a *negasystem through a system to* a negasystem; and (2) a measure of selective information from a *system through a negasystem to* a system. This reveals tremendous gaps in selective information on this model. A clear deficiency of his system is the extent to which variations in an environment are ignored. There is no recognition in the model of information measures sufficient to handle the interdependencies, interactions and transactions with an environment.

Albus' model completely misses fundamental necessary conditions of natural intelligence. It completely omits any dynamic context-interdependency and sensitivity that are necessary to any natural intelligence system. Moreover, it omits the very goal-seeking and intentionality it was purportedly designed to include, and is instead an essentially reactive system.

On the other hand, artificial *neural networks* are supposed to be designed to model the way the brain performs a particular task or function. The basic principle behind neuromorphic computational systems is that building such systems true to the representations and architectures used by neurobiology will permit the production of capable, autonomous robots while also suggesting testable hypotheses as to how biological systems accomplish sensory and motor tasks.

However, a major difficulty for neural network programs is the fact that most of the current theoretical work done on brain dynamics is still based on the original McCulloch-Pitts (1943) single switch model. Moreover, it is yet unclear whether or not newer models of real neuronal activity have made much of an impact on artificial models of intelligence. Though biological systems such as us are remarkably adaptive in most environments, including highly cluttered and noisy ones, the most advanced computer architectures available are "brittle" and virtually unable to adapt at all.

I presented analysis of the Cocktail Party so as to demonstrate the brittleness of computer programs in comparison to relatively ordinary intelligent abilities of human beings. The Cocktail Party Problem demonstrates that our real-world, real-time environment is very "noisy" and cluttered, yet, assuming a healthy organism, human beings *know how* to "track" single, unique objects such as voices in all that noise and clutter without much effort and without awareness that we are even doing so.

Realistic environments require reasoning systems capable of identifying and implementing effective actions in the face of incompleteness in knowledge, knowing, and awareness of the world. Yet in spite of burgeoning research and development in the neurosciences, we do not know the mechanisms of our natural intelligence which enable us to *know how* to isolate and excise objects from their background on the basis of many things, including the *timing* of the occurrences of things.

I argued that Artificial Intelligence (AI) and Artificial Life (AL) approaches cannot handle indexically-functioning words or gestures, images, and patterns of action. That is primarily because indexicals are highly context-interdependent; but it is also because such indexicals cannot be conflated with mathematical functions. This is largely due to the first-person nature of their occurrence and meaning.

We also briefly examined findings in computer pattern recognition, noting that it is in fact discrete feature pattern matching *classification*. The concept "recognition" is reduced to the concept "feature classification", and underlying this reduction is the further reduction of cognition in general to classificatory functions. Only quantitative (QN) intelligence is acknowledged. Moreover, the enormously varied and highly complicated and dynamic nature of neural networks is not built into pattern recognition in part because pattern recognition is still based upon the original McCullough-Pitts single-switch model.

We noted that pattern recognition broadly construed must include objects remembered as well as mathematical objects (functions) just mentioned, as well as (for example) patterns of set theoretical objects. Pattern recognition should also include linguistic and nonlinguistic indexicals that are woven throughout natural intelligence. These patterns disclose much of the breadth and depth of any natural intelligence system, including the intelligence of human beings. Yet machine pattern recognition to date has little to say about either indexicals or their patterns in intelligence.

Assessment of current computer recognition programs apparently show that machine face recognition architectures and algorithms are not based upon the human face recognition system. Though human beings can recognize faces in extremely low resolution images, the temptation of machine face recognition systems is to increase the amount of detail in face images. However, high resolution images are not always available. Future research might turn to face recognition with low resolution imagery and more theoretical work done on human recognition to determine how humans are able to recognize faces under extremely low resolution.

Also, human beings have the ability to tolerate degradations of face images if they are familiar with target subjects. Yet, the authors note, we do not know what aspect of the increased experience with a given individual leads to an increase in the robustness of the encoding permitting face recognition under conditions of degradation. At the very least, the authors propose that researchers draw a distinction between familiar and unfamiliar face recognition.

We suggest that a thorough analysis and theoretical understanding of immediate awareness intelligence as outlined above and earlier (Estep 2003) would provide the research basis for examining this ability of human beings. Immediate awareness is that kind of natural intelligence found within the person and is usually revealed or disclosed in their *knowing how*.

Additionally, facial features are processed holistically. Contrary to most machine face recognition algorithms which process features, evidence shows that facial recognition is dependent on "holistic" processes involving an interdependency between featural and configural information. Thus conventional machine recognition approaches cannot work (at least by themselves) to simulate natural intelligence face recognition. The actual interdependency of feature and configuration information processing is evidence that conventional and neural network (neuromorphic) programs and architectures are necessary.

As noted, neuromorphic architectures require large numbers of parallel sensors and like numbers of redundant actuators. This poses complexity problems that are unlikely to go away, especially given the increasing inclusion of computational systems into more devices. This requires more attention to complexity and problems posed by the classical theory.

Complexity problems also have implications for kinds of natural intelligence that I have grouped as *rule-governed* and *rule-bound.* Only rule-governed natural intelligence is recursively enumerable. In many respects, where all of natural intelligence is correlated with recursive enumerability this correlation amounts to the use of conceptual tools of classical computational theory and formal logical systems to address the nature of and to resolve dynamic, fine-grain intelligence problems. Paraphrasing Mandelbrot (1983) it is similar to attempts to use Euclidean geometry to describe the shape of a cloud, a mountain, or a tree. It cannot work. As he succinctly pointed out, clouds are not spheres, mountains are not cones, and bark is not smooth.

Rule-governed *knowledge that* (QN) is what we call a recursively enumerable set. That means that it is a set of things we can count. It is a computable set, on the standard digital computer. It is a set of problems or function instances that can be defined over discrete, countable domains. As such, there are effective algorithms we can use to address those problems. Algorithms are sequential decision procedures to generate an answer to a problem. *Knowledge that* (QN) problems are those requiring "yes" or "no" decisions.

On the other hand, rule-bound know*ing*, which includes *knowing how* (PF) and immediate awareness (QL), requires a different concept or approach to computability altogether, if they are indeed computable at all. Minimally, *knowing how* requires an approach that generates dynamic self-organizing patterns of interactions among very large numbers of components or elements in the way something is done. That is, a *knowing how* problem requires that we look at the dynamic patterns in the actual doing of something. Computationally, *knowing how* requires a massively parallel and distributed approach. But any given instance of knowing how to do something may not be entirely computable at all due to immediate awareness embedded within it.

Confusions about the reducibility of *knowing how* (PF) to *knowledge that* (QN) are still pervasive in the artificial intelligence research community. However, the top down sequential approach is intrinsically not suited to *knowing how* and immediate awareness. Classical computer architectures and programs fail as approaches to mapping natural intelligence to machine space. Neural network models are likewise insufficient.

8 Summary and Conclusions of Self-Organizing Natural Intelligence

We sought to investigate natural intelligence to broadly carve the universe of discourse in terms of actual human and animal experience. We also sought to investigate and critically evaluate current leading scientific theories of intelligence and methods of research, along with standardized tests.

The Western world has inherited a single-capacity, top-down additive view of intelligence limited to human beings, as the unfolding of logico-linguistic abilities genetically determined at birth and largely unmodifiable by education and experience. This view is substantially at odds with the facts of natural intelligence as found in actual human and animal experience and evidence from a variety of disciplines including biology and neuroscience.

We found major problems with this theory that is supported by classical methods of behavioral science. Since the late nineteenth century, psychologists have analyzed human behavior in terms of what are called theoretical "constructs" and single causal chains of certain kinds. In some cases, American psychologists took a theoretical construct meant to explain a small segment of human behavior, reified and expanded it to explain everything in human behavior. 'Intelligence quotient' is one such construct.

Among many other problems, the classical approach to intelligence research is based upon a misleading heritage of inductivism evidenced by an anti-theory, narrow data collection methodology based upon mechanistic models, faulty reductionism, and causal-correlation confusions. Additionally, this approach is characterized by neglect of theory construction and concept formation along with narrowing the domain of inquiry, intelligence, to suit the classical data collection methods at hand.

Results based upon this approach include the single-capacity theory founded upon unexamined assumptions, inadequate concepts, and numerous fallacies. It has resulted in an excessively narrow view of human intelligence generally, limited largely to verbal knowledge verified by verbal test instruments. This reflects an underlying bankrupt theory of knowing based upon an equally narrow view of the scope of cognition.

Even more broadly, the classical approach to intelligence research has supported a view of intelligence that neglects multiple signs and disclosure of kinds of intelligence in a larger domain of human and animal experience.

8.1 A History of Biased Intelligence Space

Historically, theories of intelligence developed based largely upon faulty reasoning, inadequate empirical methods, and obedience to religious and state authority. The very meaning of 'intelligence' and the carving of intelligence space has largely been the result of the inertia of false dichotomies written into unquestioned dogma and tradition instead of the result of objective reason and science.

Questions about intelligence seem to have revolved around "who and what has it," "who and what does not," "who has more of it than anyone else," and, (by virtue of having more of it than anyone else) "who should rule" and "who should be ruled."

In most philosophic theories during the classical and medieval periods, mankind--but more specifically the human male--was regarded as the center of the universe. According to Aristotelian doctrine, the highest levels of intelligence are found only in some human males. This was adopted as dogma by the Church, largely making human males God's centerpiece.

As such, the human male became regarded as the highest point in the hierarchy of all being in the entire universe under God. The human male was regarded as a reflection of God, having been made in His image by virtue of his superior intelligence. His was held to be a superior life, based upon this reflected divinity, based in large part on his assumed superior intelligence.

Theories of intelligence were used to rationalize slavery and war against those considered inferior. Before the period known as the Enlightenment, monarchies headed by kings together with religious leaders used the same theories embedded within so-called sacred texts, written in accordance with Aristotelian doctrine, and legal systems, to make women and others legally dependent upon and in some cases owned by husbands and masters.

Theories of intelligence are not mere academic exercises. Such theories can and do have far-reaching, sometimes unimaginably horrible consequences to the value, quality, and length of life itself.

Over many centuries, intelligence space has been cut on the human bias because until recently lower animals have been largely excluded as not having any intelligence. Showing his obedience to Church authority, Descartes had argued that animals do not have any intelligence because, he argued, they have no souls. More recently, lower animals have been excluded by some intelligence researchers because it is claimed (in spite of growing evidence to the contrary) they do not have recursive language capabilities.

Intelligence space has also over many centuries been cut on the language, knowledge *cum* propositional bias since it is only there, some argue, that we can have true sentences. Due in large part to the enormous influence of nominalism over many centuries, verifiable sentences became the gold standard used to determine whether or not someone is intelligent. Hence, intelligence space has been cut on the public language and verificationist bias ruling out references to internal states such as kinds of sensation, imagery, and sensorimotor awareness, as well as any *knowing how* to do anything at all evidenced in physical practices.

Concurrently, it was also cut on the paper and pencil test bias, excluding those parts of intentional, purposeful behavior that are not amenable to existing (verbal) data collection instruments.

Additionally, intelligence space has been cut on the brain bias excluding organic life that has no brain or spinal cord. As astonishing as it may seem to some readers, there may be very good evidential reasons to expand our understanding of intelligence beyond animals with brains to include living things that do not.

More specifically, intelligence space has been cut on the language-centers-in-the-cerebral-cortex bias, excluding other parts of the brain that we have known for some time play a large part in intelligent activity and practices.

In spite of mounting evidence to the contrary, many still believe that mathematics (often thought associated solely with the highest levels of verbal intelligence) is the result of language ability, that it is made possible by those language centers in the brain. Yet we know that mathematical thought and doing cannot be reduced to language ability, it cannot be reduced to activity in the Broca and Wernicke areas of the brain. Evidence even shows that mathematical thought and doing are often *at odds* with language ability.

Moreover, intelligence space has also for centuries been cut on gender and racial biases that still hold sway in spite of all evidence to the contrary. In some respects, reasoning on these matters may as well be occurring around 384-323 B.C., alongside Aristotle. These biases are often based on spurious economic and other arguments by many intelligence research professionals who ought to know better. At the very least, they should be willing to reexamine the mountain of unquestioned and uncritically accepted assumptions upon which their arguments are based, along with the fallacies liberally found throughout.

The domain of intelligence generally, but natural intelligence in particular, has been demarcated far too narrowly and too many of the cuts have clearly been biased to serve arbitrary traditions of power or prejudices of the day. Of course not all of these biased cuts of intelligence space are accepted or implicitly assumed by all intelligence researchers. Nonetheless, some or all of the above crooked cuts are still found liberally peppered throughout the research literature and current research strategies in the field.

8.2 Natural Intelligence as Self-Organizing and Emerging

The single-capacity, top-down additive theory of intelligence holds that it is the unfolding of logico-linguistic abilities genetically determined and given at birth and largely unmodifiable by education or experience. Their theory places 'g' for general intelligence, at the very top of the intelligence hierarchy, controlling everything else beneath it. It is largely this theory that provides the basis for public policies related to education in the United States and is the basis for standardized intelligence tests.

However, evidence from empirical research in a variety of disciplines studying human and animal learning and intelligence, as well as actual human and animal experience, show that natural intelligence generally is a kind of self-organizing, emergent nonlinear phenomenon. Contrary to the single-capacity theory based upon classical methods, natural intelligence phenomena cannot be understood by breaking it up into constituent parts at lower levels of development and analyzing these parts in isolation. These "parts" include verbal ability, mathematical reasoning, spatial visualization and memory. According to the theory these parts are analyzed independently in isolation, scores are added together, then according to the standardized test requirements, one's intelligence quotient score is determined.

But actual natural intelligence cannot be understood this way. Properties of intelligence are not found in parts and their analysis in isolation. Indeed, natural intelligence is found as properties emerging from the *interactions between parts* rather than being properties of the parts themselves. Moreover, the single-capacity theory does not even have an adequate theory of the fundamental parts involved in natural intelligence.

Natural intelligence is not a static hierarchical stack of "blocks" with '*g*' at the top of the stack, controlling everything below it. On the contrary, it is one of nature's enormously complex sets of patterns of certain kinds revealed and disclosed sometimes in very complicated, subtle and interesting ways in action and thought by members of the animal kingdom.

There is enormous empirical evidence showing that human intelligence is a self-organizing dynamic phenomenon that emerges from an "ocean of complexity" and is not a *direct* consequence of simplicities of natural laws, such as genetic laws, at levels of biological development. Single-capacity theories have no way to account for growth and emergence of intelligence since it is viewed on the classical model as largely a mechanical consequence of genetic conditions in parts of the brain.

Intelligence emerges at higher structural levels made possible by the interactive activity of sometimes immense numbers of elements across multiple domains. That interactive activity involves large numbers of structures and relations that combine in highly complex ways at lower levels, progressively leading to higher ones, still interacting and combining all along the way in even more highly complex ways.

The model for natural intelligence must be an organismic model of a living thing. It must be organizational, not isolated parts or "building blocks" that get added together by a controlling central processor as the top block on the stack. The interaction-based properties are not there *in the parts* when they are studied independently but are in the interactions themselves.

Moreover, we fundamentally reject the genetic determinism of current theories of intelligence. The DNA of intelligent beings such as ourselves and our nearest relatives in the animal kingdom, chimpanzees, cannot be studied independently to find our intelligence. Our DNA is not us. Our genes are not us nor are they our intelligence.

In contrast to the biological determinism of the single-capacity theory, the view developed here is that a theory of natural intelligence cannot be reduced to the biology of the brain. It must look more broadly at the context and experience within which intelligence emerges. There are at least two major categories of phenomena to which a science of intelligence (where we are specifically addressing human intelligence) must attend. One category includes the brain and its context; the other category includes the person and the person's experience in the objective world. The former major category is solely from an objective science point of view; the latter category is addressed with both an objective science point of view as well as the phenomenal view of the person in relation to an object.

Moreover, the latter category requires viewing the phenomenal experience of the person within a set or matrix of relations obtaining between Subject (S) and Object(s) (O). Not all objects within such a set will be physical; many will be artifacts, including ideas. For example, the accumulated knowledge of a discipline may constitute an artifact which becomes an object in the relation with a Subject. As such, these require a different level and kind of inquiry than object-level science as performed in a clinical laboratory.

8.2.1 Multidimensional and Multilayered Intelligence

Given the enormous empirical evidence of the self-organizing and emergent view of natural intelligence, we approached the intelligence domain as multidimensional and multi-layered. We demonstrated that the scope of intelligence is extended far beyond the single capacity *g*-theory to reflect the facts of natural intelligence found in human and animal experience and in enormous evidence from research studies in a variety of disciplines.

An extended domain of intelligence includes many kinds of knowing, intentional rational behavior. These kinds of knowing extend beyond the traditional verbal language-based propositional knowing *that* (QN) and beyond applied knowing *that* in action or behavior.

For decades, it has been both logically and empirically demonstrated that knowing how (PF) to do something is not reducible to language-based propositional *knowledge that.* Though recognized by Classical Greek philosophers, the nature of *praxis* as a rational activity was virtually eliminated as a kind of intelligence, particularly after the Church-approved mind-body split by Descartes.

Contrary to the Cartesian split, *knowing how* (PF) is a different kind of intelligence from *knowledge that* (QN) that must be included in any exhaustive classification of natural intelligence.

Moreover, empirical studies of the nature of *cognitive* immediate awareness (QL), even under surgical anesthesia, and its integration in the sensory, somatosensory, and sensorimotor systems related to intentional moving and touching (among other kinds of behavior) demonstrates the need to include immediate awareness as a natural intelligence category as well. Moreover, because of the part immediate awareness plays in making *knowing how* possible, there is a necessity for far more research to more accurately determine the parameters of this category of knowing.

8.2.2 Three Major Kinds of Natural Intelligence

These three major categories of knowing, knowledge that (QN) or "quantitative knowing), knowing how (PF or "performative knowing"), and immediate awareness (QL or "qualitative knowing) comprise the major categories of the universe of natural intelligence. Because of the dynamical nature of these extended kinds of knowing, I argued for an appropriate and comparable extension of theoretical and empirical methods including mathematical models for defining the problems of that domain.

8.3 Nonlinear Methods for a Science of Intelligence

For reasons cited above and more we explored throughout, nonlinear methods and models are superior to classical methods and models in intelligence research. Other sciences across the board from physics to most of the life sciences have since moved far beyond the classical methods still used in intelligence research by many in the behavioral sciences.

There are good reasons for intelligence research to likewise move into the future. These include at least the following:

1. We demonstrated that nonlinear theory models approach to natural intelligence permits a more complete and exhaustive classification of kinds of knowing; it does not limit the scope of cognition to only one kind or category of knowing, verbal (linguistic) intelligence. In fact, we have shown that there are minimally three major categories of kinds of natural intelligence. These include *knowledge that* or quantitative intelligence (QN), emphasizing its classificatory structures; *immediate awareness* or qualitative intelligence (QL), emphasizing primary and secondary qualities as objects of this kind of intelligence; and *knowing how* or performative intelligence (PF) obviously emphasizing its structures in doing.
2. We also demonstrated how more advanced nonlinear theory models incorporate the strengths of classical statistical and nonstatistical models and methods, permitting characterizations of organized simplicity and unorganized complexity, but goes beyond these to permit characterizations of organized complexity. In effect, these permit configurations of self-organizing, emergent dynamics of natural intelligence. This is a geometric bottom-up approach as opposed to the classical top-down symbolic approach.
3. We also demonstrated that a more advanced nonlinear theory models approach is not reductionist as are classical models. We demonstrated the use of retroduction, a source for generating ideas, concepts, and hypotheses instead of assuming that ideas, concepts, and hypotheses are already made.

The use of nonlinear theory modeling in natural intelligence research can profoundly enlarge our understanding of natural intelligence. It can change the current single or limited multiple capacity view of intelligence as a top-down, knowledge-based rule-governed static phenomenon to one of the dynamic emergence of self-organizing intelligence and growth in possibilities.

Classical approach to intelligence research has from the beginning embraced and continues to embrace underlying theory, assumptions, and methodology that bear little or no demonstrable relationship to the method by which intelligence is actually generated in natural systems. It is an approach which has focused upon outputting intelligent *solutions*, *knowledge that,* rather than intelligent behavior, *knowing*, thus missing most of the domain of intelligence altogether.

A nonlinear scientific theory model of self-organizing complexity is an organismic point of view permitting representations of organized complexity in natural intelligence. This cannot be done with classical methods Living systems of organizing complexity such as natural intelligence require *geometric* theory models that permit analysis of properties and relations of entire ensembles of coupled elements in dynamic interaction with one another and their environment. It is by understanding the structure and active, dynamic behavior of ensembles of coupled epistemic elements that we can understand the emergence of natural intelligence.

Among some of our findings, based upon a nonlinear theory models approach involving the SIGGS theory model, we included the following:

- Human intelligence is sorted from infra-animal intelligence by the fact that it exhibits or discloses itself on a spectrum from a purely *quantitative* intelligence (QN), 'knowledge *that*', to a purely *qualitative* intelligence (QL), 'immediate awareness', with *performative* intelligence (PF), 'knowing how', interwoven throughout. The scope of infra-animal intelligence on that same spectrum is not yet determined, in part due to a regrettable lack of sufficient research.

- The approach to intelligence found in single capacity theories largely rests on principles that define a narrow scope of *quant*itative intelligence (QN), reduced to neural, genetic, or rule-governed linguistic *cum* logico-mathematical structures. As noted, this conceptualization of intelligence is largely defined in terms of the operations of class logic. Their methods of data collection and verification, hence, follow classical science approaches.

- A pivotal distinction between quantitative and qualitative intelligence rests in part on the notion of *individuation* and also on the differences between logical and non-logical indexicals. Since quantitative intelligence operates or proceeds by means of class logic, with a quantification and identity apparatus, it aims at identifying classes, universals or instances, members of classes and universals. It aims at identifying extended objects by means of logical or class operators and logical indexicals.

- Intelligence that operates according to *non*-logical indexicals is ignored, *based on the assumption that the cognitive is isomorphic solely with class (extensional) logic.* I argue that this assumption is false, that it contradicts the facts of actual natural intelligence experience. In sum, I argue that cognition extends beyond class extensional logic to include *immediate awareness* cognition of individuals, the *unique,* and includes performative, *knowing how* structures as well.

- The dynamic view of intelligence I argue for is that bodiliness [physicality] of intelligence is manifested and disclosed in the world in part by the use of indexical signs. It is indexicality that forms the foundations of a theory of multiple signs of intelligence. Those multiple signs are especially found in behavior that is comprised of the conjunction of *knowing how* and immediate awareness.

- But I also do not sanction any reduction of these kinds of cognition to physical (neural) mechanisms or sensation, as found in many materialist views. I support a self-organizing, emerging organismic model for natural intelligence. Where intelligence is reduced to physical mechanisms or sensation, that characterization is too narrow because they cannot account for actual facts of either human or animal intentional behavior. Moreover, as should by now be obvious, I do not support those theories of intelligence that are restricted to *mentalness*, such as linguistic ordering or use; they are also entirely too narrow, ignoring major facets of actual intelligence found in structures of *knowing how* and experience.

There is *nonlinguistic* (nonverbal) intelligence, nonlinguistic knowing, and there are nonlinguistic indexicals by which humans and animals manifest that broad array of natural intelligence. Where indexicality is extended beyond linguistic signs to include nonlinguistic ones, we have a basis for an extended notion of representation of knowing, and for a concept of intelligence as a naturally emergent, self-organizing and adaptive system.

8.4 Some Issues Left Unresolved

Inevitably, there remain a number of unresolved issues. Certain of these are directly related to problems identified in "Mapping Natural Intelligence to Machine Space," but they are found across the entire natural intelligence space. We will touch upon only a few here as we have dealt with each of these in greater detail throughout.

The Problem of Universals

In spite of claims by some neuroscientists, cognitive scientists, and philosophers to the contrary, the problem of universals persists. This most ancient of philosophical questions is at the core of issues surrounding our lack of understanding of mind, intelligence, and consciousness. That problem may be stated as follows: "Why do universal concepts or words have universal or common meaning applied to many things?" Some concepts and words can apply in principle to an *uncountable* infinite number of things. How is this so? More to the point: how do we *know* that this is so? What are the structures of our intelligence that permit our knowing to extend so far?

The problem of universals is not resolved by efforts to collapse concepts into percepts, as found in Edelman's (2003) theory and elsewhere (Lakoff and Núñez 2000; Scott 2002). It is not resolved by spurious biological and language reductionist or analytic arguments intended to define the problem away.

With respect to theories of intelligence, the problem of universals is central to resolving other issues related, for example, to mathematical intelligence, including knowing how to prove theorems or perform transformations on abstract objects.

In my opinion, the problem of universals is directly related to coming to terms with the intelligence of *knowing how.* As a culture we still largely either deny or disparage practical intelligence due to centuries of cumulative prejudice and poor reasoning. Currently, though some standardized intelligence tests claim to measure what they refer to as 'procedural knowledge' it is neither identical nor equivalent to *knowing how*. The underlying assumption on those tests is that *knowing how* is reducible to rule-governed *knowledge that* when it is not; and neither is mathematical intelligence including mathematical knowing how.

The Problem of Indexicals

The intelligence research community has not adequately addressed the larger set of ways intelligence is disclosed or exhibited in the world. Indeed, it barely addresses disclosure at all. It has almost exclusively focused upon verbal intelligence that is amenable to and measurable with the data collection instruments they have at hand.

I have throughout addressed the role that linguistic and nonlinguistic indexicals play in intelligence. Patterns of linguistic indexicals such as the use of 'this', 'that', 'now', 'I', and others, along with patterns of nonlinguistic indexicals such as gestures and other "pointings" to physical as well as abstract objects and ideas, are woven throughout natural intelligence. These patterns disclose much of the breadth and depth of any natural intelligence system. Yet theories of intelligence have little to say about them.

Though I addressed immediate awareness and the role nonlinguistic indexicals play at that level of intelligence, I also argued (2003) that Artificial Intelligence (AI) and Artificial Life (AL) approaches cannot handle either indexically-functioning words or gestures, images, and patterns of action. That is primarily because indexicals are highly context-interdependent; but it is also because such indexicals cannot be conflated with mathematical functions. This is largely due to the first-person nature of their occurrence and meaning.

In order to fully develop a robust theory of natural intelligence, a broad theory of indexicality must also be developed. This must be interwoven with a broad theory of signs that clearly identify the means humans and animals use to exhibit and disclose their intelligence. This will include taking seriously the first-person, phenomenal experience of Subjects with Objects.

The Problem of Awareness

Many fundamental questions remain about the scope and depth of awareness in general, but immediate awareness in particular. The research literature on preattentive and attentive processes shows that there is some dispute on when and where the preattentive processes make the transition to attention. Preattentive orientation proceeds *subconsciously,* interpreted as the absence of "consciousness *that*" such and such is the case, at the level of the nervous system. Many researchers hold that it is only when sensory perception is attained that *attention* can then focus upon information as an object with which it can operate. Some appear to argue that only at the attention phase, interpreted as "consciousness that" or "awareness that" (aligned with language) does a subject's neurophysical activity become cognitive.

However, there is substantial evidence in multiple experiments of cognitive immediate awareness *below* the threshold of attention. We cited a number of experiments showing some deeper level of cognitive awareness, below the attention system threshold, not aligned with language that *correctly* affected subjects' overall behavioral responses.

This evidence shows that the circle of cognition is larger and deeper than previously thought and this should be aligned with the scope of natural intelligence generally. Minimally, the evidence supports our position that intelligence begins with immediate awareness; this is prior to the level of attention and is not aligned with language. This is so not only as it pertains to vision, but also the psychomotor and entire sensory motor parts of the brain.

Thus I argued that not only do we need to revise our understanding of the scope and depth of the cognitive domain and the place where we enter it, we must also revise our understanding of a network of related concepts, minimally including cognition itself, natural intelligence and learning. If the empirical findings and our interpretations of them are correct, natural intelligence begins with cognitive immediate awareness in the preattentive phase.

Additionally, we looked at experiments involving patients under surgical anaesthesia. Researchers investigating cognitive awareness under surgical anaesthesia recommend a broader spectrum of the concept of awareness. The general concept of awareness is not viewed by many researchers in terms of two mutually exclusive states, awareness or unawareness, but is viewed as a continuum of states ranging from unaware through an infinite number of partially aware states, to complete awareness. This continuum also distinguishes between "*awareness that*" such and such is the case [tying awareness to "that" clauses or linguistic reports] and "*immediate awareness*" which is not tied to such reports.

In spite of the growing research into the nature of awareness, however, we still do not have a clear understanding of its scope and depth even under normal conditions. Nor do we have an adequate understanding of all the variables involved under what are admittedly highly variable conditions under the best of circumstances. There is a need to pursue far more research, especially given the interrelatedness of awareness with all other intentional behavior.

The Problem of Autonomy

Moreover, the issue of autonomy, a necessary condition to self-direction and self-organizing behavior at higher levels of natural intelligence, is one of the conditions separating reactive mechanisms from kinds of living intelligent organisms such as ourselves. In principle, it also separates human beings from animals. Yet it is usually noticeably missing in theories of intelligence.

The issue of autonomy is sometimes confused with the issue of intentionality. In philosophic circles, 'intentionality' is often taken to mean that one "knows what one is doing," where the meaning of 'know' is limited to *knowledge that* (Searle 1992, 1995). That means that one can provide a verbal or written description of what one is doing. In my opinion this definition or understanding of intentionality falls into the nominalist trap.

As I have argued and made clear with evidence throughout, much that we know extends far beyond the verbal and is not reducible to it. As noted, I have used 'intentionality' to mean deliberativeness or purposiveness in behavior, not limited to the verbal category of knowing.

However, the concept of 'autonomy' is neither identical nor equivalent to the concept 'intentional.' The usual definition of 'autonomy' is either "self-ruling" or "self-governing;" it is central to the notion of independence and freedom. In Kant's words (1785), however, autonomy is the "giving of law to one's self," specifically the moral law, a categorical imperative.

While I do not wish to enter into arguments about Kant's moral theory, it appears to me that his concept of autonomy as the *giving of law to oneself* is more appropriate to an understanding of human reason and intelligence, specifically the independence of reason, thought, and human freedom.

In contrast, the notion of "self-ruling" does not necessarily include the notion that one is ruled by a law or rule *that one has given to one's self.* It may be a rule that has been externally imposed upon a person, even in an otherwise benign fashion. It could be that someone or some agency taught or otherwise handed the rule to a person who then unquestioningly accepted it and used it to rule oneself. We may rightly ask in that case, however, whether or not that person is autonomous and free. But Kant's notion of "giving law to oneself" is clearly a free and rational independent choice.

Autonomy in Kant's sense is found throughout higher levels of reason; it is not found solely in moral or ethical reasoning (as Kant may have surmised) but is found even in more esoteric areas of reason and thought in mathematics, science, philosophy, and elsewhere.

At present, I do not see many intelligence researchers pursuing theories of autonomy, in Kant's sense, in human intelligence and reason. Ironically, though these questions, along with questions about the nature of intentionality, are found in *artificial* intelligence, I do not believe they have made much recent impact in studies of natural intelligence. At least not since Kant asked them.

References

Acar W (1988) Theory Versus Model Further Comments. Systems Research, 5, pp 172–173

Aceto P, Valente A, Gorgoglione M, Adducci E, De Cosmo G, (2003) Relationship between awareness and middle latency auditory evoked responses during surgical anaesthesia. British Journal of Anaesthesia 90 (5) pp 630–5

Adleman L (1995) A Boom in Plans For DNA Computing. Science 268, pp 498–499

Albus JS (1981) Brains, Behavior, and Robotics. Peterborough, NH, BYTE

Albus JS (1991) Outline for a Theory of Intelligence. IEEE Transactions on Systems, Man and Cybernetics 21 (3) May/June

Allen C, Bekoff MA (1997) Species of Mind: The Philosophy and Biology of Cognitive Ethology. MIT Press, Cambridge

Allgood K, Yorke J (1989) Fractal Basin Boundaries and Chaotic Attractors. Proceedings of Symposia in Applied Mathematics: Chaos and Fractals, 39, American Mathematical Society

Almog J, Perry J, Wettstein H (eds) (1989) Themes From Kaplan. Oxford, New York

American Heritage College Dictionary (1993) Third Edition, Houghton Mifflin, Boston, New York

Anderson AK, Phelps EA (2001) Lesions of the human amygdala impair enhanced perception of emotionally salient events. Nature 411, 17, pp 305–309

Anderson PB, Emmeche C, Finnemann NO, Christiansen PV (2000) Downward Causation: Minds, Bodies and Matter. Aarhus Univ Press, Aarhus Denmark

Anscombe GEM (1959) An Introduction to Wittgenstein's Tractatus. Hutchinson Univerity Library, London

Antognetti P, Milutinovic V (eds) (1991) Neural Networks: Concepts, Applications, and Implementations Volume III. Prentice-Hall, New Jersey

Arbib MA, Hanson AR (eds) (1987) Vision, Brain, and Cooperative Computation. MIT Press, Cambridge

Armstrong DF, Stokoe WC, Wilcox S (1995) Gesture and the nature of language. Cambridge Univ Press, New York

Ashby WR (1960) Design for a Brain. Chapman and Hall, London

Ayres FJr (1952) Theory and Problems of Differential Equations. Schaum Publishing, New York

Baars BJ (1998) A Cognitive Theory of Consciousness. Oxford, New York

Bacon F (1960) The New Organon and Related Writings. Anderson F (ed) Liberal Arts Press, New York

Bak P, Tang C, Wiesenfeld K (1988) Self-Organized Criticality. Phys Rev, A, 38, p 364
Bartlett FC (1958) Thinking. Basic Books, New York
Barwise J, Perry J (1983) Situations and Attitudes. MIT Press, Cambridge.
Baumgartner P, Payr S (1995) Speaking Minds: Interviews with Twenty Eminent Cognitive Scientists. Princeton University Press, New Jersey
Baumgartner T, Burns TR, et al (1976) Open Systems and Multilevel Processes: Implications for Social Research. International Journal of General Systems 3 (1) pp 25–42
Becker S (1991) Unsupervised Learning Procedures for Neural Networks. International Journal of Neural Systems 2, pp 17–33
Begley S (1999) Shaped by life in the womb. Newsweek 13
Bekoff A, Marc A (1997) Species of Mind: The Philosophy and Biology of Cognitive Ethology. MIT Press, Cambridge
Bell DJ (1990) Mathematics of Linear and Nonlinear Systems. Oxford,, New York
Bell ET (1952) Mathematics: Queen and Servant of Science. G Bell & Sons Ltd, London
Benacerraf P, Putnam H (eds) (1964) Philosophy of Mathematics. Prentice-Hall, New Jersey
Bertalanffy L von (1968) General System Theory: Foundations, Development, Applications. Braziller, New York
Berthoz A, Israël I, François PG, Grasso R, Tsuzuku T (1995) Spatial Memory of Body Linear Displacement: What is Being Stored. Science 269 (7), pp 95–98
Binet A (1898) Historique des recherches sur les rapports de l'intelligence avec la grandeur et la forme de la tête. L'Année psychologique 5: 245–298
Binet A (1898) Recherches sur la technique de la mensuration de la tête vivante, plus 4 other memoirs on cephalometry. L'Année psychologique 7: 314–429
Binet A (1909) Les idées modernes sur les enfants. Flammarion, Paris
Binet A, Simon T (1912) A Method of Measuring the Development of the Intelligence of Young Children. Courier Co, Illinois
Bjorkman M, Juslin P, Winman A (1993) Realism of confidence in sensory discrimination: The Underconfidence Phenomenon. Perception & Psychophysics 54, pp 75–81
Black M (1964) A Companion to Wittgenstein's Tractatus. Cornell Univ Press, Ithica
Block N J, Dworkin G (1976) The IQ Controversy. Pantheon, New York
Block N (1995) On a Confusion About a Function of Consciousness. Behavioral and Brain Sciences 18, pp 227–287
Bloom P (1994) Generativity within language and other cognitive domains. Cognition 51, pp 177–189
Blum A (1992) Neural Networks in C++, An Object-Oriented Framework for Building Connectionist Systems. Wiley & Sons, New York
Blum L (1990a) Lectures on a Theory of Computation and Complexity over the Reals (or an Arbitrary Ring). Lectures in the Sciences of Complexity II, Jen (ed) Addison-Wesley, Massachusetts

Blum L (1990b) A Theory of Computation and Complexity Defined over the Real Numbers. Technical Report International Computer Science Inst, Berkeley

Blum L, Shub M, Smale S (1989) On a Theory of Computation and Complexity over the Real Numbers: NP Completeness, Recursive Functions and Universal Machines. Bulletin of the American Mathematical Society 21 (1), pp 1–46

Blum L, Smale S (1990) The Gödel Incompleteness Theorem and Decidability Over a Ring. Technical Report International Computer Science Inst, Berkeley

Boolos G (1995) Gödel *1951: Introductory Note. Feferman et al (eds) Kurt Gödel Collected Works Vol III Unpublished Essays and Lectures. Oxford, New York

Bower TGR (1972) The Visual World of Infants. In: Perception Mechanisms and Models. WH Freeman, San Francisco, pp 349–357

Bradley MC (1969) Comments and Criticism: How Never to Know What You Mean. J Phil LXVI (5) March 13

Braine MDS, O'Brien DP (eds) (1998) Mental Logic. Earlbaum, Mahwah, NJ

Breazeal C, Fitzpatrick P, Scassellati B (2005) An Active Vision System for a Social Robot. Artificial Intelligence Laboratory, MIT Cambridge.

Britannica Online (1994–1997) The History of Epistemology: Ancient Philosophy. Retrieved 1997 from http://www.britannica.com

Britannica Online (1994–1997) Physiological Psychology. Retrieved 1997 from http://www.britannica.com

Brooks R, Maes P (eds) (1994) Artificial Life IV. MIT Press, Cambridge

Brooks R (1996) Challenge Problems for Artificial Intelligence. In: Proceedings of AAAI–96

Brown V, Huey D, Findlay J M (1997) Face detection in peripheral vision: Do faces pop out? In: Perception 26: 1555–1570

Bruner J, Olver, Greenfield, et al (1966) Studies in Cognitive Growth. John Wiley & Sons, New York

Burch R (2001) Charles Sanders Peirce. In: The Stanford Encyclopedia of Philosophy (Fall 2001 Edition), Edward N. Zalta (ed) http://plato.stanford.edu/archives/fall2001/entries/peirce/

Burt C (1909) Experimental Tests of General Intelligence. Brit J Psy 3: 94–177

Burt C (1912) The Inheritance of Mental Characters. Eugenics Review 4: 168–200

Burt C (1914) The Measurement of Intelligence by the Binet Tests. Eugenics Review 6: pp 36–50, 140–152

Burt C (1937) The Backward Child. Appleton, New York

Burt C (1940) The Factors of Mind. Univ of London Press, London

Burt C (1943) Ability and Income. Brit J Ed Psy 13: 83–98

Burt C (1949) The Structure of the Mind. Brit J Ed Psy 19: 100–111, 176–199

Burt C (1955) The Evidence for the Concept of Intelligence. Brit J Ed Psy 25: 158–177

Burt C (1972) The Inheritance of General Intelligence. Am Psy 27: 175–190

Cahmi JM (1984) Neuroethology: Nerve Cells and the Natural Behavior of Animals. Sinauer Associates, Mass

Camazine S, Deneubourg JL, Nigel R, Franks JS, Theraulaz G, Bonabeau E (2001) Self-Organization in Biological Systems. Princeton University Press, New Jersey

Cantor G (1932) Gesammelte Abhandlungen. Fraenkel A, Zermelo E (eds) Springer-Verlag, Berlin
Cantor G (1955) Contributions to the Founding of the Theory of Transfinite Numbers. Dover, New York
Cartmill M (1990) Human uniqueness and theoretical content in paleoanthropology. In: Int J Primat (3), pp 173–192
Cartwright RL (1967) Classes and Attributes. In: Noûs (1) pp 231–242
Castañeda H-Ñ (1967) Indicators and Quasi-indicators. Am Phil Quarterly (4), pp 85–100
Castañeda H-Ñ (1975) Individuation and Non-Identity: A New Look. Am Phil Quarterly (12), pp 131–140
Castañeda H-Ñ (1977) Perception, Belief, and the Structure of Physical Objects and Consciousness. Synthese (35), pp 285–351
Castañeda H-Ñ (1981). "The Semiotic Profile of Indexical (Experiential) Reference," in Synthese, Vol. 49, pp. 275–316.
Castañeda H-Ñ (1987) Self-Consciousness, Demonstrative Reference, and the Self-Ascription View of Believing. In: Philosophical Perspectives I, Metaphysics. Tomberlin J (ed) Ridgeview, Atascadero, California, pp 405–459
Castañeda H-Ñ (1989) Direct Reference, The Semantics of Thinking, and Guise Theory: Constructive Reflections on David Kaplan's Theory of Indexical Reference. In Themes from Kaplan. Almog J, Perry J, Wettstein H (eds) Oxford, New York
Castañeda H-Ñ (1989) The Reflexivity of Self-Consciousness: Sameness/Identity, Data for Artificial Intelligence. In: Philosophical Topics XVII (1) Spring, pp 27–58
Castañeda H-Ñ (1990) Indexicality: The Transparent Subjective Mechanism for Encountering a World. In: Noûs, XXIV (5), pp 735–749
Castañeda H-Ñ (1990) Philosophy as a Science and as a Worldview. The Institution of Philosophy. Cohen A, Dascal M (eds) Nous Publications, Bloomington, Indiana
Catani M, Jones DK, Ffytche DH (2005) Perisylvian language networks of the human brain. Ann Neurol 57(1): 8–16
Ceci SJ, Williams WM (1997) Schooling, intelligence, and income. Am Psy 52:1051–1058
Chabris CF (1998) IQ since 'The Bell Curve'. Commentary 106, pp 33–40
Chaitin GJ (1966) On the Length of Programs for Computing Binary Sequences. J Assoc Comp Mach 13: 547–569
Chaitin GJ (1974) Information Theoretical Limitations of Formal Systems. J Assoc Comp Mach 21: 403–424
Chaitin GJ (1987) Algorithmic Information Theory. Cambridge University Press, Cambridge.
Chalmers D (1996) The Conscious Mind: In Search of a Fundamental Theory, Oxford University Press, New York
Chau T (2001) A review of analytical techniques for gait data Part 1: fuzzy, statistical and fractal methods. Gait and Posture 13: 49–66

Cherry EC (1953) Some experiments in the recognition of speech, with one and two ears. J Acoust Soc Am 25: 975–979.
Cherry C (1957) On Human Communication. MIT Press, Cambridge
Chevalier-Skolnikoff S (1981)The Clever Hans Phenomenon, Cuing, and Ape Signing: A Piagetian Analysis of Methods for Instructing Animals. In: Sebeok T, Rosenthal R (eds) Annals NY Acad Sci, 364: 60–93
Chisholm RM (1977) Theory of Knowledge, 2nd edn. Prentice-Hall, New York
Chomsky N (1972) Language and Mind. Harcourt, New York
Churchland P, Churchland P (1990) Could a Machine Think? Sci Am January
Churchland PS, Sejnowski TJ (1992) The Computational Brain. MIT Press, Cambridge
Churchland P (1995) Interview. In: Baumgartner P, Payr S (eds) Speaking Minds: Interviews with Twenty Eminent Cognitive Scientists. Princeton Univ Press, New Jersey
Cipolotti L, Harskamp N (2001) Disturbances of number processing and calculation. In: Berndt RS (eds) Handbook of Neuropsychology. Elsevier, Amsterdam
Clark A (1989) Microcognition. MIT Press, Cambridge
Clayes G (2001) Introducing Francis Galton 'Kantsaywhere' and 'The Donoghues of Dunno Weir. Utopian Studies 12(2): 188–190
Cohen J, Stewart I (1994) The Collapse of Chaos: Discovering Simplicity in a Complex World. Viking: Penguin Group, New York
Cohen J, Stewart I (1994) Our Genes Aren’t Us. Discover April pp 78–83
Cohen J (1998) The Human Mind as an Emergent Phenomenon: The Complicit Coevolution of Intelligence and Extelligence. Keynote Address OACES
Colombo J, Ryther JS, Frick J, Gifford J (1995) Visual Pop-out in Infants: Evidence for Preattentive Search in 3- and 4-month-olds. Psych Bul & Rev 2 (2): 266–268
Collishaw SM. Hole GJ (2000) Featural and configurational processes in the recognition of faces of different familiarity. In Perception 2000 Vol 29 (8): 893–909
Corballis MC (1999) The Gestural Origins of Language. Am Sci 87 (2) March-April
Coren S (1994) The Intelligence of Dogs: Canine Consciousness and Capabilities. The Free Press, New York
Cormen TH, Leiserson CE, et al (1992) (eds) Algorithms. MIT Press, Cambridge
Crick F, Koch C (1990) Towards a Neurobiological Theory of Consciousness: Seminars in the Neurosciences. Vol 2 pp 263–275
Crick F, Koch C (1992) The Problem of Consciousness. Sci Am Vol 267 (110)
Crick F (1994) The Astonishing Hypothesis: The Scientific Search for the Soul. Simon and Schuster, New York
Culham JC, Kanwisher NG (2001) Neuroimaging of Cognitive Functions in Human Parietal Cortex. Current Opinion in Neurobiology Vol 11, pp 157–163
Cullen FT, Gendreau P, Jarjoura GR, Wright JP (1997) Crime and the Bell Curve: Lessons from intelligent criminology. Crime and Delinquency (43): 387–411

Cutland NJ (1980) Computability, An Introduction to Recursive Function Theory. Cambridge University Press, Cambridge
Darwin F (ed) (1887) The Life and Letters of Charles Darwin. Kessinger Publishing, London
Darwin F, Seward AC (eds) (1903) More Letters of Charles Darwin. Kessinger Publishing, London
Davenport JH, Heintz J (1988) Real Quantifier Elimination is Doubly Exponential. In: Algorithms in Real Algebraic Geometry. J Symb Comp Vol 5 (1)
Davis M (1993) How Subtle is Gödel's Theorem? More on Roger Penrose. Beh Br Sci 16 (3): p 612
Damasio A (1994) Descartes' Error: Emotion, Reason, and the Human Brain. Putnam, New York
Dawson G, Fischer K (eds) (1994) Human Behavior and the Developing Brain. Guilfor, New York
De Becker G (1997) The Gift of Fear and Other Survival Signals That Protect Us From Violence. Dell, New York
Dehaene S (1992) Varieties of Numerical Abilities. Cognition 44 (1–2) pp 1–42
Dehaene S, Cohen L (1995) Towards an anatomical and functional model of number processing. Math Cogn 1: 83–120
Dehaene S, Spelke E, Pinel P, Stanescu R, Tsivkin S (1999) Sources of Mathematical Thinking: Behavioral and Brain-Imaging Evidence. Science Vol 284, pp 970–974
De Lange F, Hagoort P, Toni I (2005) Neural Topography and Content of Movement Representations. J Cog Neur (17): 97–112
Dennett D (1991) Consciousness Explained. Little, Brown and Co, New York
Dennett D (1994) The Role of Language in Intelligence. In Khalfa J (ed) What Is Intelligence? Cambridge University Press, Cambridge
Descartes R (1960) Discourse on Method and Meditations. Lafleur LJ (trans). Bobbs-Merrill, Indianapolis
Descartes R (1984–1985) The Philosophical Writings of Descartes, Volumes I and II. Cottingham J, Stoothoff R, Murdoch D (trans). Cambridge Univ Press, Cambridge
Devaney RL (1989) An Introduction to Chaotic Dynamical Systems 2nd edn. Addison-Wesley, Massachusetts
Devaney RL (1989) Dynamics of Simple Maps. In Proceedings of Symposia in Applied Mathematics: Chaos and Fractals Vol 39. AMA, Washington DC
Devlin B, Daniels M, Roeder K (1997) The heritability of IQ. Nature (388): 468–471
Devlin K (1994) Mathematics the Science of Patterns. Scientific American Library, Washington
Devlin K (1997) The logical structure of computer-aided mathematical reasoning. Am Math Monthly Vol 104 (7), pp 632–646
Dickason A (1976) Anatomy and Destiny: The Role of Biology in Plato's Views of Women. In: Gould CC, Wartofsky MW (eds) Women and Philosophy: Toward a Theory of Liberation. Penn State Press, PA

Dreyfus HL (1992) What Computers Still Can't Do: A Critique of Artificial Reason. MIT Press, Cambridge

Dreyfus HL (1995) Cognitivism Abandoned. In: Baumgartner P, Payr S (eds) Speaking Minds: Interviews with Twenty Eminent Cognitive Scientists. Princeton University Press, New Jersey

Dubois D, Prade H (1991) Fuzzy Labels, Imprecision and Contextual Dependency: Comments on Milan Zeleny's 'Cognitive Equilibrium: a Knowledge-Based Theory of Fuzziness and Fuzzy Sets. In Intl J Gen Sys Vol 19: 383–386

Dukas R (ed) (1998) Cognitive Ecology: The Evolutionary Ecology of Information Processing and Decision Making. University Chicago Press, Chicago

Eccles J (2002) The Effect of Silent Thinking on the Cerebral Cortex. In TruthJournal, Leadership University

Edman I (ed) (1928) Theaetetus. In The Philosophy of Plato. Modern Lib, New York

Edelman G (2004) Wider than the Sky: The Phenomenal Gift of Consciousness. Yale University Press, New Haven

Egner RE, Denonn LE (1961) (eds) The Basic Writings of Bertrand Russell: 1903–1959. Simon & Schuster, New York

Ekman P (1980) The Face of Man: Expressions of Universal Emotions in a New Guinea Village. Garland STPM Press, New York

Ekman P, Rosenberg EL (1997) (eds) What the Face Reveals: Basic and Applied Studies of Spontaneous Expression Using the Facial Action Coding System (FACS). Oxford, New York

Elman JL (1990) Finding Structure in Time. Cog Sci Vol 14 pp 179–211

Elman JL (2003) Development: It's About Time. Dev Sci 6:4 pp 430–433

Elman J (2005) Connectionist Models of Cognitive Development: Where Next? Trends Cog Sci Vol 9: 3 March

Elmund A, Melin L, Knorring AL, Proos L, Tuvemo T (2004) Cognitive and neuropsychological functioning. In Acta Paediatrica Vol 93 (11) pp 1507–1513

Elsasser WM (1966) Atom and Organism: A New Approach to Theoretical Biology. Princeton University Press, New Jersey

Enard W, Khaitovich P, Klose J, Zollner S, Heissig F, Giavalisco P, Niselt-Struwe K, Muchmore E, Varki A, Ravid R, Doxiadis G M, Bontrop R E, Paabo S (2000) Intra- and Interspecific Variation in Primate Gene Expression Patterns. Science Vol 296: 340–344

Engel AK, Singer W (2001) Temporal binding and the neural correlates of sensory awareness. Trends Cog Sci Vol 5 (1) pp16–25

Esch H, Zhang S, Srinivasan MV, Tautz J (2001) Honeybee dances communicate distances measured by optic flow. Nature 411 pp 581–583

Estep M (1978a) The Concept of Understanding. SISTM Quarterly Vol 1 (3)

Estep M (1978b) Toward a SIGGS Characterization of Epistemic Properties of Educational Design. In Applied General Systems Research: Recent Developments and Trends. Klir G (ed) NATO Conference Series. Plenum Press, New York

Estep M (1978c) Pragmatics of the SIGGS Theory Model Relative to Pedagogical Epistemological Inquiry. In Avoiding Social Catastrophes and Maximizing Social Opportunities. SGSR, AAAS, Washington, DC

Estep M (1978d) A SIGGS Information Theoretic Characterization of Qualitative Knowing: Cybernetic and SIGGS Theory Models. In Sociocybernetics Vol 2 Martinus Nijhoff Social Sciences Division, Leiden, Boston, London

Estep M (1979) Open Systems Characterizations of Epistemic Properties: Implications for Inquiry in the Human Sciences. In Improving the Human Condition: Quality and Stability in Social Systems. SGSR and Springer-Verlag, Berlin, Heidelberg, New York, London

Estep M (1981) Ways of Qualitative Worldmaking: The Nature of Qualitative Knowing. In Applied Systems and Cybernetics Vol. II: Systems Concepts, Models, and Methodology. Lasker G (ed). Pergamon Press, New York

Estep M (1984) Toward Alternative Methods in Systems Analysis: The Case of Qualitative Knowing. In Cybernetics and Systems Research Vol 2 Trappl R (ed). Elsevier Science Publishers B.V. (North-Holland), Netherlands

Estep M (1986) The Concept of Power and Systems Models for Developing Countries. In: Cybernetics and Systems Research: An International Journal, Hemisphere Publishing Corp of Harper and Row, Washington, DC

Estep M (1987) Systems Analysis and Power. In Problems of Constancy and Change: The Complementarity of Systems Approaches to Complexity. Hungarian Academy of Sciences, Budapest

Estep M (1992) On Models and Retroductive Inference. In Cybernetics and Systems Research Vol 1.Trappl R (ed).World Scientific, Singapore, New Jersey, London, Hong Kong

Estep M (1993) On Qualitative Logical and Epistemological Aspects of Fuzzy Set Theory and Test-Score Semantics: Indexicality and Natural Language Discourse. In First European Congress on Fuzzy and Intelligent Technologies Proceedings Vol 2 Zimmerman HJ (ed). Verlag der Augustinus Buchhandlung, Aachen, Germany

Estep M (1996) Critique of James' Neutral Monism: Consequences for the New Science of Consciousness. J Cons Stdies

Estep M (1998) What Gödel Said: On the Non-algorithmic Nature of the Second Theorem. Syst Res. Vienna, EMCSR 1998

Estep M (1999) Teaching the Logical Paradoxes: On Mathematical Insight or Is there a Non-Algorithmic Element in Gödel's Second Theorem? Abstracts of Papers Presented to the Amer Math Soc, Vol 20 (1)

Estep M (2003) A Theory of Immediate Awareness: Self-organization and Adaptation in Natural Intelligence. Springer, Dordrecht

Eubank S, Farmer D (1990) An Introduction to Chaos and Prediction. In Jen E (ed). 1989 Lectures in Complex Systems, Santa Fe Institute Studies in the Sciences of Complexity. Addison-Wesley, Massachusetts

Fagot J, Wasserman EA, Young ME (2001) Discriminating the Relation Between Relations: The Role of Entropy in Abstract Conceptualization by Baboons (Papio papio) and Humans (Homo sapiens). An Behav Proc, Vol 27 (4)

Familant ME, Detweiler MC (1993) Iconic Reference: Evolving Perspectives and an Organizing Framework. In Int J Man-Machine Stdies 39: 705–728

Farmer J, Doyne (1990) A Rosetta Stone for Connectionism. In Physica D Vol 42, pp 153–187

Feferman S, Dawson JW, Kleene SC, et al (eds). (1986) Kurt Gödel Collected Works Vol I. Oxford, New York

Feferman S, Dawson JW, Kleene SC, et al (eds). (1990) Kurt Gödel Collected Works, Volume II. Oxford, New York

Feferman S, Dawson JW, Kleene SC, et al (eds). (1995) Kurt Gödel Collected Works, Volume III. Unpublished Essays and Lectures. Oxford, New York

Feldman BL, Gross J, Christensen T, Benvenuto M (2001) Knowing what you're feeling and knowing what to do about it: Mapping the relation between emotion differentiation and emotion regulation. Cognition and Emotion 15:713–724

Fine K (1989) The Problem of De Re Modality. In: Themes From Kaplan. Almog J, Perry J, Wettstein H (eds). Oxford, New York

Fischer CS, Hout M, Sanchez Jankowski M, Lucas SR, Swidler A, Voss K (1996). Inequality by Design: Cracking the Bell Curve Myth. Princeton Univ Press, New Jersey

Flanagan DP, Kaufman AS (eds) (2004) Essentials of WISC-IV Assessment. John Wiley, New York

Flynn JR (1984) The Mean IQ of Americans: Massive Gains. Psych Bul Vol 95 pp 29–51

Flynn JR (1987) Massive IQ Gains in 14 Nations: What IQ Tests Really Measure. Psych Bul, Vol 101, pp 171–191

Forrest S, Miller JH (1991) Emergent Behavior in Classifier Systems. In Emergent Computation. Forrest S (ed). MIT Press, Cambridge

Foster L, Swanson JW (1970) Experience and Theory. University of Mass Press, Amherst

Frank LM, Stanley GB, Brown EN (2004) Hippocampal plasticity across multiple days of exposure to novel environments. J Neurosci 24 (35): 7681–9

Freeman JA (1994) Simulating Neural Networks with Mathematics. Addison-Wesley, Reading, Massachusetts

Frege G (1892) Über Begriff und Gegenstand. In Vierteljahrsschrift für wissenschaftliche Philosophie 16, pp 192–205

Frege, Gottlob (1952). "Uber Sinn und Bedeutung," (On Sense and Reference), in Geach, Peter, and Max Black, (eds.), Translations From the Philosophical Writings of Gottlob Frege, New York, Oxford: Oxford University Press.

Frick T (1983) Non-metric Temporal Path Analysis (NTPA): An Alternative to the Linear Models Approach for Verification of Stochastic Educational Patterns. Indiana University, Bloomington, Indiana

Frick T (1990) Analysis of Patterns in Time: A Method of Recording and Quantifying Temporal Relations in Education. AERA J, Vol 27 (1), pp 180–204

Frith C, Perry R, Lumer E (1999) The neural correlates of conscious experience: an experimental framework. Trends in Cog Sci, Vol 3 (3), pp 105–114

Fritzke B (1993) Growing Cell Structures-A Self-Organizing Network for Unsupervised and Supervised Learning. Technical Report. International Computer Science Inst, Berkeley, California

Fritzke B (1993) Kohonen Feature Maps and Growing Cell Structures-A Performance Comparison. International Computer Science Inst, Berkeley, California

Galton F (1869/1892/1962) Hereditary Genius: An Inquiry into its Laws and Consequences. Macmillan/Fontana, London.

Galton F (1883/1907/1973). Inquiries into Human Faculty and its Development. AMS Press, New York.

Gannon P, Holloway R (1998) Similarities Found In Human, Chimp Brains. Science, AAAS January

Gardner H (1982) Art, Mind, and Brain. Basic Books, New York

Gardner H (1985) The Mind's New Science. Basic Books, New York

Gardner H (1993) Frames of Mind: The Theory of Multiple Intelligences. Basic Books, New York

Gardner H (1995) Cracking open the IQ box. In The American Prospect http://www.prospect.org/archives/20/20gard.html

Gardner H (1998) Are there additional intelligences? The case for naturalist, spiritual, and existential intelligences. In Kane J (ed). Education, information, and transformation. Merrill-Prentice-Hall, Upper Saddle River, New Jersey

Gärdenfors P (1988) Knowledge in Flux: Modeling the Dynamics of Epistemic States. MIT Press, Cambridge

Geach P (1971) Mental Acts, Their Content and Their Objects. Humanities Press, New York

Gelman R, Gallistel CR (2000) Non-Verbal Numerical Cognition: from reals to integers. Trends in Cog Sci, Vol 4 (2)

Gelman R, Butterworth B (2005) Number and Language: How Are They Related. In: Trends in Cog Sci, Vol 9 Issue 1 pp 6–10

George M (1999) What Aquinas Really Said About Women. In: First Things 98, pp 11–13

Goddard HH (1914) Feeble-mindedness: Its Causes and Consequences. Macmillan, New York

Goddard HH (1919) Psychology of the Normal and Subnormal. Dodd, Mead and Co, New York

Gödel K (1951) Some Basic Theorems on the Foundations of Mathematics and Their Implications. In: Feferman, et al (eds) (1995) Kurt Gödel Collected Works, Volume III. Unpublished Essays and Lectures. Oxford, New York

Gödel K (1964a) Russell's Mathematical Logic. In Benacerraf P, Putnam H (eds) Philosophy of Mathematics. Prentice-Hall, New Jersey

Gödel K (1964b) What is Cantor's Continuum Problem? In Benacerraf P, Putnam H (eds) Philosophy of Mathematics. Prentice-Hall, New Jersey

Goldin-Meadow S, Mylander C (1998) Spontaneous sign systems created by deaf children in two cultures. Nature 391:279–281

Gomi H, Kawato M (1996) Equilibrium-Point Control Hypothesis Examined by Measured Arm Stiffness During Multijoint Movement. Science Vol 272, 5, pp 117–120

Goodall J (1986) The Chimpanzees of Gombe. Harvard University Press, Cambridge

Goodman,N (1973) Fact, Fiction, and Forecast. Bobbs-Merrill Publishing Company, Indianapolis

Gottfredson LS (1997a) Inequality by design (book review). Personnel Psychology, 50, pp 741–746

Gottfredson LS (1997b) Why g matters: The complexity of everyday life. Intelligence, 24, pp 79–132

Gottfredson LS (1998c) The general intelligence factor. Scientific American Presents, 9, pp 24–29, 51

Gould CC, Wartofsky MW (eds) (1976) Women and Philosophy: Toward a Theory of Liberation University Park, Penn State Press, PA

Gould JL, Gould CG (1994) The Animal Mind. Scientific American Lib, New York

Gould SJ (1981) The Mismeasure of Man. WW Norton & Company, New York

Gould SJ (1994) Curveball. The New Yorker, New York

Grandin T (1995) Thinking in Pictures and Other Reports from my Life With Autism. Vintage Books, New York

Grandin T (2005) Animals in Translation: Using the Mysteries of Autism to Decode Animal Behavior. Scribner, New York

Graves K, et al (1973) Tacit Knowledge. The Journal of Philosophy, Vol. LXX, No. 11, June 7

Grewal D, Salovey P (2005) Feeling Smart: The Science of Emotional Intelligence. American Scientist, Vol. 93, No. 3, July-August, p 330

Griffin D (1984) Animal Thinking. Harvard University Press, Cambridge

Griffin D (1992) Animal Minds. University of Chicago Press, Chicago

Grimaldi R (1994) Discrete and Combinatorial Mathematics 3rd edn, Addison-Wesley, Reading, Massachusetts

Grimson WE, Patil RS (eds) (1987) AI in the 1980s and Beyond: An MIT Survey. MIT Press, Cambridge

Gruber HE, Vonëche J (1995) The Essential Piaget: An Interpretive Reference and Guide, 2nd edn. Jason Aronson, New Jersey

Guilford JP (1967) The Nature of Human Intelligence. McGraw-Hill, New York

Gupta D (2001) Computer Gesture Recognition: Using the Constellation Method. Caltech Undergraduate Research J, Vol. 1, April

Gurwitsch A (1963) On the Conceptual Consciousness. In: The Modeling of Mind. Sayre KM, Crosson FJ (eds) Notre Dame University, South Bend

Hackworth D (1997) Military Operations: Vietnam Primer, Lessons Learned. Department of the US Army, Washington, DC

Hadamard J (1945) The Psychology of Invention in the Mathematical Field. Princeton University Press, Princeton, New Jersey

Hameroff S, Rasmussen S, Mansson B (1989) Molecular Automata in Microtubules: Basic Computational Logic of the Living State. In Artificial Life Langton, C (ed), Vol. VI, Santa Fe Institute Studies in the Sciences of Complexity, Addison-Wesley Publishing Company, Reading, MA

Hanson NR (1972) Patterns of Discovery. Cambridge University Press, Cambridge
Harary F (1969) Graph Theory. Addison-Wesley, Reading, Massachusetts
Harmon GH (1977) Thought. Princeton University Press, New Jersey
Harmon LD (1973) Recognition of Faces. In Sci Am, November
Hartley RVL (1928) Transmission of Information. In The Bell Sys Tech J, Vol 7, pp 535–563
Hartshorne C, Weiss, P (eds) (1931–1958) The Collected Papers of Charles Sanders Peirce, Vols I-VI. Harvard University Press, Cambridge
Hauser MD (2000) Wild Minds: What Animals Really Think. Henry Holt, New York
Hauser MD, et al (2002) The Faculty of Language: What Is It, Who Has It, and How Did It Evolve? In Science, Vol 298, pp 1569–1579
Hauser RM, Huang MH (1997) Verbal ability and socioeconomic success: A trend analysis. In Social Science Res, 26, pp 331–376
Hausman A, Foster T (1977) Is Everything a Class? In Phil Studies, 32, pp 371–376
Haykin S (1994) Neural Networks: A Comprehensive Foundation. Macmillan, New York
Healey CG (1993) Visualization of Multivariate Data Using Preattentive Processing. Masters Thesis, University of British Columbia
Healey CG (2005) Perception in Visualization. N C State University
Hebb DO (1949) The Organization of Behavior: A Neuropsychological Theory, John Wiley, New York
Heijenoort J (ed) (1967) From Frege to Gödel. Harvard University Press, Cambridge
Hempel CG (1965) Aspects of Scientific Explanation. The Free Press, New York,
Hernegger R (1995) Wahrnehmung und Bewusstsein: Ein Diskussionsbeitrag zur Neuropsychologie. Spekrum, Heidelburg
Herrnstein RJ, Charles M (1994) The Bell Curve: Intelligence and Class Structure in American Life. The Free Press, New York
Higashi M, Klir G (1982) Measures of Uncertainty and Information Based on Possibility Distributions. In: Int J Gen Sys, 8, SGSR, Washington, DC
Higashi M, Klir G (1983) On the Notion of Distance Representing Information Closeness. In: Int J Gen Sys, 9, ISGSR, Washington, DC
Higgins C (2001) Sensory Architectures for Biologically Inspired Autonomous Robots. Biol Bull 200: 235–242, April
Hinton GE (1981) Shape Representation in Parallel Systems. In: Proc 7th International Joint Conference on Artificial Intelligence, Vancouver, British Columbia
Hinton GE, Sejnowski TJ (1986) Learning and Relearning in Boltzmann Machines. In: Parallel Distributed Processing: Explorations in Microstructure of Cognition. Rumelhart DE, McClelland JL (eds), MIT Press, Cambridge
Hinton GE, Dayan P, Frey B, Neal R (1995) The 'Wake-Sleep' Algorithm for Unsupervised Neural Networks. In: Sci Am, AAAS, Vol 268, No 5214, pp 1158–1161

Holland J (1975) Adaptation in Natural and Artificial Systems. University of Michigan Press, Ann Arbor

Holland JH, Holyoak KJ, Nisbett R, Thagard P (1986) Induction: Processes of Inference, Learning, and Discovery. MIT Press, Cambridge

Horwitz B, Rumsey JM, Donohue BC (1998) Functional connectivity of the angular gyrus in normal reading and dyslexia. Proc Natl Acad Sci USA, July 21: 95(15): 8939–44

Irvine P (1986) Sir Francis Galton (1822–1911). J Spec Ed, 20(1)

Jackson P, Reichgelt H, Harmelen F (1989) Logic-Based Knowledge Representation. MIT Press, Cambridge

Jacob F (1977) Evolution and Tinkering. Science, Vol 196, pp 1161–1166

James W (1884) Some Omissions of Introspective Psychology. Mind, 9, January, 1884, pp 1–26

James W (1890) The Principles of Psychology, Volumes I and II. Macmillan, London

James W (1976) Essays in Radical Empiricism. Harvard University Press, Cambridge

Jen E (ed) (1989) Lectures in Complex Systems, Santa Fe Institute Studies in the Sciences of Complexity. Addison-Wesley, Reading, Massachusetts

Jencks C, Phillips M. (1998) The black-white test score gap. Brookings, Washington, DC

Jensen A (1998) The g Factor: The Science of Mental Ability. Praeger, Westport, CT

Jensen AR (1999) Precis of: "The g Factor: The Science of Mental Ability" Psycoloquy, 10(023)

Johnson-Laird PN (1983) Mental Models. Harvard University Press, Cambridge

Jordan MI (1986) An Introduction to Linear Algebra in Parallel Distributed Processing. In Parallel Distributed Processing: Explorations in the Microstructure of Cognition, Vol 1: Foundations. Rumelhart DE, McClelland JL, PDP Research Group, MIT Press, Cambridge

Joseph R (2000) Neuropsychiatry, Neuropsychology, Clinical Neuroscience. Academic Press, New York

Jourdain PEB (1912) The Development of the Theories of Mathematical Logic and the Principles of Mathematics. In: The Quart J Pure and App Math, 43, pp 219–314

Jusczyk PW (1997) The Discovery of Spoken Language. MIT Press, Cambridge

Kalish D, Montague R, Mar G (1980) Logic: Techniques of Formal Reasoning. Harcourt, New York

Kandel ER, Schwartz JH (1991) Principles of Neural Science, 3rd edn. Elsevier, New York

Kane J (ed) (1998) Education, information, and transformation. Merrill-Prentice Hall, Upper Saddle River, NJ

Kant I (1783) Prolegomena to any Future Metaphysics that will be able to present itself As a Science. Johann Friedrich Hartknoch, Riga.

Kant I (1785; 1998) Groundwork for the Metaphysics of Morals (Grundlegung zur Metaphysik der Sitten) Cambridge University Press, Cambridge

Kant I (1788; 1996) Critique of Practical Reason (Kritik der practischen Vernunft) Prometheus Books, New York

Kant I (1929) Critique of Pure Reason. (Smith NK, trans), Macmillan, Toronto

Kaplan D (1989) Demonstratives. In Themes From Kaplan. Almog J, Perry J, Wettstein H (eds) Oxford, New York

Kauffman SA (1990) Requirements for Evolvability in Complex Systems: Orderly Dynamics and Frozen Components. Physica D, Vol 42, pp 135–152

Kauffman SA (1991) Antichaos and Adaptation. Sci Am, Vol 265, No 2, August, pp 78–84

Kauffman SA (1993) The Origins of Order: Self-Organization and Selection in Evolution. Oxford, New York

Kauffman SA (1995) At Home in the Universe: The Search for the Laws of Self-Organization and Complexity. Oxford, New York

Keen L (1989) Julia Sets. In Proc of Symposia in Applied Mathematics: Chaos and Fractals, Vol 39, Am Math Soc

Kellert SH, Mark A, Stone AF (1990) Models, Chaos, and Goodness of Fit. In: Philosophical Topics, Vol 18, No 2, Fall

Kelly K (2005) Where You Stand Determines What You See and How You Live. CommonDreams

Kerlinger F (1973) Foundations of Behavioral Science, 2nd edn. Holt, Rinehart and Winston, New York

Kessen W (1965) Child. John Wiley, New York

Kirshner D, Awtry T (2004) Visual Salience of Algebraic Transformations. JRes Math Ed, Vol 35, No 4, pp 224–257

Klima G (2004) The Medieval Problem of Universals. In: The Stanford Encyclopedia of Philosophy. Zalta EN (ed) http://plato.stanford.edu/archives/win2004/entries/universals-medieval/

Klir G (1972) Trends in General Systems Theory. John Wiley, New York

Klir G (1989) Is There More to Uncertainty Than Some Probability Theorists Might Have Us Believe? In Int J Gen Sys, Vol 15, pp 347–378

Klir GJ, Yuan B (1995) Fuzzy Sets and Fuzzy Logic: Theory and Applications. Prentice Hall, PTR, Upper Saddle River, NJ

Klir GJ, Wierman MJ (1999) Uncertainty-Based Information: Elements of Generalized Information Theory. Physica-Verlag/Springer-Verlag, Heidelberg and New York

Knauff M, Johnson-Laird PN (2000) Visual and spatial representations in spatial reasoning. In: Proc 22nd Annual Conference of the Cognitive Science Society, pp. 759-765,Earlbaum, Mahwah, NJ

Knauff M, Mulack T, Kassubek J, Salih HR, Greenlee MW (2002) Spatial imagery in deductive reasoning: a functional MRI study. Cognitive Brain Research 13, pp 203–212

Köhler W (1973) The Mentality of Apes. Routledge and Kegan Paul, London

Kohonen T (1982) Self-organized Formation of Topologically Correct Feature Maps. Biological Cybernetics, Vol 43, pp 59–69

Kohonen T (1988) An Introduction to Neural Computing. In: Neural Networks, Vol 1, pp 3–16

Kohonen T (1990) The Self-organizing Map. In: Proc IEEE, 78, pp 1464–1480
Kornblith H (1994) Naturalizing Epistemology, 2nd edn. MIT Press, Cambridge
Kornblith H (1999) In Defense of a Naturalized Epistemology. In: The Blackwell Guide to Epistemology. Greco J, Sosa E (eds). Blackwell, Oxford
Kosslyn S (198) Image and Mind. Harvard University Press, Cambridge
Kosslyn S (1996) Image and Brain. MIT Press, Cambridge
Kosslyn S, Ganis G, Thompson WL (2001) Neural Foundations of Imagery. Nature Publishing Group, Washington DC
Kosslyn S (2002) Visual Mental Images in the Brain: How Low Do They Go. Presented at a meeting of AAAS, Cognitive Neuroscience of Mental Imagery, February
Kripke S (1972) Naming and Necessity. Harvard University Press, Cambridge
Kuhn TS (1970) The Structure of Scientific Revolutions, 2nd edn. Int Enc Unif Sci, University of Chicago Press, Chicago
Kunimoto C, et al (2001) Confidence and Accuracy in Near-Threshold Discrimination Responses. In Consciousness and Cognition, Vol 10, no 3, pp 294–340
Lacourse MG, Turner J, Randolph-Orr E, Schandler S, Cohen M (2004) Cerebral and cerebellar sensorimotor plasticity following motor imagery-based mental practice of a sequential movement. J Rehab Res Dev, Vol 41, No 4, July-August, pp 505–524
Lakoff G, Johnson M (1998) Philosophy in the Flesh: The Embodied Mind and Its Challenge to Western Thought. Basic Books, New York
Lakoff G, Núñez RE (2000) Where Mathematics Comes From: How the Embodied Mind Brings Mathematics Into Being. Basic Books, New York
Lamme VAF, Roelfsema PR (2000) The Distinct Modes of Vision Offered by Feedforward and Recurrent Processing. Trends in Neuroscience, Vol 23, pp 571–579
Langton CG (ed) (1989) Artificial Life, Santa Fe Institute Studies in the Sciences of Complexity, Vol 6. Addison-Wesley, Redwood City, California
Lasswell H (1930) Psychopathology and Politics. University of Chicago, Chicago
Laughlin RB (2005) A Different Universe: Reinventing Physics from the Bottom Down. Basic Books, New York
Lay SR (1990) Analysis With An Introduction to Proof, 2nd edn. Prentice-Hall, New Jersey
Leakey R, Lewin R (1992) Origins Reconsidered: In Search of What Makes Us Human. Little, Brown, Boston
Lehrer K, Paxson T, (1968) Knowledge: Undefeated Justified True Belief. J Phil, Vol LXVI, No 8, April
Lehrer K (1974) Knowledge. Oxford, New York
Lehrer K (1980) Knowledge. In Bogdan RJ (ed), Keith Lehrer. D. Reidel Publishing Company
Lehrer K (1983) Coherence and Indexicality in Knowledge. In Tomberlin JE (ed), Agent, Language, and the Structure of the World: Essays Presented to Hector-Neri Castañeda With His Replies. Ridgeview, Atascadero, California

Leibniz GW (1703–05; 1989) Preface to the New Essays (1703–05). In G.W. Leibniz: Philosophical Essays. Ariew R, Garber D (trans). Hackett Publishing Co, Indianapolis

Lenat DB (1995) Artificial Intelligence: A Critical Storehouse of Commonsense Knowledge is Now Taking Shape. Sci Am, September, pp 80–82

Lewis D (1979) Attitudes De Dicto and De Se. Phil Rev, Vol 88, pp 513–543

Li F, Van Rullen R, Koch C, Perona P (2002) Rapid natural scene categorization in the near absence of Awareness. In Proc Nat Acad Sci, vol 99, July

Libet B (1973) Electrical Stimulation of Cortex in Human Subjects, and Conscious Memory Aspects. In Iggo A (ed), Handbook of Sensory Physiology, Vol. II. Springer-Verlag, Berlin, Heidelberg, New York

Lipton RJ (1995) DNA Solution for Hard Computational Problems. Science, AAAS, Vol 268, 28 April, pp 542–545

Livingstone M, Hubel D (1988) Segregation of Form, Color, Movement, and Depth: Anatomy, Physiology, and Perception. Science, Vol 240, pp 740–749

Lloyd JE (1983) Bioluminescence and communication in insects. Annual Review of Entomology, Vol 28, pp 131–160

Loux MJ (1970) Universals and Particulars, Readings in Ontology. Anchor Books, New York

Luria AR (1968) The Mind of the Mnemonist. Harvard University Press, Cambridge

Maccia ES (1964) Retroduction: A Way of Inquiring Through Models. Communicaciones Libres, Memorias del XIII Congreso Internacional de Filosofia, Universidad Nacional Autonoma de Mexico.

Maccia ES, Maccia G (1966). Development of Educational Theory Derived From Three Educational Theory Models, Final Report, U.S. Department of HEW. U.S. Office of Education, Bureau of Research, Washington DC

Maccia ES, Maccia G (1971) System Theory and the SIGGS Theory Model. In: Proc Int Congress of the History of Science, Moscow

Maccia ES, Maccia G (1973) Information Theoretic Extension of the Cybernetic Model and Theory of Education. Advances in Cybernetics and Systems, Rose J (ed), Gordon and Breach Science Publishers

Maccia ES, Maccia G (1976) The Logic of the SIGGS Theory Model. In: Proc AERA

Maccia ES (1976) Logical and Conceptual Analytic Techniques for Educational Researchers. In: Proc AERA

Maccia G (1973) Epistemological Considerations of Educational Objectives. Presented to The Philosophy of Education Section, XVth World Congress of Philosophy. Varna, Bulgaria, September

Maccia G (1987) Genetic Epistemology of Intelligent, Natural Systems. In: Systems Research, Vol 3

Maccia G (1989) Genetic Epistemology of Intelligent Systems: Propositional, Procedural, and Performative Intelligence. Presented at Hangzhou University, Hangzhou, Zhejiang Province, The People's Republic of China

Mach E (1959) Analysis of the Sensations. Dover Publications, New York

Mackintosh NJ (1998) IQ and human intelligence. Oxford University Press, Oxford

Mager RF (1975) Preparing Instructional Objectives, 2nd edition. Feron Publishing Co, Belmont, California

Malcolm N (1958) Ludwig Wittgenstein: A Memoir. Oxford University Press, Oxford

Mandelbrot B (1983) The Fractal Geometry of Nature. WH Freeman, New York

Marcus G (2003) The Birth of the Mind: How a Tiny Number of Genes Creates the Complexities of Human Thought. Basic Books, New York

Markie P (2004) Rationalism vs. Empiricism. The Stanford Encyclopedia of Philosophy (Fall 2004 Edition). Zalta, EN (ed). <http://plato.stanford.edu/archives/fall2004/entries/rationalism-empiricism/>

Marsh RC (ed) (1956) Bertrand Russell: Logic and Knowledge Essays 1901–1950, Capricorn Books, New York

Martin E (1973) The Intentionality of Observation. Can J Phil, Volume III, Number 1, September, pp 121–129

McClelland JL, Rumelhart DE, PDP Research Group (1986) Parallel Distributed Processing, Volumes 1 and 2. MIT Press, Cambridge

McCulloch WS,Pitts W (1943) A logical calculus of the ideas immanent in nervous activity. Bull of Math. Biophysics, 5:115–137

McGue M (1997) The democracy of the genes. Nature, 388, 417–418

McInerney JD (1999) Genes and Behavior: A Complex Relationship. Judicure: Genes and Justice: The Growing Impact of the New Genetics on the Courts, November-December, Vol 83, No 3

McKeon R (ed) (1941) The Basic Works of Aristotle. Random House, New York

McLaren C (2005) The Great White Way. Stay Free! Magazine

McNeill D (1992) Hand and Mind: What Gestures Reveal about Thought. University of Chicago Press, Chicago

Medawar P (1964) Is the Scientific Paper Fraudulent? Sat Rev, August 1, pp 42–43

Medawar P (1969) Induction and Intuition in Scientific Thought, Jayne Lectures for 1968. APA Memoirs, Philadelphia

Meinong A (1899) Über Gegenstände höherer Ordnung und deren Verhältniss zur inneren Wahrnehmung. In Zeitschrift für Psychologie des Sinnesorgane, 21, pp 182–272

Mitra S, Sankar KP (1994) Self-Organizing Neural Network as A Fuzzy Classifier. In IEEE Transactions on Systems, Man, and Cybernetics, Vol 24, No 3, March, pp 385–398

Mizrai AR (1990) Artificial Intelligence: Concepts and Applications in Engineering. Thomson Learning, London

Morgan AW, Sullivan SA, Darden C, Gregg N (1997) Measuring the intelligence of college students. J Learn Dis, Sep-Oct, Vol 30, Number 5, pp 560–565

Moss F, Wiesenfeld K (1995) The Benefits of Background Noise. Sci Am, Volume 273, Number 2, August, pp 66–69

Murray C (1995) For Whom the Bell Curve Tolls. Skeptic, Vol 3, Number 2, pp 34–41

Murray C (2000) Heritability and the Independent Causal Role. Psycoloquy, Volume 11, number 105

Myerson J, Rank MR, Raines FQ, Schnitzler MA (1998) Race and general cognitive ability: The myth of diminishing returns to education. Psych Sci, 9, 139–142

Näätanen R, Tervaniemi M, Sussman E, Paavilinen E, Winkler I (2001) 'Primitive Intelligence' in the Auditory Cortex. Trends in Neurosciences, Vol 24, number 5, pp 283–288

Nagel E, Newman J (1958) Gödel's Proof. NY University Press, New York

Neisser U (1997) Rising Scores on Intelligence Tests. Am Sci, September-October

Neurath O, Morris C, Carnap R (eds) International Encyclopedia of Unified Science, Volume 1, Nos. 1–10. University of Chicago Press, Chicago

Newman L (2000) Descartes' Epistemology. In: Zalta EN (ed) The Stanford Encyc Phil

Nicolelis MA, Luiz A, Baccala R, Lin CS, Chapin JK (1995) Sensorimotor Encoding by Synchronous Neural Ensemble Activity at Multiple Levels of the Somatosensory System. Science, AAAS, Vol 268, 2 June, pp 1353–1358

Nordenstam T (1972) Empiricism and the Analytic-Synthetic Distinction. Universitetsforlaget Stockholm

Ostwald P (1959) When People Whistle. Language and Speech, 2, pp 137–145

Ostwald P (1960) The Sounds of Human Behavior-A Survey of the Literature. Logos, 3, pp 13–24

Ostwald P (1964) How the Patient Communicates About Disease with the Doctor. Approaches to Semiotics. Sebeok TA, Hayes AS, Bateson MC (eds). Mouton & Co, The Hague

Page G (1999) Inside the Animal Mind: A Groundbreaking Exploration of Animal Intelligence. Doubleday, New York

Pant V, Higgins C (2004) A Biomemetic VLSI Architecture for Small Target Tracking, IEEE

Parsons KM (1989) God and the Burden of Proof. Prometheus, Buffalo, New York.

Pearson K (1976) The Control of Walking. Sci Am, Vol 235, pp 72–86

Peirce CS (1931–1958) Collected Papers of Charles Sanders Peirce vols. 1–6 Hartshorne C, Weiss P (eds). vols. 7–8 Burks AW (ed). Harvard University Press, Cambridge, Mass

Penrose R (1974) The Role of Aesthetics in Pure and Applied Mathematical Research. Bull Inst Math Appl, July/August, pp 266–271

Penrose R (1989) The Emperor's New Mind. Oxford, New York

Penrose R (1994) Shadows of the Mind. Oxford, New York

Perkins D (1995) Outsmarting IQ : The Emerging Science of Learnable Intelligence. The Free Press, New York

Perkins DN, Grotzer TA (1997) Teaching intelligence. Am Psych, 52, 1125–1133

Perry J (1979) The Problem of the Essential Indexical. NOÛS, 13, pp 3–21

Piaget J (1950) The Psychology of Intelligence. Percey M, Berlyne DE (transl). Routledge, London and New York

Piaget J, Inhelder B (1956) The Child's Conception of Space. Routledge, London

Piaget J (1971) Biology and Knowledge: An Essay on the Relations Between Organic Regulations and Cognitive Processes. University of Chicago Press, Chicago

Piaget J (1972) The psychology of the child. Basic Books, New York

Piaget J (1990) The child's conception of the world. Littlefield Adams, New York

Pinker S, Bloom P (1990) Natural language and natural selection. Behav Brain Sci 13(4), pp 707–784

Pinker S (1994) The Language Instinct: How the Mind Creates Language. William Morrow & Co, New York

Pinker S (1999) How the Mind Works. WW Norton, New York

Pinker S (2004) How to Think About the Mind. Newsweek, September 27

Plomin R, Petrill SA (1997) Genetics and intelligence: What's new? Intelligence, 24, 53–77

Plucker JA (ed) (2003) Human intelligence: Historical influences, current controversies, teaching resources. Retrieved September 8, 2004, from http://www.indiana.edu/~intell

Pojman LP (1995) What Can We Know, An Introduction to the Theory of Knowledge. Wadsworth, Belmont

Polanyi M (1966) The Tacit Dimension. Doubleday, New York

Polanyi M (1969) The Unaccountable Element in Science. In: Knowing and Being. University of Chicago, Chicago

Polanyi M (1969) Knowing and Being. Grene M (ed). University of Chicago Press, Chicago

Popper K (1972) Objective Knowledge. Clarendon Press, Oxford

Popper K, Eccles JC (1977) The Self and Its Brain. Springer-Verlag, New York, Heidelberg, London

Pot P (1997) Logical Structures of Young Chimpanzees' Spontaneous Object Grouping. Int J Primat, Vol 18, No. 1, pp 33–59

Premack A, Premack D (1972) Teaching Language to an Ape. Sci Am, October, 1972, pp 92–99

Premack D, Premack A (2003) Original Intelligence: Unlocking the Mystery of Who We Are. McGraw-Hill, New York

Presenti M, et al, (2000) Neuroanatomical Substrates of Arabic Number Processing, Numerical Compaison and Simple Addition: A PET Study. J Cog Neuroscience, Vol 12, 2000, pp 461–479

Putnam H (1994) The Dewey Lectures 1994: Sense, Nonsense, and the Senses: An Inquiry into the Powers of the Human Mind. The J Phil, Vol. XCI, Number 9, September, 1994

Quine WVO (1951) Two Dogmas of Empiricism. The Phil Rev, Vol. 60, 1951, pp 20–43

Quine WVO (1953) From a Logical Point of View. Harvard University Press, Cambridge

Quine WVO (1960) Word and Object. MIT Press, Cambridge

Quine WVO (1963) Set Theory and Its Logic. Harvard University Press, Cambridge

Quine WVO (1969) Ontological Relativity and Other Essays. Columbia University Press, New York
Quine WVO (1970) Grades of Theoreticity. In: Experience and Theory, Foster, Swanson (eds). University of Massachusetts Press, Amherst
Quine WVO, Ullian JS (1978) Web of Belief, 2nd edn. Random House, New York
Quine WVO (1981) Theories and Things. Harvard University Press, Cambridge
Quine WVO (1990) Norms and Aims. In: The Pursuit of Truth. Harvard University Press, Cambridge
Quiñones E (2005) New Way Of Tracking People's Mental State As They Think Back To Previous Events. Medical New Today, 24 Dec
Rizzolatti G, Arbib MA (1998) Language Within Our Grasp. Trends in Neuroscience, Volume 21, number 5, 1998, pp 188–194
Reifman A (2000) Revisiting the Bell Curve. Psycoloquy: 11(099)
Renegar J (1988) A Faster PSPACE Algorithm for Deciding the Existential Theory of the Reals. In: Proc 29th Ann Symp Comp Sci, October, IEEE Computer Society Press
Repp B (2001) Phase Correction, Phase Resetting, and Phase Shifts After Subliminal Timing Perturbations in Sensorimotor Synchronization. J Exp Psych: Human Perception and Performance, APA, Vol 27, Number 3, June, 2001
Rips LJ (1994) The Psychology of Proof. MIT Press, Cambridge
Rose N (1988) Mathematical Maxims and Minims. Raleigh NC
Rosen R (1970) Dynamical System Theory in Biology, Vol. I: Stability Theory and Its Applications. John Wiley, New York
Rosenblatt F (1958) The Perceptron: A probabilistic model for information storage and organization in the brain. Psychol Rev 65, pp 298–311
Rothstein E (2004) The Brain? It's a Jungle in There. NY Times, New York Times
Rowland T (1999) Manifold. Eric Weisstein's Math World, Wolfram Research, Inc., Chicago
Royal CDM, Dunston G (2004) Changing the Paradigm from 'Race' to Human Genome Variation. Nature Genetics, 2004, Volume 36, pp S5–S7
Rucker R (1982) Infinity and the Mind: The Science and Philosophy of the Infinite. Bantam Books, New York
Russell B (1903) Principles of Mathematics. WW Norton, New York
Russell B (1911–1912) On the Relations of Universals and Particulars. Proc Aristotelian Society
Russell B (1912) The Problems of Philosophy. Thornton Butterworth, London
Russell B (1914) Preliminary Description of Experience. The Monist, 24, (January), pp 1–16
Russell B (1915) Our Knowledge of the External World. Open Court, Chicago
Russell B (1918) Mysticism and Logic. Penguin, London
Russell B (1921) The Analysis of Mind. George Allen & Unwin, London
Russell B (1927) An Outline of Philosophy. Allen & Unwin, London
Russell B (1940) Language and Metaphysics. In: An Inquiry into Meaning and Truth. George Allen & Unwin, London
Russell B (1948) Human Knowledge. Simon and Schuster, New York

Russell B (1984) Theory of Knowledge: The 1913 Manuscript. Eames ER (ed). Allen & Unwin, London and New York
Ryle G (1949) The Concept of Mind. Barnes and Noble, New York, London
Saaty TL, Bram J (eds) (1964) Nonlinear Mathematics. Dover, New York
Sapolsky R (2000) It's Not 'All in the Genes.' Newsweek, April 10
Savage-Rumbaugh S, Lewin R (1994) Kanzi: An Ape at the Brink of the Human Mind. John Wiley, New York
Sayre K, Crosson F (1963) The Modeling of Mind. Notre Dame University Press, South Bend, Indiana
Scheffler I (1965) Conditions of Knowledge: An Introduction to Epistemology and Education. Scott, Foresman, Glenview, Illinois
Schilpp PA (ed) (1946) The Philosophy of Bertrand Russell. Library of Living Philosophers, Illinois
Schoenemann PT (1999) Syntax as an Emergent Characteristic of the Evolution of Semantic Complexity. In: Minds and Machines, Volume 9, Kluwer Academic Publishers, 1999, pp 309–346
Schwender D, Klasing S, Daunderer M, Madler C, Poppel E, Peter K (1995) Awareness during general anesthesia: Definition, incidence, clinical relevance, causes, avoidance and medicolegal aspects. Anaesthesist, Nov 44 (11): 743–54
Scott A (1995) Stairway to the Mind: The Controversial New Science of Consciousness. Springer-Verlag, New York
Scott A (2002) Neuroscience: A Mathematical Primer. Springer-Verlag, New York
Searle J (1967) Proper Names and Descriptions. Encyclopedia of Philosophy, Edwards P (ed), Volume 6. Macmillan, New York
Searle J (1992) The Rediscovery of the Mind. MIT Press, Cambridge
Searle J (1995) The Mystery of Consciousness. In The New York Review of Books, November and December, NY Times, New York
Sebeok T, Rosenthal R (1981) Clever Hans Phenomenon: Communication with Horses, Whales, Apes, and People, New York Academy of Sciences
Sebeok T (1990) The Sign Science and the Life Science. Symbolicity, Bernard J, Deely J, Voigt V, Withalm G (eds)
Seeley TD (1989) The honey bee colony as a superorganism. Am Sci Vol 77, pp 546–553
Seeley R, Trent D, Stephens, Tate, P (2002) Anatomy and Physiology, 6th edn, McGraw-Hill Science/engineering/Math.
Sejnowski T, Hinton GE (1987; 1990) Separating Figure from Ground with a Boltzmann Machine. In: Arbib M, Hanson A (eds), Vision, Brain, and Cooperative Computation. MIT Press, Cambridge
Seligman D (2002) Good breeding. National Review, 54(1), 53–54
Selverston AI (1992) Pattern generation. Current Opinion in Neurobiology, Dec., Volume 2, Number 6, pp 776–780
Shafer G (1976) A Mathematical Theory of Evidence. Princeton University Press, Princeton, New Jersey

Shannon C, Weaver W (1949) The Mathematical Theory of Communication. University of Illinois Press, Urbana
Shepard RN, Cooper LA (1982) Mental Images and Their Transformations. MIT Press, Cambridge
Simon SR (2004) Quantification of human motion: gait analysis-benefits and limitations to its application to clinical problems. J Biomech, Volume 37, no 12, pp 1869–1880
Simonton DK (2003) Francis Galton's Hereditary Genius: Its place in the history and psychology of Science. In Sternberg RJ (ed), The anatomy of impact: What makes the great works of psychology great. APA, Washington, DC
Singh S (1997) Fermat's Enigma: The Epic Quest to Solve the World's Greatest Mathematical Problem. Walker & Co, New York
Sinha P, Balas B, Ostrovsky Y, Russell R (2005), Face Recognition by Humans: 20 Results All Computer Vision Researchers Should Know About. MIT Press, Cambridge
Skinner BF (1953) Science and Human Behavior. The Free Press, New York
Sluga H (1980) Gottlob Frege. Routledge, London
Soussan T (2004) Prairie Dogs Have Own Language, Researcher Claims. AP Press Report
Steele CM, Aaronson J (1998) Stereotype threat and the test performance of academically successful African Americans. In: Jencks C, Phillips M (eds), The black-white test score gap. Brookings, Washington, DC
Steels Luc (1993) The Artificial Life Roots of Artificial Intelligence. Artificial Life, Vol 1, Number 1/2, MIT Press, Cambridge
Stein DL (1988) Lectures in the Sciences of Complexity, Santa Fe Institute Studies in the Sciences of Complexity, Addison-Wesley Publishing Company
Steiner E (1976) Logical and Conceptual Analytic Techniques for Educational Researchers. In: Proc AERA, San Francisco, Washington, DC
Steiner E (1988) Methodology of Theory Building. Educology Research Assoc, Sydney, Australia
Sternberg RJ, Wagner RK, Williams WM, Horvath JA (1995) Testing common sense. Am Psy, 50, 912–927
Sternberg RJ (2003) The Anatomy of Impact: What Makes the Great Works of Psychology Great. APA, Washington, DC
Stewart I (1995) Nature's Numbers. Basic Books, New York
Stewart I (1998) Life's Other Secret: The New Mathematics of the Living World. John Wiley, New York
Stich S, Nisbett R (1980) Justification and the Psychology of Human Reasoning. Phil Sci, Vol 47, pp 188–202
Stich S (1990) The Fragmentation of Reason. MIT Press, Cambridge
Stix G (1995) Boot Camp for Surgeons. Sci Am, September, p 24
Stout GF (1901) A Manual of Psychology, 2nd edn, University Tutorial Press, London
Tanenhaus MK, Spivey-Knowlton MJ, Eberhard K, Sedivy J (1995) Integration of Visual and Linguistic Information in Spoken Language Comprehension. Science, AAAS, Volume 268, 16 June, pp 1632–1634

Tarski A (1951) A Decision Method for Elementary Algebra and Geometry, 2nd revised edn. University of California Press, Berkeley

Terman LM (1906) Genius and Stupidity: A Study of Some of the Intellectual Processes of Seven "Bright" and Seven "Stupid" Boys." Pedagogical Seminary, 13, pp 307–373

Terman LM (1916) The Measurement of Intelligence. Houghton Mifflin, Boston

Thompson KR (2005) 'General System' Defined for Predictive Technologies of A-GSBT (Axiomatic-Genereal Systems Behavioral Theory), Raven58 Technologies

Thurstone LL (1960) The Nature of Intelligence. Littlefield Adams

Tomberlin J (ed) (1983) Agent, Language, and the Structure of the World: Essays Presented toHector-Neri Castañeda with His Replies. Ridgeview Publishing, Atascadero, California

Tononi G, Edelman G (1998) Consciousness and Complexity, Science, Vol 282, no 5395, pp 1846–1851

Trehub SE, Trainor LJ (1993) Listening Strategies in Infancy: The Roots of Music and Language Development. In: McAdams S, Bigand E (eds), Thinking in Sound: The Cognitive Psychology of Human Audition. Clarendon, Oxford, pp 278–327

Triesman A (1985) Preattentive processing in vision. In: Computer Vision, Graphics and Image Proc 31, 1985, pp 156–177

Triesman A (1991) Search, similarity, and integration of features between and within dimensions. J Exp Psych: Human Perception & Performance, Volume 17, no.3, 1991, pp 652–676

Triesman A, Gelade G (1980) A feature-integration theory of attention. Cog Psych, Vol 12, 1980, pp 97–136

Triesman A, Gormican S (1988) Feature analysis in early vision: Evidence from search asymmetries. In: Psych Rev, Vol 95, no 1, 1988, pp 15–48

Treisman A, Vieira A, Hayes A (1992) Automatic and preattentive processing. Am J Psych, 1992, Volume 105, pp 341–362

Trojano L, Grossi D, Linden DEJ, Formisano E, Hacker H, Zanella FE, Rainer Goebel, Salle GD (2000) Matching Two Imagined Clocks: the Functional Anatomy of Spatial Analysis in the Absence of Visual Stimulation, Cerebral Cortex, May, Vol 10: 473–481

Turing AM (1937) On Computable Numbers With An Application to the Entscheidungsproblem. In: Proc London Math Soc, Vol 42, pp 230–265

VanRullen R, Reddy, Lavanya,Koch C (2004) Visual Search and Dual Tasks Reveal Two Distinct Attentional Resources. J Cog Neuroscience, 16:1, pp 4–14

Vikhanski L (2001) In Search of the Lost Cord. Joseph Henry Press, Washington, DC

Vilis T (2002) The Physiology of the Senses: Transformations for Perception and Action. University of Western Ontario

Vinod VV, Santanu Chaudhury J Mukherjee, Ghose S (1994) A Connectionist Approach for Clustering with Applications in Image Analysis. In: IEEE Transactions on Systems, Man, and Cybernetics, Vol 24, No. 3, March, pp 365–383

Von Bertalanffy L (1968) General System Theory. Braziller, New York

Wadsworth SJ, DeFries JC, Fulker DW (1993) Cognitive abilities of children. J Learn Dis, November, Volume 26, No 9, pp 611–615

Webster's Encyclopedic Unabridged Dictionary (1989). Portland House, New York

Weiskrantz L (1997) Consciousness Lost and Found. Oxford, New York

Wittgenstein L (1922) Tractatus Logico-Philosophicus. Ogden CK (trans), Routledge, London

Wittgenstein L (1953) Philosophical Investigations. 3rd edn. Anscombe GEM, (trans). Macmillan, New York

Wittgenstein L (1969) Über Gewissheit: On Certainty. Anscombe GEM, Wright GH (eds). Harper and Row, New York, London

Wolfe JM (1996) Visual Search. In: Pashler H (ed), Attention. University College, London

Wolfe JM, Bennett SC (1997) Preattentive Object Files: Shapeless Bundles of Basic Features. Vision Research, Vol 37, Issue 1, January

Wolfe JM, Cave KR, Franzel SL (1989) Guided search: An alternative to the feature integration model for visual search. J Exp Psych: Human Perception and Performance, 15, 419–433

Wolfram S (1984) Computer Software in Science and Mathematics. Scientific American, September, pp 188–203

Wolfram S (ed) (1986) Theory and Applications in Cellular Automata. World Scientific, Singapore

Wolpaw JR (1997) The Complex Structure of Simple Memory. Trends in Neurosciences, Vol 20, pp 588–594

Wright C (1983) Frege's Conception of Numbers as Objects. Aberdeen University Press, Aberdeen

Wright S (1931) Evolution in Mendelian Populations. Genetics, Vol 16, number 97

Wright S (1932) The Roles of Mutation, Inbreeding, Crossbreeding and Selection in Evolution. Proc Sixth Int Cong Genetics, Vol. 1, number 356

Yerkes RM (1921) Psychological Examining in the United States Army. National Academy of Sciences, Vol 15

Zadeh LA (1971) Fuzzy Languages and Their Relation to Human and Machine Intelligence. Proc Conf on Man and Computer, Bordeaux, France, Memorandum M-302, Electronics Research Laboratory, University of California at Berkeley

Zadeh LA (1973) Outline of a New Approach to the Analysis of Complex Systems and Decision Processes. IEEE Transactions on Systems, Man, and Cybernetics, Vol SMC-3, No 1

Zadeh LA (1977a) A Theory of Approximate Reasoning, Memorandum No UCB/ERL M77/58, Electronics Research Lab, College of Engineering, University of California at Berkeley

Zadeh LA (1977b) PRUF-A Meaning Representation Language, Memorandum No ERL-M77/61, Electronics Research Lab, College of Engineering, University of California at Berkeley

Zadeh LA (1978) PRUF-A Meaning Representation Language for Natural Languages," Int J Man-Machine Studies, Vol 10, pp 395-460

Zadeh LA (1981) Test-Score Semantics for Natural Languages and Meaning Representation Via PRUF. Empirical Semantics. Rieger BB (ed), Studienverlag Dr. N. Brockmeyer, Bochum.

Zadeh LA (1983) A Fuzzy Set-Theoretic Approach to the Compositionality of Meaning: Propositions, Dispositions and Canonical Forms. Memorandum No UCB/ERL M83/24, 4 April

Zadeh LA, et al (eds) (1990) Uncertainty in Knowledge Bases. Springer-Verlag, Berlin, Heidelberg, New York

Zalta E (1999) Natural Numbers and Natural Cardinals as Abstract Objects: A Partial Reconstruction of Frege's Grundgesetze in Object Theory. Journal of Phil Logic, Vol 28, number 6, pp 619–660

Zimmerman HJ (1993) Proceedings of the First European Congress on Fuzzy and Intelligent Technologies, Volumes 1–3, European Laboratory for Intelligent Techniques Engineering, Aachen, Germany, 1993

Zuzne L (ed) (1957) Names in the history of psychology. John Wiley, New York

Index

N

O

P

Q

R

S